Industrial Electrical Systems

Industrial Electrical Systems

by

Dale R. Patrick, Professor

and

Stephen W. Fardo, Associate Professor

Department of Industrial Education and Technology
College of Applied Arts and Technology
Eastern Kentucky University

Howard W. Sams & Co., Inc.
4300 WEST 62ND ST. INDIANAPOLIS, INDIANA 46268 USA

FIRST EDITION
FIRST PRINTING—1977

International Standard Book Number: 0-672-21446-6
Library of Congress Catalog Card Number: 77-77401

Printed in the United States of America.

Preface

Industrial Electrical Systems is a unique contribution to a vast and ever-expanding body of knowledge. This book is a comprehensive collection of information related to industrial systems. The content of the book provides a broad overview of industrial electrical systems. The current "state of the art" for industrial systems is presented in a logical manner. Each chapter of the book covers a vast array of material, and it would be very easy for the authors to expand each chapter into a separate volume. However, the manner of presentation allows the reader to become familiar with the basics of many electrical and associated systems that are commonly used in industry.

This book is intended to be either a vocational/technical or college-level text, or a self-study vehicle for those aspiring to become familiar with industrial electrical systems. The authors feel that an industrial technician should be a generalist in many respects. He should be aware of the broad spectrum of industrial systems that might be encountered in a typical industrial environment. With the basic knowledge that will be acquired from this book, the technician will be better able to understand industrial system operation. The mathematical applications used in the book are limited to basic operations, simple algebraic functions, and simple trigonometric functions.

The organization of many technical textbooks is diametrically opposite to the way in which learning takes place. The student must fit seemingly unrelated pieces of information together, similar to putting a jigsaw puzzle together. The "systems" approach, followed in this book, avoids the use of disconnected or unrelated information. Industrial systems are discussed in the following format: (1) electrical *energy sources* for industry; (2) *distribution* systems for industry; (3) industrial *control* systems; (4) industrial electrical *loads;* and (5) *indicating* systems for industry. This format is then expanded to include the *applications* of each of the parts of an industrial system. The basic organization includes both electromechanical and fluid power systems. The authors feel that these areas should be studied together, rather than separately, as they traditionally are in technical programs.

The text is organized into seven chapters as follows: Chapter 1 is an introduction to the basic systems concept with a detailed discussion of how this concept is applied to various industrial systems; Chapter 2 discusses the various codes, standards, and symbols that govern industrial systems; Chapter 3 describes the various electrical energy sources used by industry; Chapter 4 covers the distribution of electrical power to and within industrial plants and equipment; Chapter 5 describes the various types of industrial loads; Chapter 6 discusses industrial indicating devices and how they are used; and Chapter 7 covers various types of electromechanical and fluid power systems and devices.

Included at the back of the book are four appendixes of reference material: Appendix 1 is a comparison chart showing U.S. and metric equivalent units;

Appendix 2 contains most of the standard graphic symbols used in industrial electrical/electronic and fluid power system diagrams; Appendix 3 is a brief discussion of trigonometric functions as applied in ac circuits, and includes a table of natural sines, cosines, tangents, and cotangents; while Appendix 4 shows three typical industrial electrical diagrams.

We would like to express our sincere thanks to the many industries who have provided technical data and photographs for the development of the manuscript. With their cooperation, much up-to-date material was provided for the book. Thanks also is extended to our wives for their patience and understanding during the manuscript preparation.

DALE R. PATRICK
STEPHEN W. FARDO

To our wives, Helen and Kay

Contents

1

Introduction

The advancement of science and technology has brought about a number of very important changes in the manufacture of industrial products. This includes such things as product inspection techniques, quality control, data transmission, detection procedures, automatic processing, sequence timing, warning signal sensors, and microprocessors. The term *industrial electronics* is commonly used to describe this broad field of study today.

At one time, the field of industrial electronics was limited to only motor systems and devices that achieved some form of control. Gaseous tubes, magnetic contactors, and electrical switchgear served as the nucleus of the field at this time. Recent developments in solid-state electronics and microminiaturization have brought a number of significant changes to the field of industrial electronics. In addition to this, electrical, mechanical, optoelectronic, hydraulic, and pneumatic systems are often combined in the control of a single machine. A person working in this field must be knowledgeable of these systems and of numerous control devices in order to be successful today.

THE SYSTEMS CONCEPT

For a number of years, people have worked with jigsaw puzzles as a source of recreation. This type of puzzle contains a number of discrete parts that must be properly placed together to produce a picture. Each part then plays a specific role in the finished product. When a puzzle is first started, it is difficult to imagine the finished product unless one sees a representative picture.

When one studies a complex field such as industrial electronics by discrete parts, it poses a problem that is somewhat similar to the jigsaw puzzle. In this case, it is difficult to determine initially the role that a discrete part plays in the operation of a complex machine. A picture of the machine divided into its essential parts, therefore, becomes an extremely important aid in understanding its operation.

The systems concept will serve as our "big picture" in the study of industrial electronics. In this approach, we will initially divide a system into a number of essential blocks. The role played by each block will then become more meaningful in the operation of the overall system. After the location of each block has been established, discrete component operation related to each block will then become more relevant. Through this approach, one should soon be able to see how some of the "pieces" of the industrial electronics field fit together in a more meaningful order.

BASIC SYSTEM FUNCTIONS

The word *system* is commonly defined as, "an organization of parts that are connected together to form a complete unit." In this respect, there are a wide variety of different systems used in industry today. An electrical power system, for example, is needed to produce electrical energy and distribute it to each industrial site. Hydraulic and pneumatic systems are used in industry to accomplish forming operations and to control other system functions. Optoelectronic systems are commonly found in inspection applications and for data production and transmission, while mechanical systems are needed to hold objects for machining operations and to move them physically in a production line.

Each industrial system obviously has a number of unique features or characteristics that distinguish it from other systems. More importantly, however, there is a common set of parts found in each system. These parts play the same basic role in all systems. The terms *energy source, transmission path, control, load,* and *indicator* will be used in this study to describe the various system parts. A block diagram of these basic parts of the system is shown in Fig. 1-1.

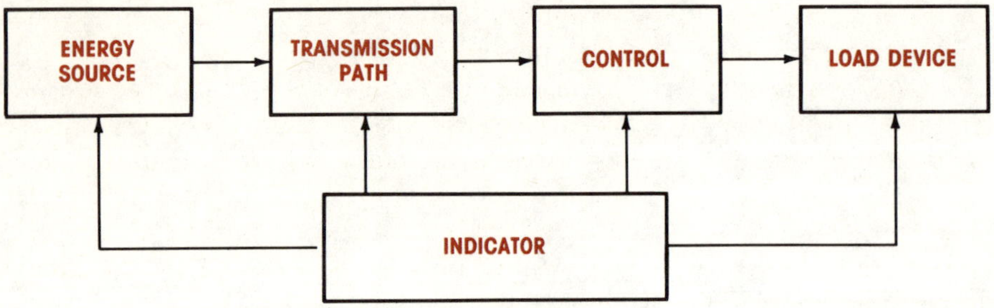

Fig. 1-1. Block diagram of a basic system.

Each block of a basic system has a specific role to play in the overall operation of the system. This role becomes extremely important when a detailed analysis of the system is to take place. Hundreds and even thousands of discrete components are sometimes needed to achieve a specific block function. Regardless of the complexity of the system, each block must still achieve its function in order for the system to be operational. Being familiar with these functions and being able to locate them within a complete system is a big step in understanding the operation of the system.

The *energy source* of an industrial system is responsible for converting energy of one form into something more useful. Heat, light, sound, chemical, nuclear, and mechanical energy are considered as primary sources of energy. A primary energy source usually goes through an energy transformation before it can be used in an operating system.

The *transmission path* of a system is somewhat simplified when compared with other system functions. This part of the system simply provides a path for the transfer of energy. It starts with the energy source and continues through the system to the load device. In some cases, this path may be a single feed line, electrical conductor, light beam, or pipe connected between the source and the load. In other systems, there may be a supply line between the source and the load, and a return line from the load to the source. There may also be a number of alternate or auxiliary paths within a complete system. These paths may be series connected to a number of small load devices, or parallel connected to many independent devices.

The *control* section of a system is by far the most complex part of the entire system. In its simplest form, control is achieved when a system is turned on or off. Control of this type can take place anywhere between the source and the load device. The term *full control* is commonly used to describe this operation. In addition to this type of control, a system may also employ some type of *partial control.* Partial control usually causes some type of an operational change in the system other than an on or off condition. Changes in electric current, hydraulic pressure, light intensity, and air flow are some of the system alterations achieved by partial control.

The *load* of a system refers to a specific part or number of parts designed to produce some form of work. The term *work,* in this case, occurs when energy goes through a transformation or change. Heat, light, chemical action, sound, and mechanical motion are some of the common forms of work produced by a load device. As a general rule, a very large portion of all energy produced by the source is consumed by the load device during its operation. The load is typically the most prevalent part of the entire system because of its obvious work function.

The *indicator* of a system is primarily designed to display certain operating conditions at various points throughout the system. In some systems the indicator is an optional part, while in others it is an essential part in the operation of the system. In the latter case, system operations and adjustments are usually critical and are dependent upon specific indicator readings. The term *operational indicator* is commonly used to describe this application. *Test indicators* are also needed to determine different operating values. In this role, the indicator is only temporarily attached to the system to make measurements. Test lights, hand-deflection meters, oscilloscopes, chart recorders, digital display instruments, and pressure gauges are some of the common indicators used in this capacity. As a general rule, an indicator will add to the total load of the system to some extent.

INDUSTRIAL SYSTEMS

The number of different systems used in industry today is quite large when we consider the wide variety of different products that are being manufactured. Each industry, for example, has a number of unique system applications and manufacturing processes that require classification. To have some degree of order, all industrial systems are first divided into a classification that describes the type of energy needed to make the system operational. In this classification, only a single energy source such as heat, light, sound, and mechanical energy is needed to power a system. This type of grouping is considered as the *hybrid* classification of industrial systems. Hybrid systems are rather rare and extremely unusual today.

A more practical classification of the industrial system places two or more of the hybrid classifications into a single system. This classification is described as a *synthesized* system. Electrical, hydraulic, pneumatic, optoelectronic, timing, digital (numerical), and environmental systems represent some of the major synthesized industrial systems in use today. This classification (synthesized system) will serve as the basis of further discussion in our study of industrial electronic systems.

ELECTRICAL SYSTEMS

Nearly everyone today has had an opportunity sometime in their life to use a flashlight. This device is designed to serve as a light source in an emergency or to provide light to unusual places. In a strict sense, flashlights can be classified as portable electrical systems. They contain the four essential parts needed to make this classification. Fig. 1-2 is a cutaway drawing of a flashlight with each component part shown associated with its appropriate system block.

The battery of a flashlight serves as the primary energy source of the system. Chemical energy of the battery must be changed into electrical energy before the system becomes operational. The flashlight is a synthesized system because it utilizes two distinct forms of energy in its operation. The energy source of a flashlight is an expendable item. It must be replaced periodically when it loses its ability to produce electrical energy.

The transmission path of a flashlight is commonly achieved by a metal case or through an electrical conductor strip. Copper, brass, and plated steel are frequently used to achieve this function.

The control of electrical energy in a flashlight is achieved by a slide switch or a push-button switch. This type of control simply interrupts the transmission path between the source and the load device. Flashlights are primarily designed to have full control capabilities. This type of control is achieved manually by the person operating the system.

The load of a flashlight is a small incandescent lamp. When electrical energy from the source is forced to pass through the filament of the lamp, the lamp will produce a bright glow. Electrical energy is first changed into heat and then into light energy. A certain amount of work is achieved by the lamp when this energy change takes place.

The energy transformation process of a flashlight is irreversible. It starts at the battery when chemical energy is changed into electrical energy. Electrical energy is then changed into heat and eventually into light energy by the load device. This flow of energy is in a single direction. When light is eventually produced, it consumes a large portion of the electrical energy coming from the source. When this energy is exhausted, the system becomes inoperative. The battery cells of a flashlight require periodic replacement in order to maintain a satisfactory operating condition.

Flashlights do not ordinarily employ a specific indicator as part of the system. Operation is indicated, however, when the lamp produces light. In a strict sense, we could say that the load of this system also doubles as an indicator. In some electrical systems the indicator is an optional system part.

ELECTRICAL POWER SYSTEMS

The electrical power system that supplies energy to our homes and industrial buildings is classified as a fixed system. This type of system is completely stationary and its parts cannot be relocated without making drastic physical changes. Fig. 1-3 shows a sketch of a simple electrical power system.

The energy source of an electrical power system is much more complex than that of a portable electrical system. The primary source of energy may be derived

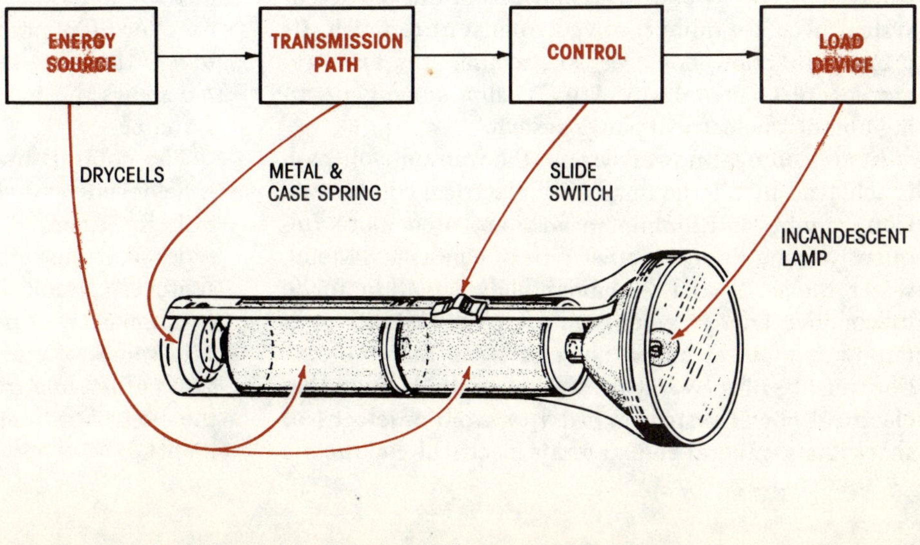

Fig. 1-2. Cutaway drawing of a flashlight illustrating the correlation of its component parts to the system blocks of Fig. 1-1.

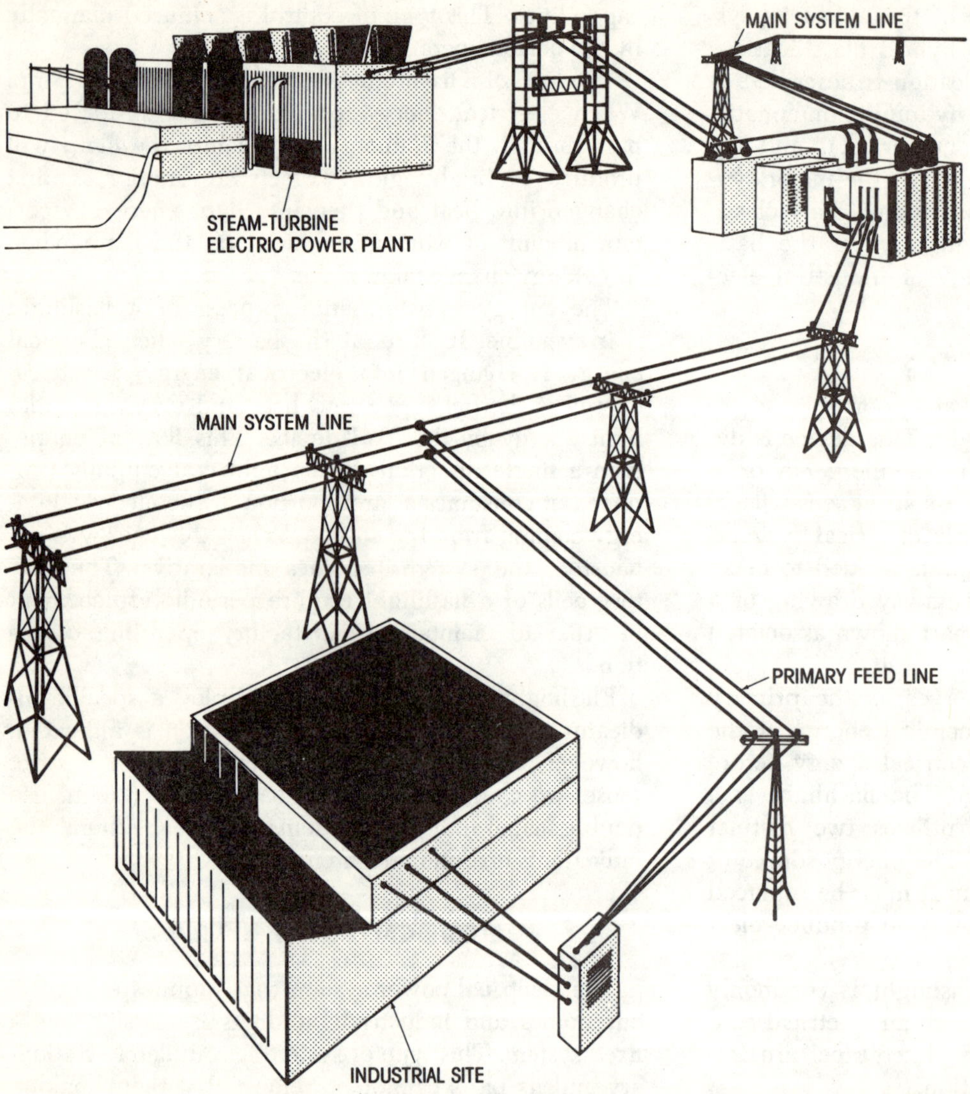

Fig. 1-3. Sketch of a simple electrical power system.

from coal, oil, natural gas, atomic fuel, or moving water. This type of energy is needed to produce mechanical energy which in turn develops the rotary motion of a turbine. Massive alternators are then rotated by the turbine to produce alternating-current electricity. The energy conversion process of this particular system becomes quite involved from start to finish. Its primary function, however, still remains the same regardless of its complexity. Fig. 1-4 shows one generating unit of an electrical power system.

In an electrical power system, the transmission path is achieved by a large number of electrical conductors. Copper wire and aluminum wire are used more frequently today than any other type of conductor. Metal, water, the earth, and the human body can all be made to conduct energy when contact is made with certain parts of an electrical power system. Extreme caution must be observed when working with an operating electrical power system in order to avoid an electrical shock. As a general rule, interior electrical conductors are coated with an insulating material in order to prevent electrical hazards.

The transmission path of an electrical power system often becomes so complex that it is considered as a separate system instead of a system part. The term *distribution system* is commonly used to describe this function. The primary function of the distribution system is the same as that of the transmission path. Fig. 1-5 shows the transmission line of an electrical power system.

The control function of an electrical power system is achieved by a variety of different devices. Full control, for example, is accomplished by three distinct types of circuit-interrupting equipment. These include switches, circuit breakers, and fuses. Each piece of equipment must be designed to pass and interrupt specific values of current. Making and breaking transmission path connections with a minimum of arc is one of the most important characteristics of this type of equipment.

Courtesy Tennessee Valley Authority

Fig. 1-4. A 1.15 megawatt steam turbine electric generator unit at Paradise, Kentucky.

Partial control of an electrical power system is achieved primarily by circuit resistance, inductance (transformers), and capacitance. The resistance of a long-distance transmission line is not a desirable feature; it must be minimized in order to avoid unnecessary system losses. Any power loss that occurs in the transmission line is wasted energy that cannot be utilized by the load device.

Fig. 1-5. Transmission line of an electrical power system.

In order to minimize power losses in an electrical system, transformers are used at strategic locations throughout the system. These partial-control devices are designed to initially *step up* the source voltage to a high value. Through this process, the source current is proportionally reduced in value. Since the power loss of a transmission line is based on the square of the current times the line resistance, losses can be reduced to a reasonable value through this process.

Transformers are also used to lower system voltages to usable values near the load device. This action of a transformer is described as its *step-down* function. When the source voltage is reduced to a lower value, the source current is proportionally increased in value. Through the use of transformers, transmission-line losses can be held to a minimum, thus causing increased system efficiency.

Capacitors are used in electrical power systems primarily as power-factor correcting devices. In this function, a capacitor can be used to cause the current to lead the voltage by a certain amount. This effect is used to reduce the current lagging condition caused by transformer inductance and large motor loads. When the alternating current and voltage of an electrical power system are kept in phase, the operating efficiency of the system is at its best.

Courtesy Philadelphia Electric Co.

Fig. 1-6. The control center of an electrical power generating plant.

The load of an electrical power system is usually quite complex. As a composite, it includes everything that utilizes electrical energy from the source. As a general rule, the load is divided into four distinct parts. This includes residential, commercial, industrial, and other uses such as street lighting and government installations. The composite load of an electrical power system is subjected to change hourly, daily, and seasonally.

The average person is probably more familiar with the load part of an electrical power system than any of its other parts. This represents the part of the system that actually does work for us. Motors, lamps, electric ovens, welders, and power tools are some of the common load devices used by industry today. Loads are frequently classified according to the type of work they produce. This includes light, heat, sound, electromechanical changes, and chemical action. Loads are also classified according to the predominant electrical characteristics involved. This includes resistive-, capacitive-, and inductive-type loads.

The indicator of an electrical power system is designed to show the presence of electrical energy at various places or to measure different electrical quantities. Hand-deflection meters, oscilloscopes, chart recording instruments, and digital display devices are some of the indicators used in this type of system today. Fig. 1-6 shows the control center of a power generating plant. Voltage indicators, line-current indicators, power-factor indicators, temperature gauges, water-level meters, pressure gauges, and chart recorders are found in this installation. Indicators of this type are designed to provide an abundance of system operating information.

The temperature gauge of Fig. 1-7 is an indicator that is commonly found attached to different types of industrial systems. Indicators of this type are designed to display the amount of temperature sensed by the thermocouple probe. This type of indicator is actually an electrical instrument. Thermal energy applied to the probe is converted into electrical energy, which in turn produces a corresponding meter deflection. Indicators play a very important role in the overall operation of a system.

FLUID POWER SYSTEMS

Systems that are designed to transfer power through fluids are a very important part of the industrial electronics field today. The term *fluid,* in this case, is a general classification for systems that use air, fluid, or a combination of air over oil. The term *hydraulic* is normally used to describe systems that use only liquids or fluids. Systems operated with air only are

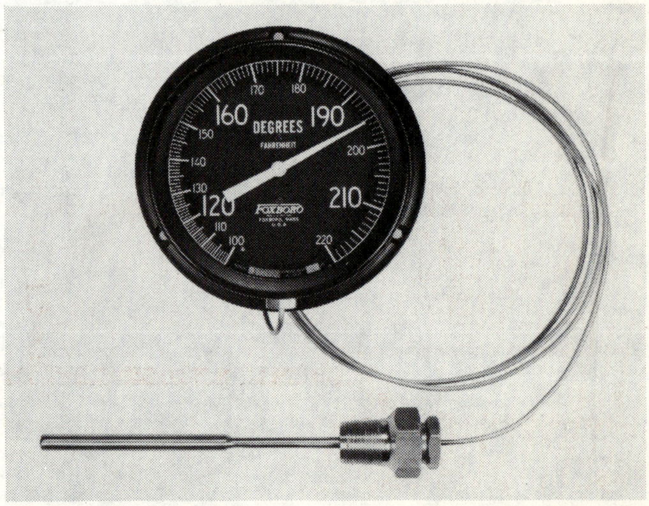

Courtesy The Foxboro Co.

Fig. 1-7. An industrial temperature indicator.

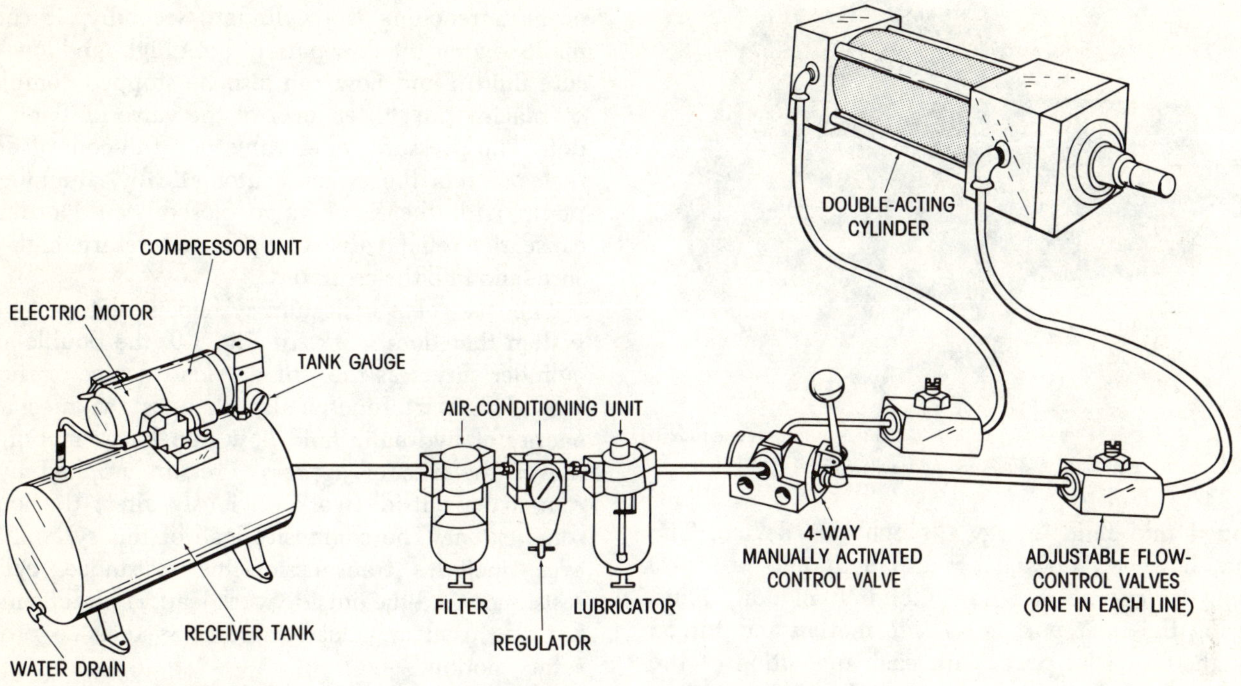

Fig. 1-8. A pneumatic type of fluid power system.

usually described by the term *pneumatic*. The operating principles associated with these systems are similar in many respects. Fig. 1-8 shows a fluid system with the basic parts labeled.

Fluid systems are primarily classified as synthesized systems. Electric energy or some form of rotary motion is needed to energize the fluid pump. Through the conversion process, electrical energy or mechanical motion is changed into the mechanical energy of a flowing fluid. This type of system is quite reliable and has a great deal of flexibility. Power can be transferred to any location where a pipe, hose, or piece of tubing can be placed. This type of system is generally closed and kept free of outside contamination.

Hydraulic Systems

Hydraulics represents a major part of the fluid classification of power systems today. This type of system is commonly found in automatic machinery control and in material handling equipment. The popularity of this type of system can be attributed to such things as operational simplicity, smoothness of operation, reliability, and adaptability. Fig. 1-9 shows a double-acting hydraulic system used to control a punch-press ram.

The primary energy source of the hydraulic system in Fig. 1-9 is an electric-motor-driven pump and the reservoir. Rotary mechanical energy of the motor is

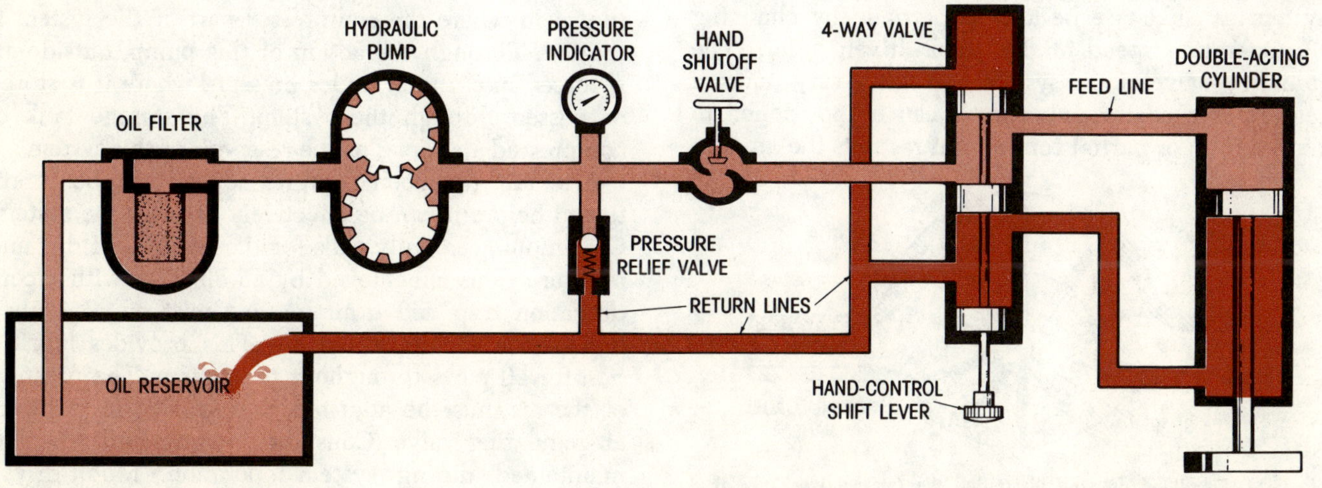

Fig. 1-9. A double-acting hydraulic system.

Courtesy Parker Hannifin Corp.

Fig. 1-10. A dual-vane hydraulic pump.

changed into fluid energy through this device. The pump, however, does not pump hydraulic pressure. During its operation fluid is set into motion. Fluid entering the inlet port is set into motion and forced through the outlet port. With each revolution of the pump rotor blade, a fixed amount of fluid is forced into the system. Fluid entering the system then encounters resistance to its flow, which creates hydraulic pressure. Fig. 1-10 shows a dual-vane hydraulic pump, and Fig. 1-11 shows a simplification of its internal workings.

The transmission path of a hydraulic system is typically solid pipes or some form of flexible tubing. Hydraulic fluid forced to pass through the transmission path encounters a resistance to its flow, which builds up system pressure. Both single-pressure and high/low-pressure systems are available today. Fig. 1-9 shows a double-acting hydraulic system with a high/low-pressure transmission path.

The hydraulic system of Fig. 1-9 has both full and partial control of system fluid. The hand shutoff valve permits full control of the system by stopping the fluid flow through the transmission line. The pressure of the system can likewise be altered partially by changing the operating speed of the motor-driven pump. The four-way valve of the system also has a very unusual feature as a control function. It can be positioned to have a form of partial control that restricts the amount

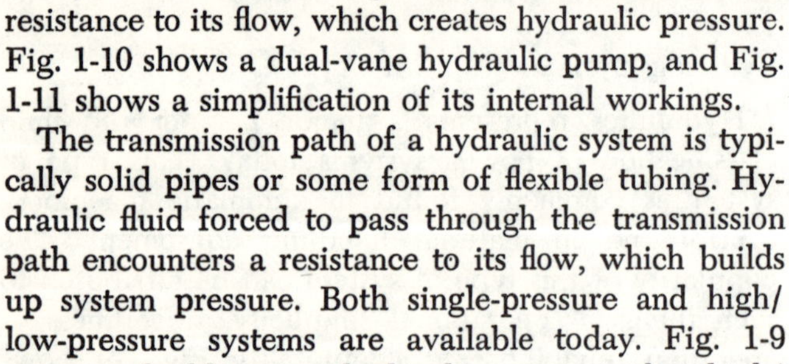

Fig. 1-11. Simplified drawing illustrating the internal workings of a dual-vane hydraulic pump.

of fluid reaching the cylinder. Secondly, it can be made to alter the flow path of both high- and low-pressure fluid. Fluid flow can also be stopped completely by placing the shifter lever of the valve in its off position. The pressure relief valve is a full-control device that protects the system automatically. Running the pump with the hand valve closed would ordinarily cause the relief valve to open and return high-pressure fluid into the reservoir.

The load of a hydraulic system is that part of the system that does work. In Fig. 1-9, the double-acting cylinder serves as the primary load device of the system. This part functionally changes the mechanical energy of hydraulic fluid flow into linear motion that moves the ram of a press. Pressure applied by the ram to its outside work obviously alters the load to some extent. The composite load of this system, however, includes transmission-line resistance, cylinder resistance, and the outside work load. Hydraulic motors are also used as a load in some systems to produce rotary motion.

The indicator of the hydraulic system in Fig. 1-9 is an optional item, as it is in other systems. In this application, the indicator is used to show system pressure under normal load conditions. Monitoring the fluid pressure and maintaining it at a constant level assures consistent operating action of the system.

Pneumatic Systems

Pneumatic systems are used in industry to power hand tools and to lift and clamp products during machining operations. The energy source of this type of system is a pump or a compressor and a storage tank to hold the air. The pump of a compressor may be driven electrically by a motor or by a portable internal-combustion engine. Fig. 1-12 shows a simple pneumatic system driving a double-acting cylinder ram.

Pneumatic systems are designed to use the air from the room where the compressor part of the system is located. Through the action of this pump, outside air is forced into a tank under pressure where it is stored or passed through the system. The storage tank of compressed air serves as the reservoir of this system.

After air has been compressed, it must be conditioned before it can be effectively used by the system. Conditioning initially calls for the removal of dirt and moisture. This is achieved by an air filter with a condensation trap and drain. A fine mist of oil is then added to the compressed air. This provides lubrication for all parts throughout the system. The pressure of the air must be adjusted to a specific level by an air-regulating valve. Constant pressure must then be maintained during system operation. Motor-driven air compressors are designed to operate only when the

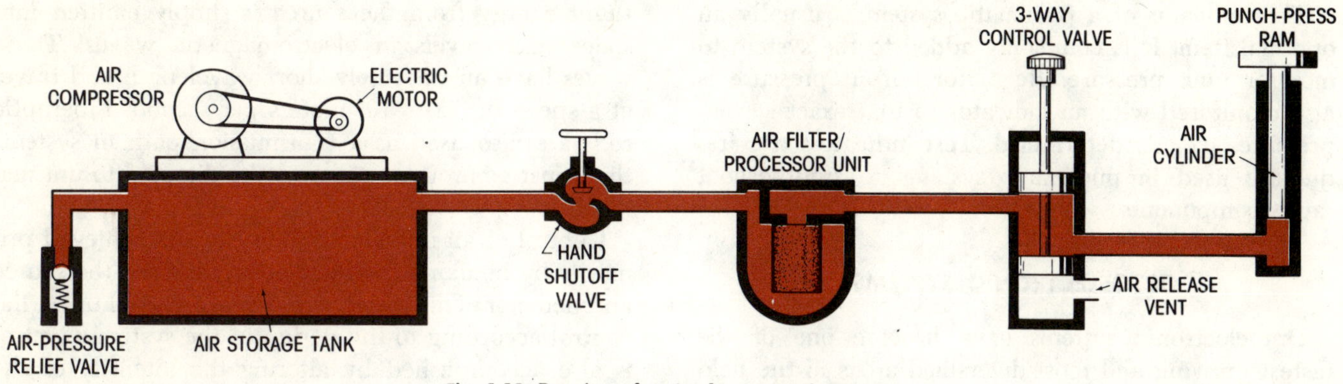

Fig. 1-12. Drawing of a simple pneumatic system.

system pressure of the storage tank drops below a predetermined level. Fig. 1-13 shows a typical industrial compressor, tank, and air conditioner in a single unit.

The transmission path of a pneumatic system represents one of its most unique parts. Solid pipes, tubing, and flexible hoses are generally used as feed lines from the compressor to different parts of the system. Return lines to the storage tank, however, are not used in pneumatic systems. Air is simply dumped from the system into the atmosphere. Pneumatic systems are somewhat simplified because of this feature.

The pneumatic system of Fig. 1-12 has both full and partial control of system air. The hand shutoff valve and pressure-relief valve provide full control of air circulating through the transmission path. Air flow can also be altered by the regulator and the three-way control valve. Pneumatic and hydraulic controls are very similar in all respects.

The load of a pneumatic system, as in other systems, is designed to perform a work function. The primary load device of Fig. 1-12 is the pneumatic cylinder. This part functionally changes the mechanical energy of air into linear motion that drives a press ram. Pressure applied by the ram to outside work also influences the primary load to some extent. The composite load of the entire system includes transmission-line resistance, control resistance, and the outside work load. Pneumatic load devices can also be used to produce rotary motion and striking blows for hammering.

Fig. 1-13. An industrial type of air compressor, storage tank, and air conditioner in a single unit.

Courtesy Gardner-Denver Co.

The indicator of a pneumatic system is usually an optional item. It is commonly added to the system to monitor tank pressure. Regulator output pressure is also monitored with an indicator so that exact system pressures can be determined. Test indicators are frequently used in pneumatic systems to troubleshoot faulty components.

OPTOELECTRONIC SYSTEMS

Optoelectronic systems have become one of the fastest growing and most diversified areas in the field of industrial electronics. The development of the laser in 1960, and numerous integrated-circuit designs shortly thereafter, have practically revolutionized this part of the field. New developments that combine optics, light, and electronics are often announced on a daily basis. System applications and operating principles of this type play a very important role in industry today.

Optoelectronic systems are classified as synthesized systems because they respond to two or more forms of energy. Light energy and electrical energy are both widely used in this type of system. A light source, transmission path, control, and load device are the essential parts of this system. Fig. 1-14 shows the basic parts of a simple optoelectronic system.

The light source of Fig. 1-14 is produced by electrical energy. Discrete bundles of light energy called *photons* are released by the source when it is placed into operation. This energy is then directed away from the source in an intense beam. The term *radiant energy* is often used to describe light energy. Incandescent lamps, flames, glow lamps, electric arcs, solid-state light, and laser energy are used today as artificial light sources.

The transmission path of an optoelectronic system is undoubtedly the most unique part of this system.

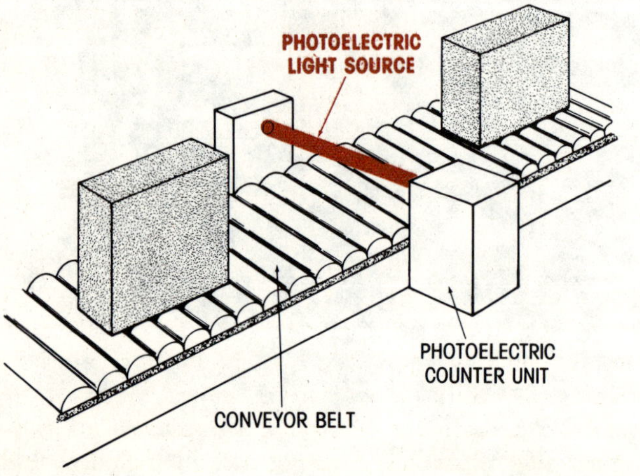

Fig. 1-14. Sketch showing the basic parts of an optoelectronic system.

Light energy from the source is simply emitted into space and travels in electromagnetic waves. These waves have an extremely short wavelength and travel at a speed of 300×10^6 meters per second. Fiberoptic rods are also used as a transmission path in systems that must go around corners or be directed to unusual locations.

Control of an optoelectronic system is achieved primarily by breaking the light beam between the source and detector. This action can cause both full or partial control according to the design of the system. Control is also accomplished by altering the intensity of the light source, its focus, shape, and wavelength. In the detector part of the light system, control is achieved by altering the sensitivity of the detector. Adjustments of this type are purposely built into the system so that it can be adapted to specific operating conditions.

The detector of an optoelectronic system is that part of the system that is designed to respond to light energy from the source. Photovoltaic, photoemissive, and photoconductive devices are commonly used to achieve this function. The output of the detector is primarily used to control the load device. The load device, as in other systems, produces a form of work. In some systems, the load may be controlled directly by light-source energy, whereas in other systems, light energy from the source must be detected and amplified before it can be effectively used to control the load. In some applications, the entire detector unit may serve as the load. As a general rule, however, the detector should be considered as a control element for the system load.

TIMING SYSTEMS

Turning a device on or off at a specific time or in step with an operating sequence is an indispensable function in automatic industrial processing. Systems designed to achieve this type of operation are called *timing systems* or simply *timers*. This type of system represents a very important part of the total industrial electronics field today.

There are three major classifications of timing systems presently being used in industry. These are primarily based on the work function that the system is designed to perform. Included in this classification are delay timers, interval timers, and cycle timers. Delay timing is designed to precede the energizing of the system load. When a certain system is placed into operation, there may be a required delay time before the load device actually becomes energized. Interval timing, by comparison, occurs after the load has been energized. For example, it may cause the load to remain energized only for a certain period of time. Cycle timing systems are typically more complex than delay and interval timers. They are designed to provide some type of energizing action in an operational sequence. This type of timing may include both interval and delay timing.

Timing systems are also classified according to the method by which they achieve control. This includes such things as thermal devices, dashpot units, motor-driven mechanisms, mechanical, electrical, electronic, and electrochemical devices.

Timing systems include all of the necessary parts that are typical of a basic system. A large portion of these systems, however, is energized by electricity. The control section is by far the most unique part of this system. Hydraulic, pneumatic, mechanical, heat, and electrical energy are often combined to achieve various types of control. The uniqueness of this system part serves as the basis for its classification as a distinct system type.

DIGITAL SYSTEMS

One of the most significant developments that has taken place in the manufacturing of industrial products today is automation. Automatic fabrication methods, packaging, printing equipment, machining operations, and drafting equipment are all an outgrowth of this principle. In systems that utilize this principle, coded instructions are supplied by perforated paper tape, punched cards, magnetic tape, or various physical changes such as pressure, temperature, or electricity. This information is then changed into digital signals and applied to the digital processor part of the system. Ultimately this information is decoded and directed to specific machines or machine parts, which will then perform the necessary operations automatically.

A large portion of the automatic machinery being used by industry today receives instructions through digital signals. The term *digital* or *numerical* system is therefore a very important part of the industrial electronics field. In a strict sense, numerical signals are used primarily to perform the control function of a digital system. The term *numerical control* or the abbreviation N/C is also used to describe a specific type of digital system.

A majority of the numerical systems in operation today are powered by electricity. This source of energy is primarily used to energize the load device, which in turn performs the work function of the system. The control function of the system must then be designed to respond to digital information. Tape machines and card readers develop this information in the first operational step. Logic circuits, which provide full or partial control of the load device, are then activated.

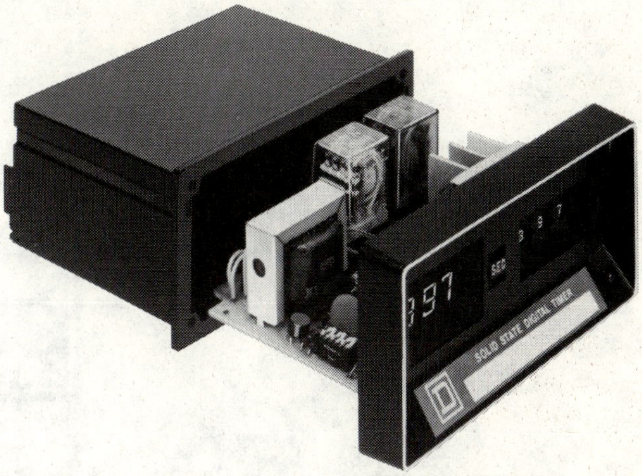

Courtesy Square D Co.

Fig. 1-15. A solid-state digital timer.

Fig. 1-15 shows a solid-state digital timer that is representative of one type of digital system.

A numerically controlled milling machine is a unique type of digital system. In the N/C milling machine system of Fig. 1-16, operating instructions are first punched on a paper tape by the perforating typewriter. This tape is then fed into the tape reader. Numerical information from the tape perforations is then translated into electrical signals by the tape reader. These signals are processed by different logic gates housed in the console unit to the right of the machine. As an end result, these signals are used to control various physical machine operations automatically.

The load of an N/C system is typically electrical actuating motors or fluid-power cylinders designed to move the physical parts of a machine. When appropriate signals from the control unit are applied to the load, they will move the two table axes or the cutting tool to a specific location. Machining operations can then be performed according to the information programmed on the tape. Cutting speed, position location, clamping operations, and material flow can all be controlled through this system. Hydraulic, pneumatic, and electrical loads are all easily controlled by programming information into a system of this type.

INDUSTRIAL ENVIRONMENTAL SYSTEMS

Industrial environmental systems represent a rather new division of the industrial electronics field. This type of system is being used today to describe a variety of systems that have been used by industry for a long time. Such things as detecting devices, pollution control units, humidity modifying equipment, alarm systems, and ultrasonic cleaning are all included under this heading. Systems of this type play an extremely

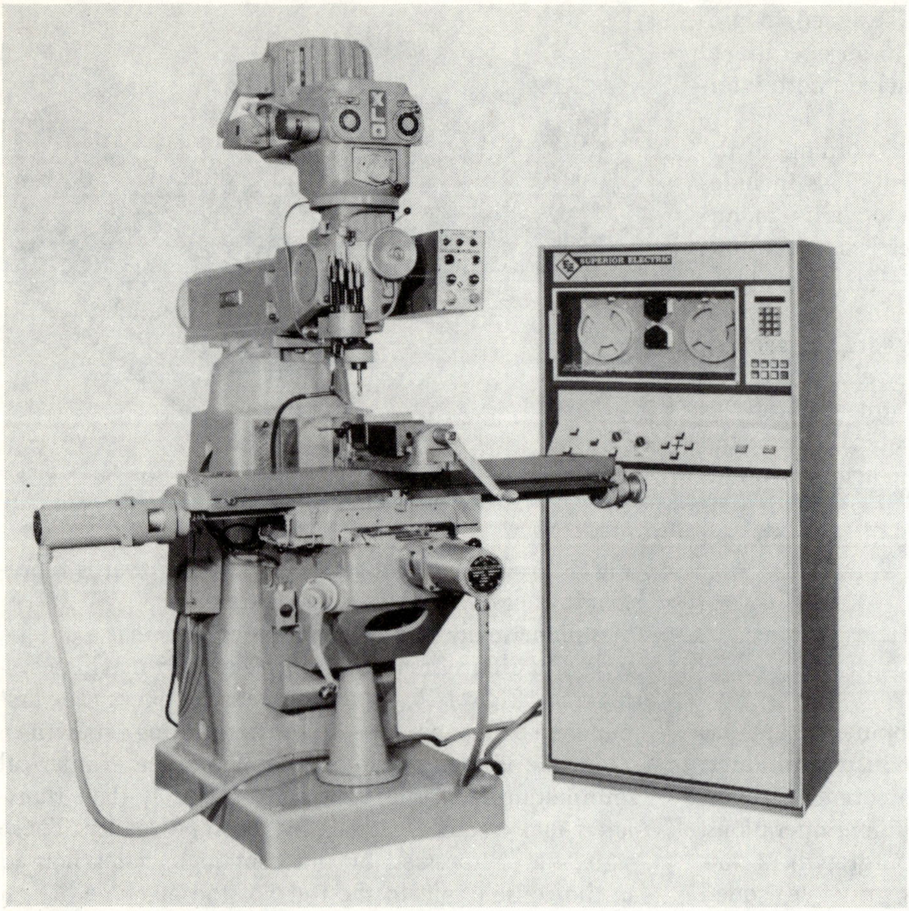

Fig. 1-16. A numerically controlled milling machine.

Courtesy The Superior Electric Co.

important role in the operation of an industrial complex today.

There is probably more concern for industrial environmental systems today than there has ever been in the past. Safety and antipollution regulations have become very strict and they are readily enforced. It has also been proved that favorable working conditions tend to produce better product quality with an increase in production output. Systems that help to achieve this will obviously become even more important in the future.

The number of different environmental systems used in industry today is so large that it requires further classification. In this regard, systems are grouped according to conditioning equipment, pollution control, alarm units, and protection equipment. Conditioning equipment primarily refers to systems that alter the surroundings of a building or machine. This includes lighting, temperature change, and humidity control. Pollution control systems are designed to respond to changes in sound level, air quality, water purity, radiation levels, etc. Alarm systems are used to detect equipment malfunctions, smoke, vapor, gas, broken windows, intruders, etc., and some type of warning signal is developed by these systems. Protection equipment is often thought of as the output or load of an alarm system. Water sprinkling units, door closing, electrical ground faults, and locking devices are included in this classification.

A large number of the environmental systems used in industry today employ electricity as the primary source of energy. This means that the system has a typical energy source, transmission path, control, a load device, and an indicator in its physical makeup. Environmental systems and electrical systems could therefore be placed in the same classification. The uniqueness of the control section and the load device of an environmental system, however, demands that it be placed in a separate classification. In a strict sense, environmental systems are often thought of as smaller parts or subsystems of the larger electrical power system. In this role, the environmental system serves only as an addition to the electrical load of the power system.

Sensors, detectors, comparators, and transducers are some of the unique devices that are used to control an environmental system. These devices achieve both full and partial control of the load device according to its needs. In addition to this, hydraulic, pneumatic, mechanical energy, heat, light, and sound may also be used to activate different system loads. Environmental systems obviously cut across a variety of different system classifications.

SUMMARY

The systems concept is an orderly method used to study the field of industrial electronics. This idea describes a common organizational plan that applies to all industrial systems. Each part of the system plays a similar role in all systems. The energy source, transmission path, control, load, and indicator are basic to all systems. An understanding of the basic system plan helps to overcome some of the complexities of different industrial systems.

The energy source of a system is responsible for converting energy of one type into something more useful for system needs. Heat, light, sound, chemical, and mechanical energy are primary sources of energy.

The transmission path of a system provides a means by which energy can be passed from the source to other system parts. Light beams, electrical conductors, and pipes are typical transmission paths.

System control can be either full or partial. Full control is an on/off operation, while partial control adjusts or varies system values. Each system type usually has a number of unique control features. Temperature, light, mechanical motion, time, sound, electric current, hydraulic fluid flow, and air are controlled in various industrial systems.

The load of a basic system is responsible for changing system energy into some form of work. Work occurs in the load when it causes a change or transformation of energy.

The indicator part of a system is designed to display or show certain operating conditions. Test indicators such as gauges, oscilloscopes, and volt-ohm-milliampere (vom) meters are temporarily attached to the system to locate faulty components. Operational indicators, by comparison, are permanently attached to a system to display critical operating values. Indicators are usually classified as optional system parts.

In an electrical power system the primary energy source is moving water, fossil fuel, or atomic energy. This type of energy is then changed into mechanical energy, which is used to rotate large turbines. Massive alternators are then rotated, which generate ac electricity. Copper and aluminum wires form a path for the transmission of electrical energy. Both full and partial control is achieved by transformers, circuit breakers, resistance, and capacitance. The load of this system is a composite of all of the devices attached to it. An industrial site represents only part of the total system load.

Fluid systems operate by transferring energy through air or a fluid. The primary energy source is typically electricity, which is used to set fluid or air into motion through the action of a pump. Control can be partial or full according to the design of the system. Linear or rotary mechanical energy is primarily produced by the load device.

Optoelectronic systems are primarily electrical systems that respond to or are controlled by light energy. The transmission path of this system is typically a beam of light energy. Control is achieved by breaking the light beam between the source and the detector.

There are three distinct kinds of timing systems presently being used in industry. These include delay, interval, and cycle timing. Control is achieved by dashpots, motors, mechanical, electrical, and electrochemical devices. Timing systems are primarily electrically energized with a number of unique control features.

Digital systems are energized primarily by electrical energy and are designed to be controlled by numerical signals. Paper tape, magnetic tape, or punched cards are commonly used to supply digital information to the system. It is then processed by logic circuits, which achieve the control function that alters the load device. Automated systems rely heavily upon digital signal processing operations.

Industrial environmental systems represent a rather new part of the industrial electronics field today. Systems of this type are divided into pollution control, conditioning equipment, alarm units, detectors, and protection equipment. Electricity serves primarily as the energy source for these systems. The uniqueness of different control functions and load devices places this type of system in a separate classification.

In this chapter, we have discussed industrial systems that are both electromechanically and electronically oriented. The emphasis of our study for the remainder of this book, however, will be on electromechanically oriented systems.

Study Questions

1. *What are the basic parts of a system?*
2. *What are the primary functions of each system part?*
3. *Discuss the system concept of the following:*
   a. *A flashlight.*
   b. *Hydraulic system.*
   c. *Pneumatic system.*
   d. *Timing system.*
   e. *Optoelectronic system.*
   f. *Digital system.*
4. *Why is the systems concept an important idea in the study of industrial electronics?*
5. *Apply the systems concept to something not discussed in this chapter, such as the automobile or a radio communications system.*
6. *What is meant by the term* work *in a system?*
7. *Which system parts are considered nonessential?*

2

Codes, Standards, and Symbols for Industrial Systems

Before industrial electrical systems and applications can be investigated, it is necessary to understand the frameworks that govern industrial systems. All industries must comply with several types of codes and standards. These codes and standards are applicable to the electrical design and operation of industry, as well as to the products manufactured.

INDUSTRIAL SYSTEM SYMBOLIC LANGUAGE

In order to develop basic commonalities between all industries, various symbols are used to standardize electrical control diagrams. These symbols provide a basis for the production of schematic diagrams that may appear very complex. However, all industrial equipment and its associated circuitry is composed of many relatively simple components. It is necessary to know the function of the component parts of an industrial system in order to understand the operation of the complete system.

In Chapter 1, we looked at industrial systems as sources, paths, controls, loads, and indicators. Industrial systems can also be broken down as shown in Fig. 2-1.

An operational *system*, such as a computer, is composed of several *subsystems*. An example of a subsystem function in a computer is signal processing by logic gates. Signal processing is accomplished by several logic gate *circuits*. A gate circuit is designed for a specific function and is part of the basic subsystem. When the circuit is broken down further, it is merely a combination of various *components* such as transistors, resistors, and diodes. Therefore, it is essential to understand the symbolic representation used for industrial schematic diagrams.

THE NATIONAL ELECTRICAL CODE (NEC)

The importance of codes, standards, and symbols for industrial applications is perhaps best shown by beginning with the National Electrical Code (NEC). This code is a national standard that governs all electrical wiring for industrial as well as commercial and residential installations.

The NEC is published by the National Fire Protection Association to promote the standardization of safe electrical wiring. All industrial supervisors, particularly those involved with electrical work, should be familiar with the NEC. It is essential that all electrical work done in industry conform to the standards of the NEC.

Each state, city, county, or district may have its own local electrical code. Most of these codes are based on the NEC with specific changes to meet local requirements for electrical wiring. A copy of the local electrical code should be obtained before attempting any electrical installations in the area. The nearest electrical utility company should be able to supply a copy of the local electrical code. If not, they should be able to explain the electrical inspection procedures used in the area.

NEC Definitions

One important section included in the NEC is a series of definitions pertaining to electrical installations. In many instances of industrial wiring, definitions may be critical factors in the electrical design. For instance, the difference between a system ground and an equipment ground should be kept clearly in mind when electrical wiring is done. Some sample definitions from the NEC are shown below:

Ampacity: Current-carrying capacity of electric conductors expressed in amperes.

Automatic: Self-acting, operating by its own mecha-

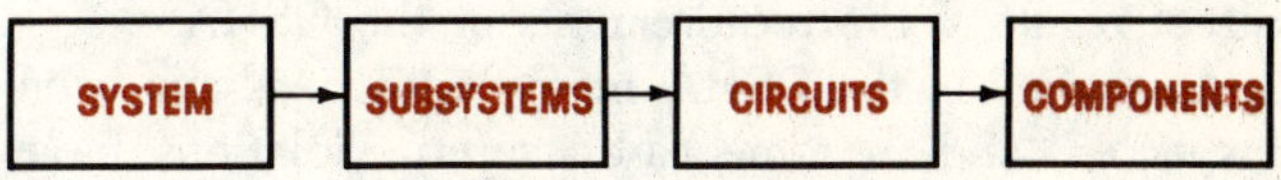

Fig. 2-1. Breakdown of industrial systems.

nism when actuated by some impersonal influence, as for example, a change in current strength, pressure, temperature, or mechanical configuration.

Branch Circuit: The circuit conductors between the final overcurrent device protecting the circuit and the outlet(s).

Branch Circuit, Appliance: A branch circuit supplying energy to one or more outlets to which appliances are to be connected; such circuits to have no permanently connected lighting fixtures not a part of an appliance.

Branch Circuit, General Purpose: A branch circuit that supplies a number of outlets for lighting and appliances.

Branch Circuit, Individual: A branch circuit that supplies only one utilization equipment.

NEC Tables

Other important sections of the NEC are the various tables pertaining to electrical installations. Included are tables for ampacity of conductors, insulation requirements, and motor protective device ratings. A table of ampacity of conductors is shown in Fig. 2-2. This table is typical of those contained in the NEC.

NEC Revisions

The purpose of the NEC is to provide a means for obtaining minimum standards necessary to prevent safety hazards due to the misuse of electrical equipment or poor equipment design. The code is ordinarily updated every three years to keep abreast of current safety requirements. Therefore, individuals in industry responsible for electrical installation or the design of electrical products should keep up to date with changes in the NEC. It should be pointed out that several other codes are used to provide standardization in areas such as plumbing and mechanical systems.

OCCUPATIONAL SAFETY AND HEALTH (OSHA) STANDARDS

The Occupational Safety and Health Act of 1970 (OSHA) has received much attention from industry in recent years. Since these safety and health standards affect both industrial operation and the products manufactured by industries, there is a need for industrial personnel to be familiar with these standards. Several changes in the 1978 edition of the NEC are a direct result of the requirements of the OSHA.

As a result of the OSHA, many industries have been inspected, and in some cases, safety violations have been cited and heavy fines imposed. When OSHA inspections began in 1971, some industries took a lackadaisical attitude; however, the number of industries inspected has increased substantially.

The purpose of the OSHA is to assure every working person in the nation safe and healthful working conditions. The importance of the safety and health of industrial workers cannot be stressed too much. We must rely upon science and technology to provide methods to cope with safety and health problems in industry.

Table 310-18. Allowable Ampacities for Insulated Conductors Rated 110 to 250°C

Not More Than Three Conductors in Raceway or Cable
Based on Ambient Temperature of 30°C (86°F).

Size	Temperature Rating of Conductor. See Table 310-13								Size
	110°C (230°F)	125°C (257°F)	150°C (302°F)	200°C (392°F)	250°C (482°F)	110°C (230°F)	125°C (257°F)	200°C (392°F)	
AWG MCM	TYPES AVA, AVL	TYPES AI, AIA	TYPE Z	TYPES A, AA, FEP, FEPB, PFA	TYPES PFAH, TFE	TYPES AVA, AVL	TYPES AI, AIA	TYPES A, AA	AWG MCM
	COPPER				NICKEL OR NICKEL-COATED COPPER	ALUMINUM OR COPPER-CLAD ALUMINUM			
14	30	30	30	30	40	....	....	....	....
12	35	40	40	40	55	25	30	30	12
10	45	50	50	55	75	35	40	45	10
8	60	65	65	70	95	45	50	55	8
6	80	85	90	95	120	60	65	75	6
4	105	115	115	120	145	80	90	95	4
3	120	130	135	145	170	95	100	115	3
2	135	145	150	165	195	105	115	130	2
1	160	170	180	190	220	125	135	150	1
0	190	200	210	225	250	150	160	180	0
00	215	230	240	250	280	170	180	200	00
000	245	265	275	285	315	195	210	225	000
0000	275	310	325	340	370	215	245	270	0000
250	315	335	....	....	....	250	270	....	250
300	345	380	....	....	....	275	305	....	300
350	390	420	....	....	....	310	335	....	350
400	420	450	....	....	....	335	360	....	400
500	470	500	....	....	....	380	405	....	500
600	525	545	....	....	....	425	440	....	600
700	560	600	....	....	....	455	485	....	700
750	580	620	....	....	....	470	500	....	750
800	600	640	....	....	....	485	520	....	800
1000	680	730	....	....	....	560	600	....	1000
1500	785	....	....	....	....	650	....	....	1500
2000	840	....	....	....	....	705	....	....	2000
CORRECTION FACTORS									
Ambient Temp. °C	For ambient temperatures over 30°C, multiply the ampacities shown above by the appropriate correction factor to determine the maximum allowable load current.								Ambient Temp. °F
31-40	.94	.95	.96	....	....	.94	.95	....	87-104
41-45	.90	.92	.94	....	....	.90	.92	....	105-113
46-50	.87	.89	.91	....	....	.87	.89	....	114-122
51-55	.83	.86	.89	....	....	.83	.86	....	123-131
56-60	.79	.83	.87	.91	.95	.79	.83	.91	132-141
61-70	.71	.76	.82	.87	.91	.71	.76	.87	142-158
71-75	.66	.72	.79	.86	.89	.66	.72	.86	159-167
76-80	.61	.68	.76	.84	.87	.61	.69	.84	168-176
81-90	.50	.61	.71	.80	.83	.50	.61	.80	177-194
91-100	....	.51	.65	.77	.80	....	.51	.77	195-212
101-120	....	....	.50	.69	.72	....	....	.69	213-248
121-140	....	....	.29	.59	.59	....	....	.59	249-284
141-160	....	....	....	....	.54	....	....	....	285-320
161-180	....	....	....	....	.50	....	....	....	321-356
181-200	....	....	....	....	.43	....	....	....	357-392
201-225	....	....	....	....	.30	....	....	....	393-437

Fig. 2-2. Typical table included in the NEC.

Many large industries employ safety engineers who are responsible for safety procedures and practices. These individuals try to detect unsafe practices and hazardous working conditions. They also recommend changes that could remove safety hazards. While some safety engineers work with industrial safety, others are responsible for product design to assure that manufactured items meet required safety standards.

THE UNDERWRITERS' LABORATORIES (UL)

Another organization that has been concerned with safety, in general, for many years is the Underwriters' Laboratories (UL). This self-supporting organization has been in existence since 1894 as a safety testing laboratory for various manufactured products and equipment. The UL now performs testing services for over 13,000 industries. Everyone is familiar with the "UL Approved" label which is affixed to many of the items we buy today. In addition to safety inspection, the UL also performs research that deals with many facets of safety.

PUBLISHED STANDARDS

There are two well-known organizations which publish standards that are of importance to industry. These organizations are the American National Standards Institute (ANSI) and the Institute of Electrical and Electronic Engineers (IEEE). The standards published by these organizations are for the use of manufacturing industries as well as consumers and, in some cases, the general public. Standards are subject to review periodically; therefore, industrial users should keep up to date with revisions.

These standards were developed through input from manufacturers, consumers, government agencies, and scientific, technical, and professional organizations. Often these published standards are used by industry as well as governmental agencies. It should be noted that the *National Electrical Code* discussed previously is an American National Standard.

MEASUREMENT STANDARDS

All industries, in one way or another, are involved in measurement and its associated instrumentation. The types of measurements performed may vary from those taken to check the tolerances of manufactured products, to the exact calibration of process control instruments in the laboratory. Regardless of how routine or how complicated the measurement may be, all industries should be aware of measurement standards for their particular applications.

Metric System

Most nations today use the metric system of measurement. In the United States, the National Bureau of Standards began a study in August, 1968, to determine the feasibility and costs of converting industry and everyday activity to the metric system. Today, this conversion is taking place.

The units of the metric system are decimal measures based on the kilogram and the meter. Although this system is very simple, several countries have been slow to adopt it. The United States has been one of these reluctant countries due to the complexity of a complete changeover of measurement systems.

Measurement Quantities

Most measurement is based on the International System of Units (SI). The basic units of this system are the meter, kilogram, second, and ampere (MKSA). These are the units of length, mass, time, and electrical current. Other systems, such as the meter-kilogram-second (MKS) and centimeter-gram-second (CGS), recognize only three base units. However, these two systems are closely associated with the metric system.

There are several derived units that are used extensively for electrical and other related measurements for industrial applications. The electrical units that we now use are part of the International System

Table 2-1. International System of Derived Units

Measurement Quantity	SI Unit
Area	square meter
Volume	cubic meter
Frequency	hertz
Density	kilogram per cubic meter
Velocity	meter per second
Acceleration	meter per second per second
Force	newton
Pressure	pascal (newton per square meter)
Work (energy), quantity of heat	joule
Power (mechanical, electrical)	watt
Electrical charge	coulomb
Permeability	henry per meter
Permittivity	farad per meter
Voltage, potential difference, electromotive force	volt
Electric flux density, displacement	coulomb per square meter
Electric field strength	volt per meter
Resistance	ohm
Capacitance	farad
Inductance	henry
Magnetic flux	weber
Magnetic flux density (magnetic induction)	tesla
Magnetic field strength (magnetic intensity)	ampere per meter
Magnetomotive force	ampere
Magnetic permeability	henry per meter
Luminous flux	lumen
Luminance	candela per square meter
Illumination	lux

of Units (SI) based on the meter-kilogram-second-ampere (MKSA) system. The International System of Derived Units is shown in Table 2-1.

The coordination necessary to develop a standard system of electrical units is very complex. An International Advisory Committee on Electricity makes recommendations to the International Committee on Weights and Measures. Final authority is held by the General Conference on Weights and Measures, which meets every few years. The laboratory associated with the International System is the International Bureau of Weights and Measures, located near Paris, France. Several laboratories in different countries cooperate in this process of standardizing units of measurement. One such laboratory is the National Bureau of Standards in the United States.

Conversion of SI Units

Sometimes it is necessary to make conversions of SI units so that very large or very small numbers may be avoided. For this reason, decimal multiples and submultiples of the basic units have been developed by using standard prefixes. These standard prefixes are shown in Table 2-2. As an example, we may express 1000 volts as 1 kilovolt or 0.001 ampere as 1 milliampere. A chart is shown in Fig. 2-3 that can be used for converting from one unit to another. To use this conversion chart, follow these simple steps:

1. Find the position of the unit as expressed in its original form.
2. Find the position of the unit to which you are converting.
3. Write the original number as a whole number or in scientific notation.
4. Shift the decimal point the appropriate number of units in the direction of the term to which you are converting, and count the difference in decimal multiples from one unit to the other.

Table 2-2. SI Standard Prefixes

Prefix	Symbol	Factor by Which the Unit is Multiplied
exa	E	1,000,000,000,000,000,000 = 10^{18}
peta	P	1,000,000,000,000,000 = 10^{15}
tera	T	1,000,000,000,000 = 10^{12}
giga	G	1,000,000,000 = 10^{9}
mega	M	1,000,000 = 10^{6}
kilo	k	1,000 = 10^{3}
hecto	h	100 = 10^{2}
deka	da	10 = 10^{1}
deci	d	0.1 = 10^{-1}
centi	c	0.01 = 10^{-2}
milli	m	0.001 = 10^{-3}
micro	μ	0.000001 = 10^{-6}
nano	n	0.000000001 = 10^{-9}
pico	p	0.000000000001 = 10^{-12}
femto	f	0.000000000000001 = 10^{-15}
atto	a	0.000000000000000001 = 10^{-18}

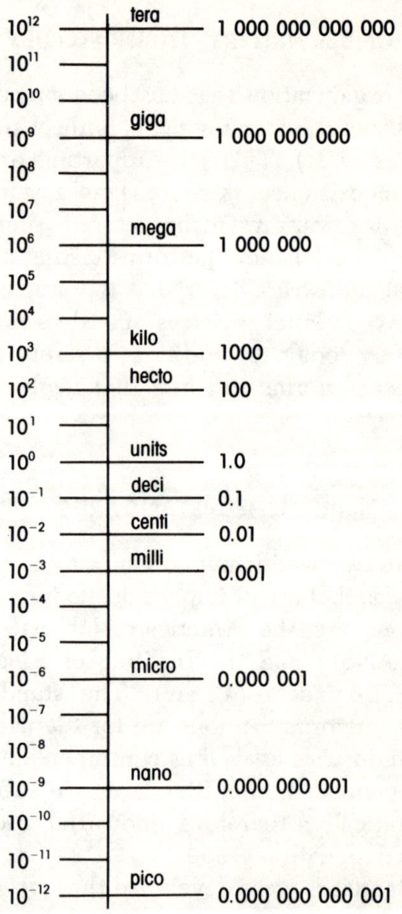

Fig. 2-3. Conversion chart for large and small numbers.

The use of this step-by-step procedure is illustrated in the following examples:

A. Convert 100 picofarads to micofarads:
   1. 100 picofarads (pF).
   2. ______ microfarads (μF).
   3. 100 pF or 100 × 10^{-12}.
   4. 0.0001 μF (decimal shifted 6 units to the left).

B. Convert 20,000 ohms to kilohms:
   1. 20,000 ohms (Ω).
   2. ______ kilohms (kΩ).
   3. 20,000 Ω or 20,000 × 10^{0}.
   4. 20 kΩ (decimal shifted 3 units to the left).

C. Convert 10 milliamperes to microamperes:
   1. 10 milliamperes (mA).
   2. ______ microamperes (μA).
   3. 10 mA or 10 × 10^{-3}.
   4. 10,000 μA (decimal shifted 3 units to the right).

Another measurement conversion method is the conversion of U.S. system units to SI units and vice versa. This type of conversion has become very prevalent in the United States today due to the changeover to the metric system. A partial conversion table is contained in Appendix 1.

COMMON INDUSTRIAL SYMBOLS

The symbols that are used by industry are somewhat varied; however, standard symbols are widely used for schematic diagrams. The symbols used to represent various components and devices in industry vary from one company to another. However, some standardization is needed in order to produce symbolic diagrams that have the same meaning to all individuals involved. Troubleshooting and maintenance is also simplified through the use of easily understood schematic diagrams.

In the discussions of industrial systems and circuit applications that follow, standard symbols will be used. The symbols shown in Appendix 2 are commonly used in industry for electrical, electronic, hydraulic, pneumatic and associated equipment control diagrams.

SUMMARY

The frameworks for much of the activity associated with industrial electrical systems and applications are the various codes, standards, and symbols that are used. Standardization of equipment and manufactured products is possible due to various codes and standards. Symbolic diagrams provide a common language for understanding complex equipment. Industrial technicians can recognize and diagnose problems that might exist in equipment through a knowledge of system, subsystem, circuit, and component functions.

The Occupational Safety and Health Act (OSHA) has added a new dimension of safety awareness to industry. Even though automated production through assembly line procedures is stressed by most industries, individual safety and health must be given top priority. The OSHA provides for the safety of industrial workers through inspections which assure that industries meet minimum safety standards. Another organization involved in safety activities is the Underwriters' Laboratories (UL). This organization tests many manufactured products to assure safe operation.

Several common standards, such as the National Electrical Code (NEC) and other published standards, provide valuable references for industry. The American National Standards Institute (ANSI) and the Institute of Electrical and Electronic Engineers (IEEE) are responsible for publishing national standards that are acceptable for industrial usage. For example, the National Electrical Code serves as a standard for all industrial wiring and for the electrical products manufactured by industry.

Other important standards are those involving measurement. The United States is now in the process of transforming industry from the U.S. system of measurement to the metric system. Most measurement is now based on the International System of Units (SI). Electrical measurements are made by using derived units of the SI system. It is important to be able to convert U.S. units to SI units, due to the changeover to the metric system. Soon, we will be using the metric system in industry almost without exception.

Industrial operations would be much more complex if it were not for the standardization of symbols and diagrams that are used for equipment. A common language is needed so that technicians and engineers can troubleshoot and maintain industrial equipment. Such a language is provided by the symbolic control diagrams used in industry. Industrial technicians should be familiar with the common component symbols that are used in order to understand the function of circuits and systems.

Study Questions

1. *Why are symbols used for industrial control diagrams?*
2. *What are two ways in which industrial systems may be subdivided or simplified?*
3. *Why is the National Electrical Code (NEC) of importance to industry?*
4. *In what ways has the Occupational Safety and Health Act (OSHA) affected industry?*
5. *How do the Underwriters' Laboratories (UL) work with industries?*
6. *What are two organizations that publish industrial standards?*
7. *What are industrial standards?*
8. *What are the basic units of the SI measurement system?*

Problems

1. Convert 3 meters to inches.
2. Convert 1850 grams to kilograms.
3. Convert 880 millimeters to meters.
4. Convert 1/60 second to milliseconds.
5. Convert 0.008 kilometer to meters.
6. Express the following numbers as powers of ten:
   a. 0.00001
   b. 0.00000001
   c. 10,000,000
   d. 1000
   e. 10
   f. 0.01
   g. 10,000
   h. 0.0001
   i. 1.0
7. Express the following numbers in scientific notation:
   a. 0.00128
   b. 1520
   c. 0.000632
   d. 0.0030
   e. 28.2
   f. 7,300,000,000
8. Convert the following units:
   a. 0.12 microfarad to picofarads
   b. 0.2 mile to meters
   c. 1500 centimeters to feet
   d. 6 feet to meters
   e. 620 inches to meters
   f. 0.215 millihenry to microhenrys
   g. 8200 centimeters to kilometers
   h. 0.0000005 farad to microfarads

3

Electrical Energy Sources for Industry

Industries use over 40% of the electrical energy produced in the United States. Sources of electrical energy, therefore, are very important for industrial operation. Without sources of electrical energy, industry could not continue to produce the goods and services upon which we have become so dependent. It is important for those involved with industrial operation to be familiar with electrical energy sources. A knowledge of how electrical energy is produced and the sources of energy available are fundamentals of industrial system operation.

Industry depends on electrical energy for heat, light, machine operation, and electrochemical processes. In addition, many industries produce electrical equipment and appliances and, therefore, depend on electrical energy for their product marketing. A substantial problem for industry is that our capacity to produce electrical energy is not increasing as rapidly as is our usage of it.

OVERVIEW OF ELECTRICAL ENERGY SOURCES

There are several methods that are presently used to produce electrical energy. The most frequently used methods in the United States are fossil-fuel (coal, oil, and gas) systems, hydroelectric systems, and nuclear-fission systems. In addition, various other methods, some of which are in the experimental stages, may be used as future energy sources. Such systems are solar cells, geothermal systems, wind-powered systems, magnetohydrodynamic (MHD) systems, nuclear fusion, and fuel cells.

Fossil-Fuel Sources

Millions of years ago, large deposits of organic materials were formed under the earth's surface. These deposits, which produced our coal, oil, and gas, are known as fossil-fuel sources. The most abundant fossil fuel is coal. Coal-fired electrical energy sources produce nearly one-half of the electrical power used in the United States. Natural gas-fired systems are used for about one-fourth of our electrical power, while oil-fired systems produce over 10%. These relative contributions of each system to the total electrical power produced in the United States are subject to rapid change due to the addition of new power generation facilities and fuel availability. At the present time, over 80% of our electrical energy is produced by fossil-fuel systems.

The basic fossil-fuel power system is illustrated in Fig. 3-1. In this system, a fossil fuel (coal, oil, or gas) is burned. The heat from the combustion process is concentrated around a boiler where circulating water is converted to steam. The high-pressure steam is used to rotate a turbine. The turbine shaft is connected directly to the electrical generator and provides the necessary mechanical energy to rotate the generator. The generator then converts the mechanical energy into electrical energy. A commercial steam turbine/generator system is shown in Fig. 3-2. Fossil-fuel systems are by far the most used electrical power sources.

Hydroelectric Sources

The energy from stored water can be used to generate electrical energy. This method of energy conversion is used in hydroelectric power systems. A basic hydroelectric power system is illustrated in Fig. 3-3. Water, which is confined in a large reservoir, is channeled through a control gate that adjusts the flow rate. The flowing water passes through the blades and control vanes of a reaction-type turbine, which produces rotation. This mechanical energy is used to rotate a generator that is connected directly to the turbine shaft. Rotation of the alternator then causes electrical energy to be produced.

Hydroelectric systems are limited by the availability of large water supplies. Many hydroelectric systems are part of multipurpose facilities. For instance, a hydro-

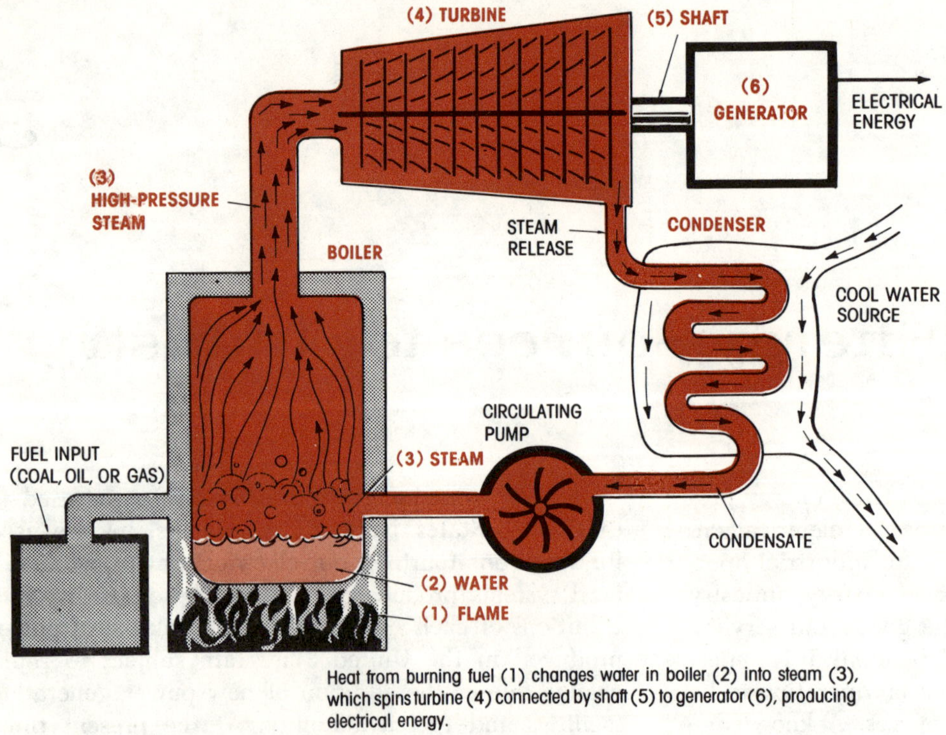

Fig. 3-1. Simplified drawing illustrating the basic fossil-fuel power system.

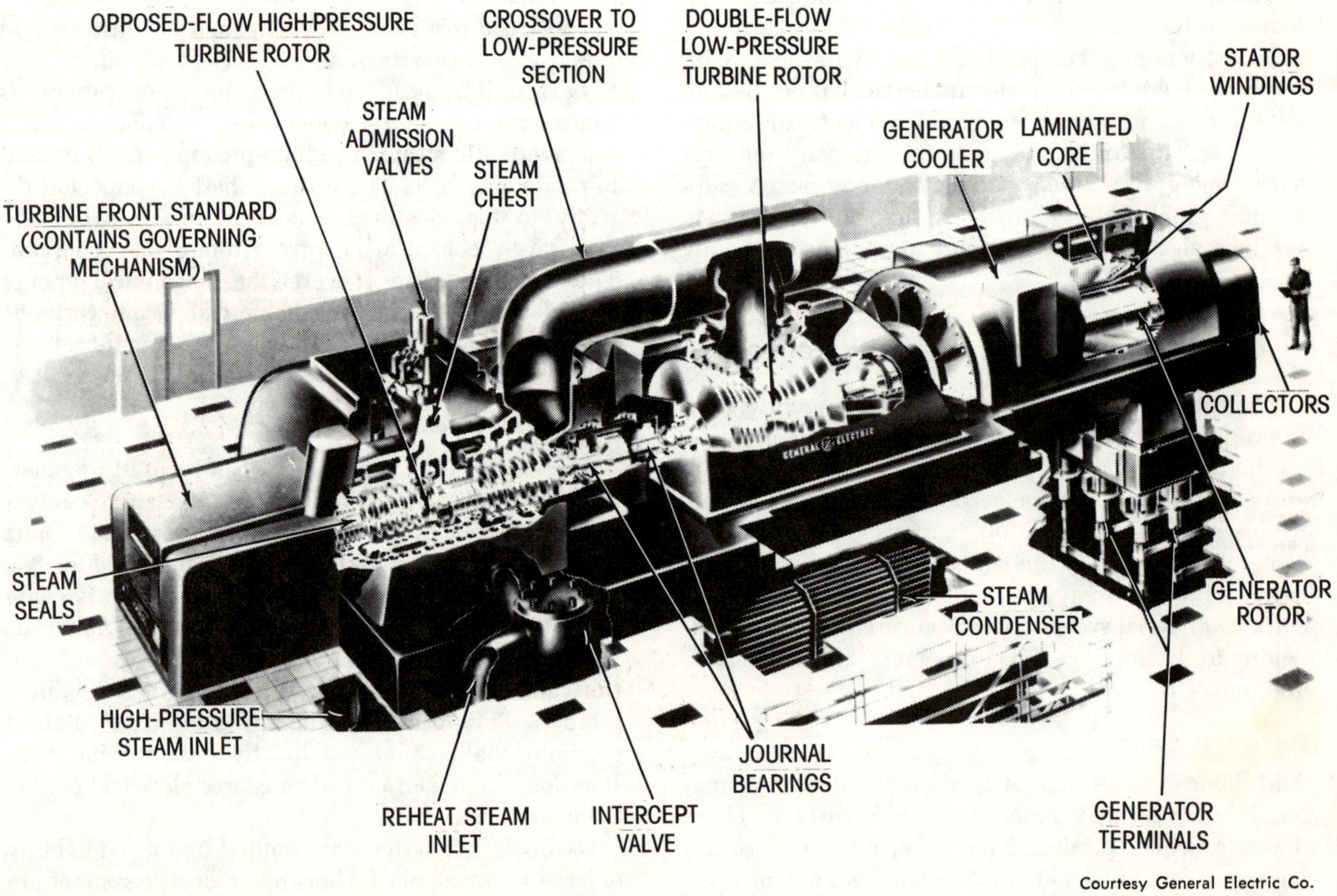

Courtesy General Electric Co.

Fig. 3-2. A commercial steam turbine/generator system.

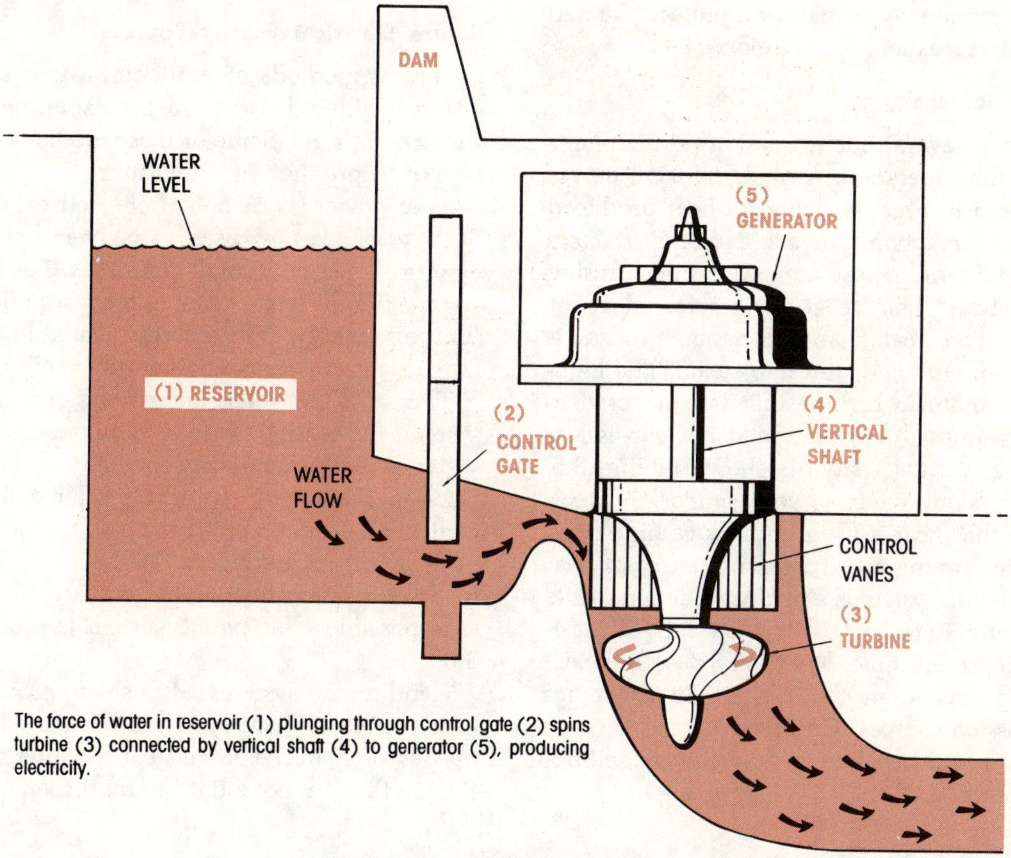

Fig. 3-3. Drawing of a basic hydroelectric power system.

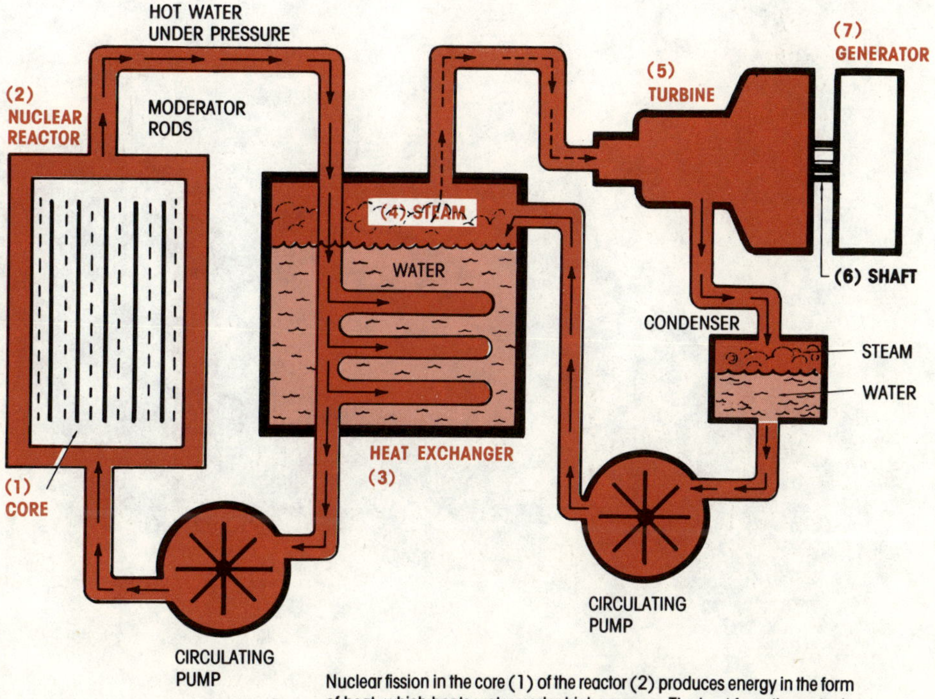

Fig. 3-4. Drawing illustrating the principles of a nuclear-fission power system.

electric power system may be part of a project planned for flood control, recreation, or irrigation.

Nuclear-Fission Sources

In recent years, several nuclear-fission power plants have been put into operation. A nuclear-fission power system, illustrated in Fig. 3-4, relies on heat produced during a nuclear reaction process. Nuclear reactors "burn" nuclear material whose atoms are split, causing the release of heat. This reaction is referred to as *nuclear fission*. The heat from the fission process is used to change circulating water into steam. The high-pressure steam rotates a turbine that is connected to an electrical generator. A commercial nuclear-fission, steam turbine/generator system is shown in Fig. 3-5.

The nuclear-fission system is very similar to fossil-fuel systems in that heat is used to produce high-pressure steam which rotates a turbine. The source of heat in the nuclear-fission system is a nuclear reaction, while in the fossil-fuel system heat is developed by a burning fuel. At the present time, less than 5% of the electrical energy produced in the United States comes from nuclear-fission sources. However, this percentage is also subject to rapid change as new power facilities are put into operation.

Future Electrical Energy Sources

Several methods of producing electrical energy are either in limited use or in the experimental stages at the present time. Some of these methods show promise for power production in the future.

Solar power is one potential electrical energy source. Solar cells are now used to convert light energy into electrical energy in small quantities. The largest energy source of all is the sun, which supplies practically limitless energy. The energy available from the sun far exceeds any foreseeable future need.

The possible solar energy systems include home systems for heating or power production, orbiting space systems, and steam-driven electrical power systems. Each of these systems utilizes solar collectors to concentrate the light from the sun so that a large quantity of heat will be produced. The heat could then be used to drive a steam turbine to generate electrical energy. One possible solar power system is illustrated in Fig. 3-6.

Geothermal systems also show promise as future energy sources. This system utilizes the heat of molten masses of material in the earth's interior. Heat from the earth is a potential source of energy for power

Courtesy General Electric Co.

Fig. 3-5. A commercial nuclear-fission, steam turbine/generator system.

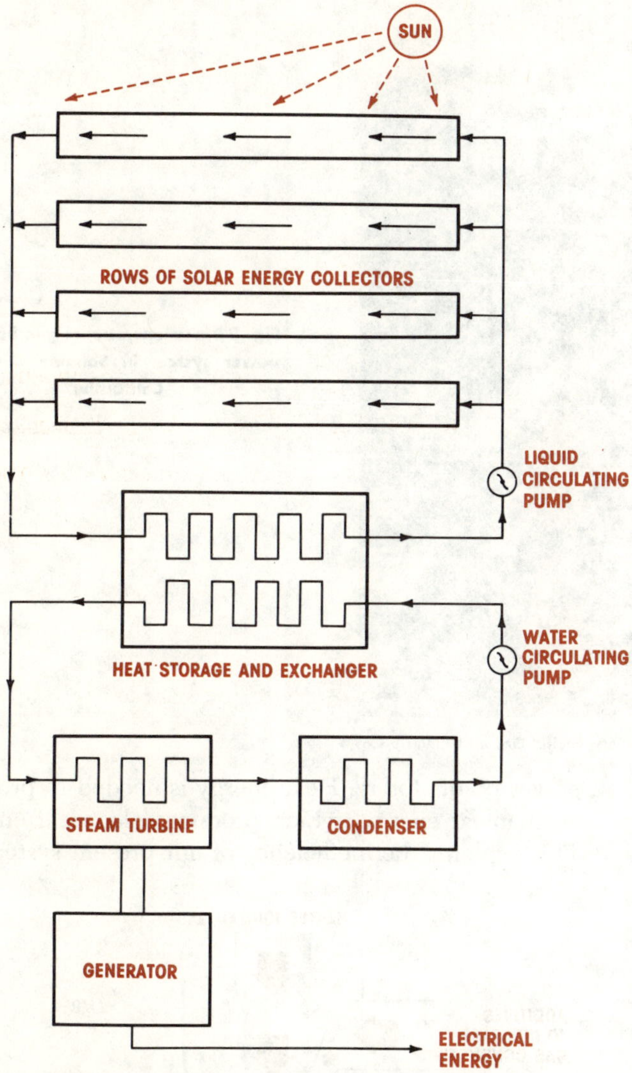

Fig. 3-6. Block diagram of a solar power system.

generation in many parts of the world. The principle of geothermal systems is similar to other steam turbine/generator systems. However, the source of steam is the heat from within the earth obtained through wells drilled up to two miles into the earth. Geothermal sources are used to produce electrical energy in certain regions of the Western United States. A basic geothermal power system is illustrated in Fig. 3-7 and an operational system is shown in Fig. 3-8.

Wind systems have also been considered for producing electrical energy. However, winds are variable in most parts of our country. This fact causes wind systems to be confined to use with storage systems, such as batteries, for use at a later time. Wind machines can be used to rotate small generators that could be located at a home. Large amounts of power would be difficult to produce by this method.

Another energy source that has some potential for future use is magnetohydrodynamics (MHD), illustrated in Fig. 3-9. The operation of an MHD system relies on the flow of a conductive gas through a magnetic field, causing a direct-current voltage to be generated. The electrical power developed depends on the strength of the magnetic field that surrounds the conductive gas, and the speed and conductivity of the gas. At the present time, only small quantities of electrical energy have been generated with the MHD principle; however, it does have some potential as a future source of electrical energy.

Still another possible future energy source is nuclear fusion. This process has not been fully developed due to the extremely high temperatures produced as the fusion of atoms takes place. A fusion reactor could use tritium or deuterium (heavy hydrogen) as fuels.

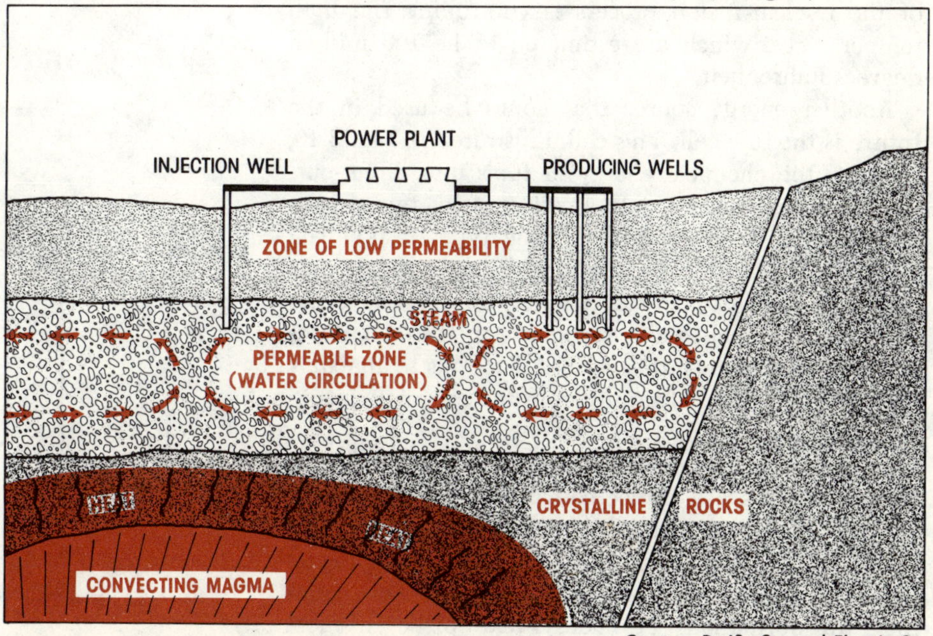

Courtesy Pacific Gas and Electric Co.

Fig. 3-7. Drawing illustrating a basic geothermal power system.

Courtesy Pacific Gas and Electric Co.

Fig. 3-8. An operational geothermal power system in Sonoma County, California.

These fuels can be found in sea water in large quantities, thus reducing the scarcity of nuclear fuel. It is estimated that there is enough deuterium in the oceans to supply all the energy the world would ever need.

If nuclear-fusion reactors could be used in the production of electrical energy, the process would be similar to the nuclear-fission plants that are now in operation. The only difference would be the nuclear reaction that takes place to change circulating water into steam to drive the turbines. The major problem of the nuclear-fusion process is controlling the high temperatures, which are estimated to be 100 million degrees Fahrenheit.

Another energy source that could be used in the future is the fuel cell. This cell, illustrated in Fig. 3-10, converts the chemical energy of fuels into direct-current electrical energy. A fuel cell contains two porous electrodes and an electrolyte. One type of fuel cell operates as hydrogen gas passes through one porous electrode, and oxygen gas passes through the other electrode. The chemical reactions of the electrodes with the electrolyte either release electrons to an external circuit or draw electrons from the external circuit, thus producing an electrical current.

It should be pointed out that many of the future energy sources are direct conversion processes. For example, the fuel cell converts chemical energy directly to electrical energy, and the solar cell converts light energy directly to electrical energy. More complex transformations of energy take place in most power plants today. Heat energy is needed to produce mechanical energy, which produces electrical energy. This explains the inefficiency of our present systems of

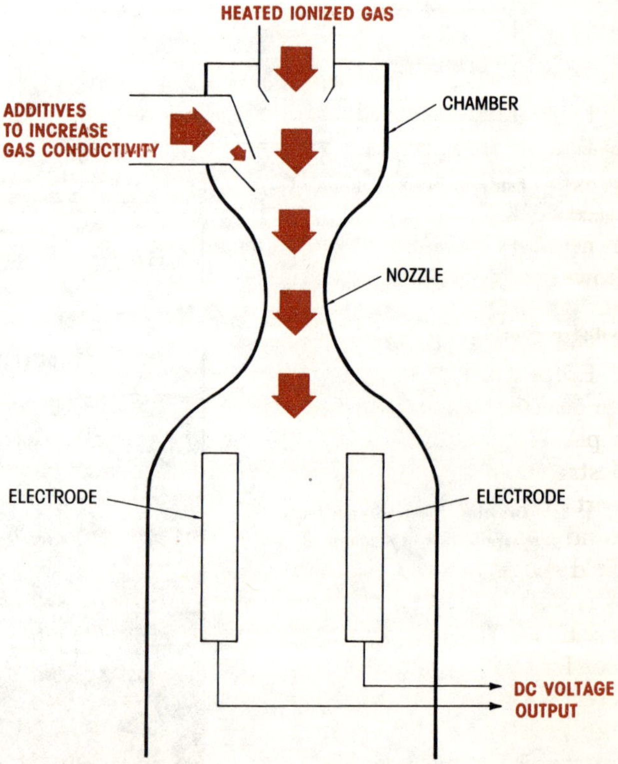

Fig. 3-9. Simplified drawing of a magnetohydrodynamic (MHD) generator system.

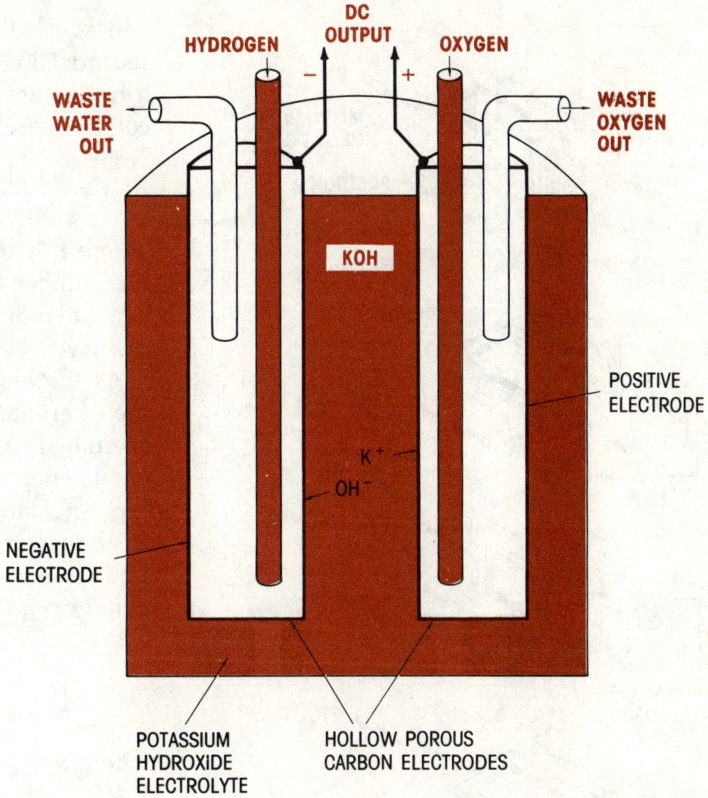

Fig. 3-10. Simplified drawing of an oxygen-hydrogen fuel cell.

producing electrical energy. Perhaps advances in our electrical power technology will bring about new and more efficient methods of producing electrical energy.

ALTERNATING-CURRENT SOURCES

Most of the electrical energy produced in the United States is alternating current. Massive generators at power plants throughout our country provide the necessary power to sustain industrial operations. Most generators produce three-phase alternating current; however, single-phase generators are also used.

Single-Phase AC Generators

Electrical energy can be produced by single-phase ac generators, commonly called *alternators.* The principle of operation of a single-phase alternator is illustrated in Fig. 3-11. In order for a generator to convert mechanical energy into electrical energy, three conditions must exist: (1) there must be a magnetic field developed; (2) there must be a group of conductors within the magnetic field; and (3) there must be relative motion between the magnetic field and the conductors. Two methods can be used to meet these conditions for electrical energy production. One is the *rotating armature* method and the other is the *rotating field* method. These methods are illustrated in Fig. 3-12.

In the rotating armature method, shown in Fig. 3-12A, ac voltage is induced into conductors of the rotating portion of the machine. The magnetic field is developed by a set of stationary pole pieces. Relative motion between the conductors and the magnetic field is provided by a prime mover or mechanical energy source connected to the shaft, which is part of the rotor assembly. Prime movers may be gasoline engines, diesel engines, steam turbines, or possibly electric motors. Generators convert mechanical energy into electrical energy.

The electrical energy produced by the rotating armature method is in the form of a sinusoidal waveform. This waveform is referred to as a *sine wave* due to its trigonometric origin. The voltage induced into the conductors of the armature varies as the sine of the angle of rotation between the conductors and the magnetic field (see Fig. 3-13). The instantaneous voltage induced into an armature conductor can be expressed as:

$$V_i = V_{max} \times \sin \theta$$

where,

V_i is the instantaneous induced voltage,
V_{max} is the maximum induced voltage,
θ is the angle of conductor rotation from the zero reference.

For example, at the 30° position, if the maximum voltage is 100 volts, then

$$V_i = 100 \times \sin \theta.$$

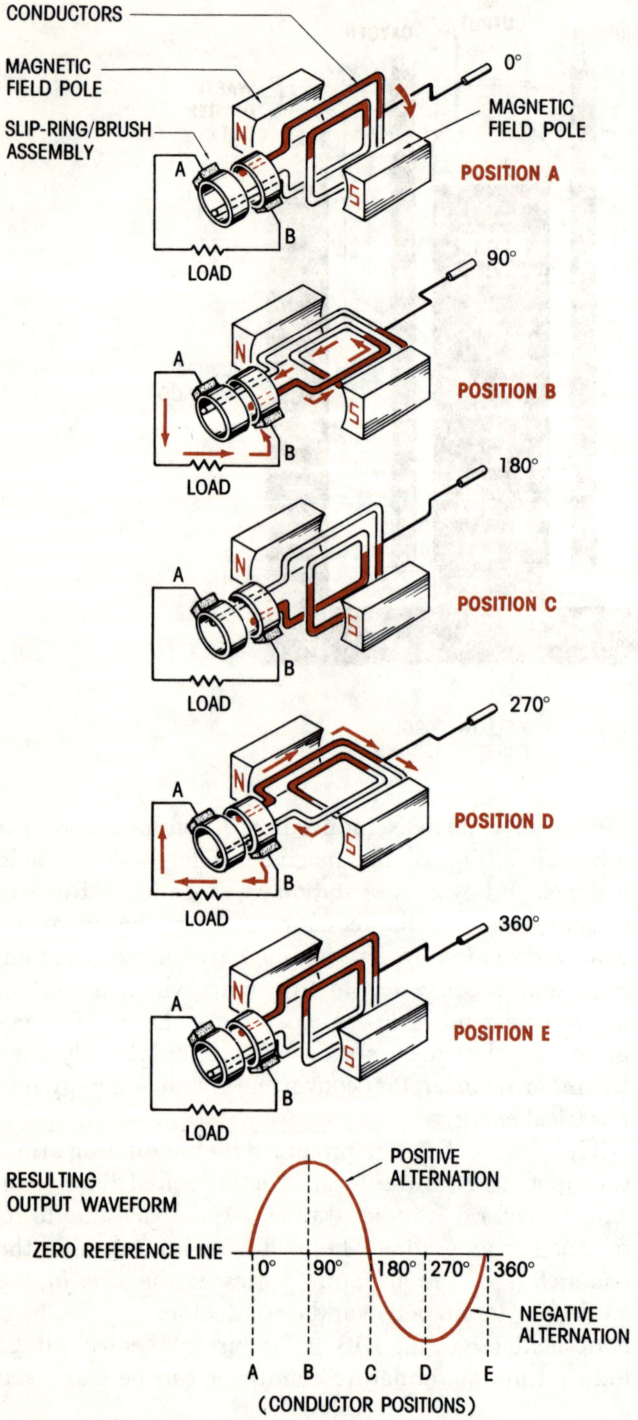

Fig. 3-11. Illustration of the basic principle of operation of a single-phase alternator.

Refer to Appendix 3 to find that the sine of 30° = 0.5, then

$$V_i = 100 \times 0.5$$
$$= 50 \text{ volts.}$$

The frequency of the sinusoidal waveforms produced by an alternator is usually 60 Hz. Since a cycle of ac is generated as the armature passes a set of north and south field poles, a speed of 60 revolutions per second (3600 r/min) must be maintained to produce a frequency of 60 Hz. The frequency of an alternator can be expressed as:

$$f = \frac{\text{number of poles} \times \text{speed of rotation (r/min)}}{120}$$

where f is the frequency in hertz (Hz). Note that if the number of poles is increased, the speed of rotation can be reduced while still maintaining a 60-Hz frequency.

As the electrical load of an alternator is increased, the alternator tends to slow down and the generated (terminal) voltage tends to decrease. The amount of change depends on the design of the alternator and the type of load connected to its terminals. The amount of change in generated voltage from a no-load condition to a rated full-load condition is referred to as *voltage regulation.* Voltage regulation can be expressed as:

$$VR = \frac{V_{NL} - V_{FL}}{V_{FL}} \times 100$$

where,

VR is the voltage regulation in percent,
V_{NL} is the no-load terminal voltage,
V_{FL} is the rated full-load terminal voltage.

Alternator efficiency is the ratio of the power output in watts and the power input in horsepower. The efficiency of an alternator can be expressed as:

$$\text{Efficiency } (\%) = \frac{P_{out}}{P_{in}} \times 100$$

where,

P_{in} is the power input in horsepower,
P_{out} is the power output in watts.

To convert horsepower to watts, remember that 1 horsepower = 746 watts. Alternator efficiencies generally range from 70% to 85%.

The basic operation of single-phase alternators was explained by referring to the rotating armature method. However, this type of alternator can only be used to produce relatively small amounts of power. The major disadvantage is that the ac voltage is extracted from the slip-ring/brush assembly (see Fig. 3-12A). A high voltage could produce tremendous sparking or arcover between the brushes and slip rings. The maintenance involved to replace the brushes and repair the slip-ring assembly would be very time consuming and expensive. Therefore this method is used only for alternators with low power ratings.

The rotating field method, illustrated in Fig. 3-12B, is used for alternators capable of producing larger amounts of power. The dc excitation voltage, which

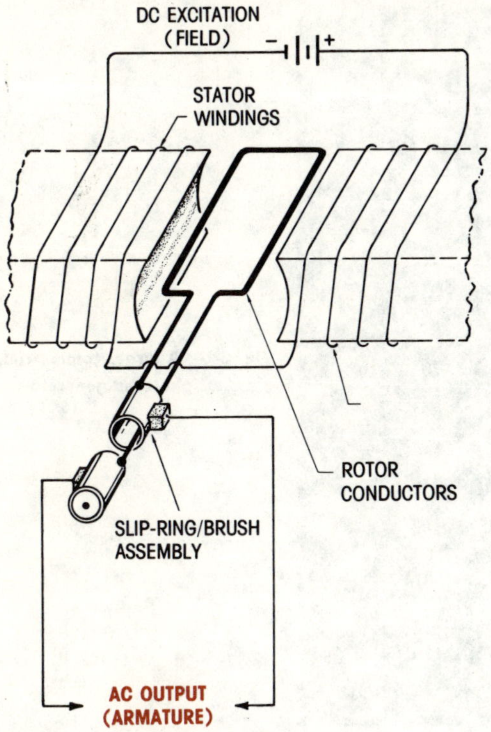

(A) Rotating armature method.

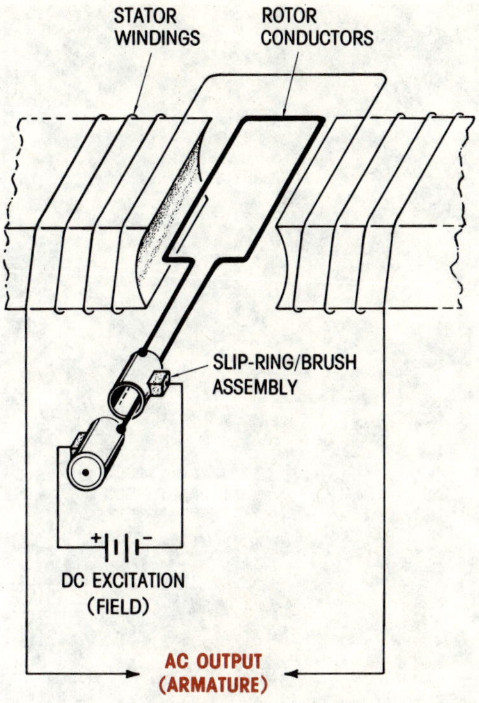

(B) Rotating field method.

Fig. 3-12. Illustration of the two basic methods of generating single-phase alternating current.

develops the magnetic field, is applied to the rotating portion of the machine. The ac voltage is induced into the stationary conductors of the machine. Since the dc excitation voltage is a much lower value than the ac voltage produced, the maintenance problems associated with the slip-ring/brush assembly are minimized. The conductors of the stationary portion of the machine can be made larger to handle more current since they do not rotate.

TRIGONOMETRIC SINE VALUES

Degrees	Sine	Degrees	Sine
0/360	0.000	180	0.000
30	0.500	210	0.500
60	0.866	240	0.866
90	1.000	270	1.000
120	0.866	300	0.866
150	0.500	330	0.500

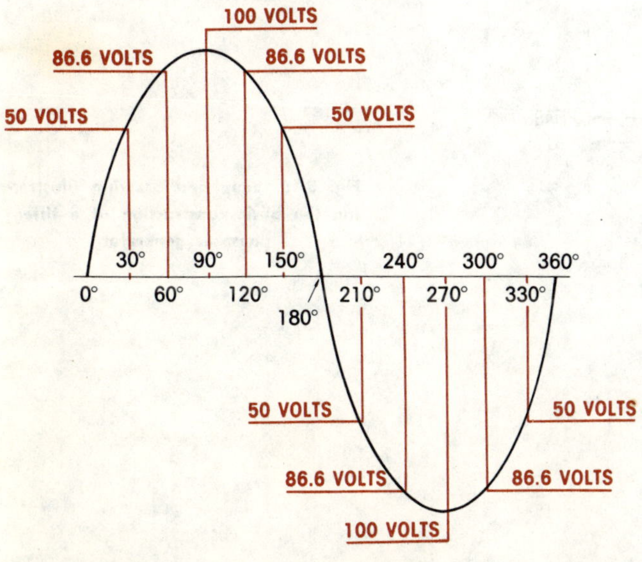

Fig. 3-13. Illustration of the trigonometric origin of the ac sine wave.

Three-Phase AC Generators

A large, commercial, three-phase ac generator is shown in Fig. 3-14. Most large commercial generators in the United States produce a three-phase voltage. Due to their large power ratings, three-phase generators utilize the rotating field method. A typical three-phase generator in a power plant might have 250 volts dc excitation applied to the rotating field through a slip-ring/brush assembly, while 13.8 kilovolts ac is induced into the stationary conductors.

The basic construction of a three-phase generator is illustrated in Fig. 3-15, and its resulting output waveform is illustrated in Fig. 3-16. Note that three-phase generators must have at least six stationary poles, or two poles per phase. In Fig. 3-15, poles A′, B′, and C′ represent the beginnings of each of the phase windings. Poles A, B, and C represent the ends of each of the windings. There are two methods that can be used to connect these windings together. These methods, called *wye* (or *star*) and *delta* connections, are illustrated in Fig. 3-17.

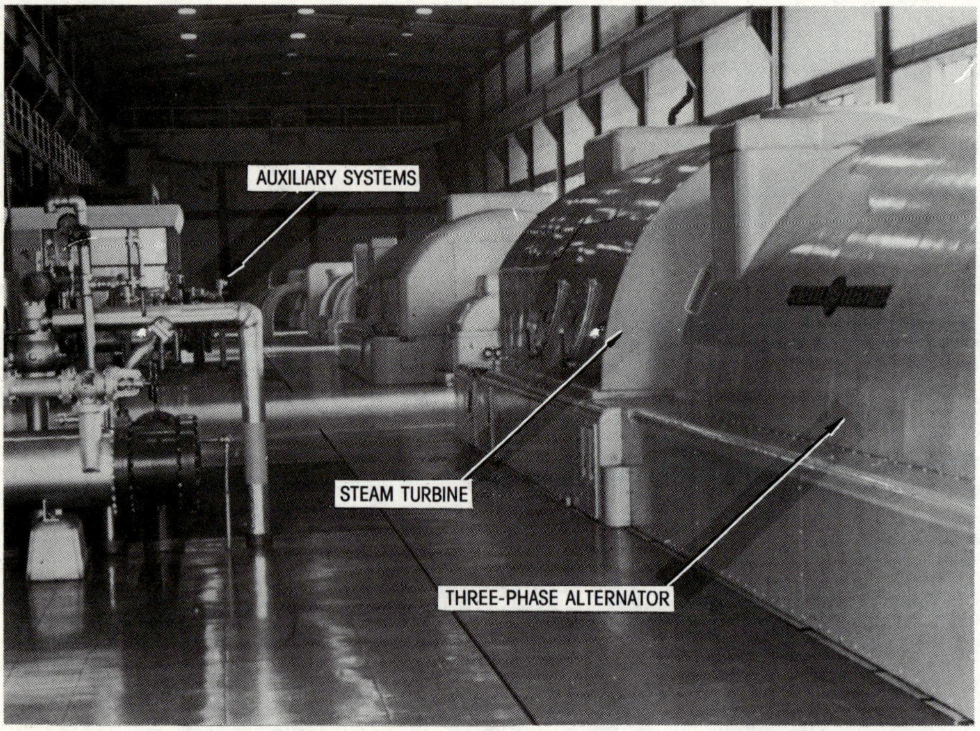

Courtesy General Electric Co.

Fig. 3-14. A large, commercial, three-phase ac generator.

In the wye connection of Fig. 3-17A, the beginnings or the ends of each winding are connected together. The other sides of the windings become the ac lines from the generator. The voltage across the ac lines (V_L) is equal to the square root of 3 (1.73) multiplied by the voltage across the phase windings (V_P), or

$$V_L = V_P \times 1.73.$$

The line currents (I_L) are equal to the phase currents (I_P), or

$$I_L = I_P.$$

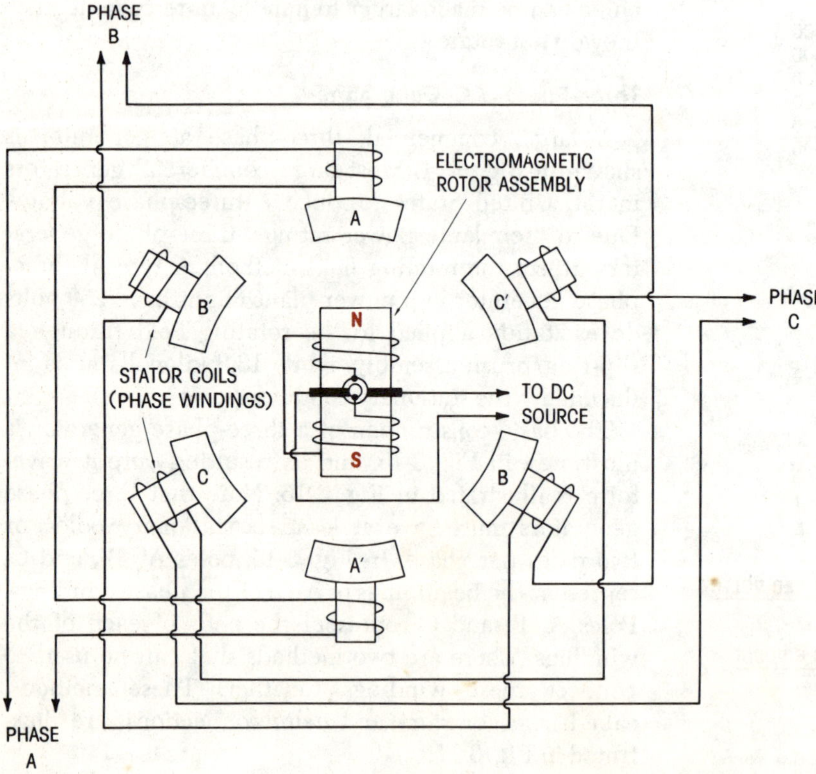

Fig. 3-15. Simplified drawing illustrating the basic construction of a three-phase ac generator.

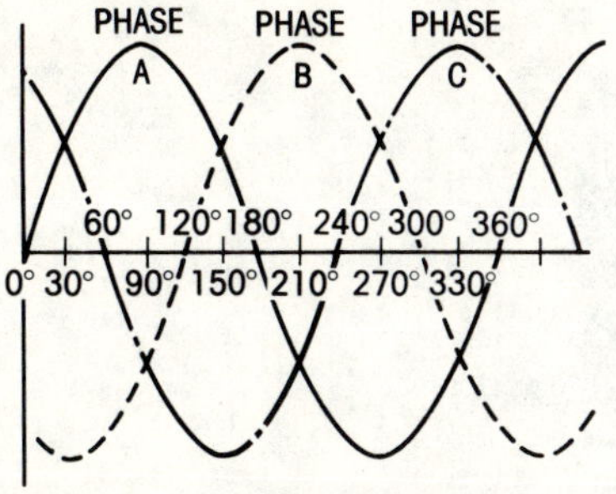

Fig. 3-16. Output waveform of a three-phase ac generator.

In the delta connection of Fig. 3-17B, the end of one phase winding is connected to the beginning of the adjacent phase winding. The line voltages (V_L) are equal to the phase voltages (V_P). The line currents (I_L) are equal to the phase currents (I_P) multiplied by 1.73.

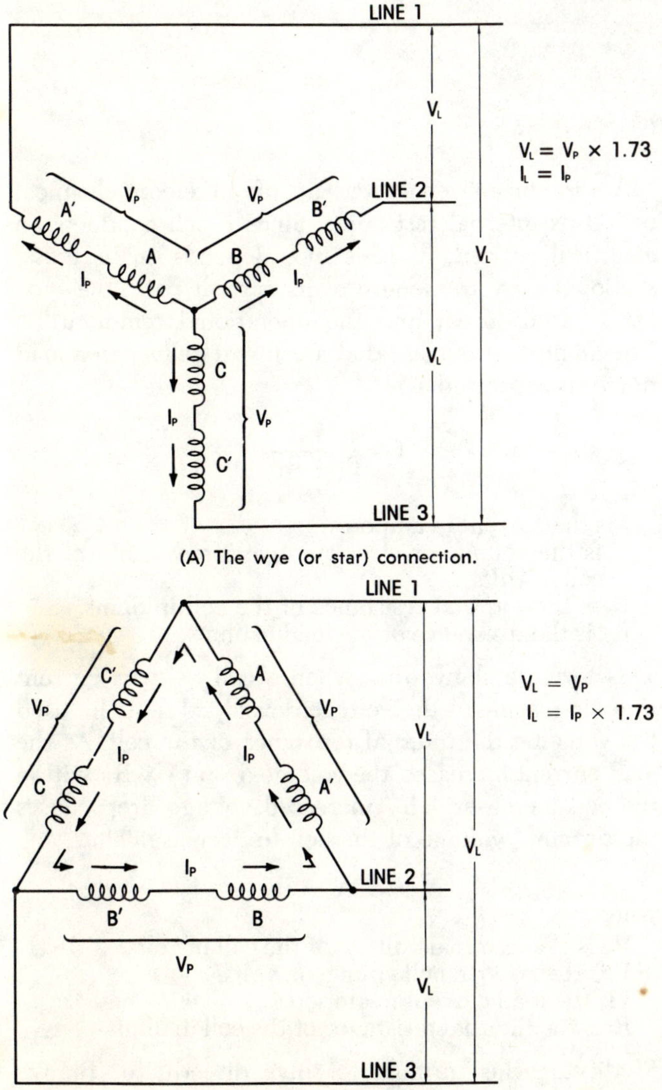

(A) The wye (or star) connection.

(B) The delta connection.

Fig. 3-17. The two methods of connecting three-phase stator coils together.

The power developed in each phase (P_P) for either a wye or a delta circuit is expressed as:

$$P_P = V_P \times I_P \times pf,$$

where pf is the power factor (phase angle between voltage and current) of the load. The total power developed by all three phases of a three-phase generator (P_T) is expressed as:

$$\begin{aligned} P_T &= 3 \times P_P \\ &= 3 \times V_P \times I_P \times pf \\ &= 1.73 \times V_L \times I_L \times pf. \end{aligned}$$

Commercial power systems use many three-phase alternators connected in parallel to supply regional load requirements. Industrial loads represent a large portion of the total load on our power systems. Due to the vast loads to be met by power systems, generators have high power ratings. Nameplate data for a commercial three-phase alternator is shown in Fig. 3-18. Note the large quantities for each of the nameplate ratings.

DIRECT-CURRENT SOURCES

Industries use direct-current sources for many of their specialized processes. Electroplating and dc variable-speed motor drives are two examples that show the need for direct-current energy to sustain industrial operations. Although most of the electrical power produced in the United States is three-phase alternating current, several methods are available to convert alternating current to direct current for industrial use. Direct current is also made available through primary and secondary chemical cells, which are used extensively in industry. In addition, direct-current generators are used by several industries to supply power for specialized applications.

Chemical Cells

The conversion of chemical energy into electrical energy is accomplished by electrochemical cells. When two or more cells are connected in series or parallel (or a combination of both), they form a battery. A cell is composed of two dissimilar metals immersed in a conductive liquid or paste called an *electrolyte*. Chemical cells can be either the primary or secondary type. Primary cells are ordinarily not usable after a certain time period when they can no longer produce electrical energy. Secondary cells can be renewed after use by reactivating the chemical process used to produce electrical energy. This reactivation is known as *charging*. Both primary and secondary cells have many applications.

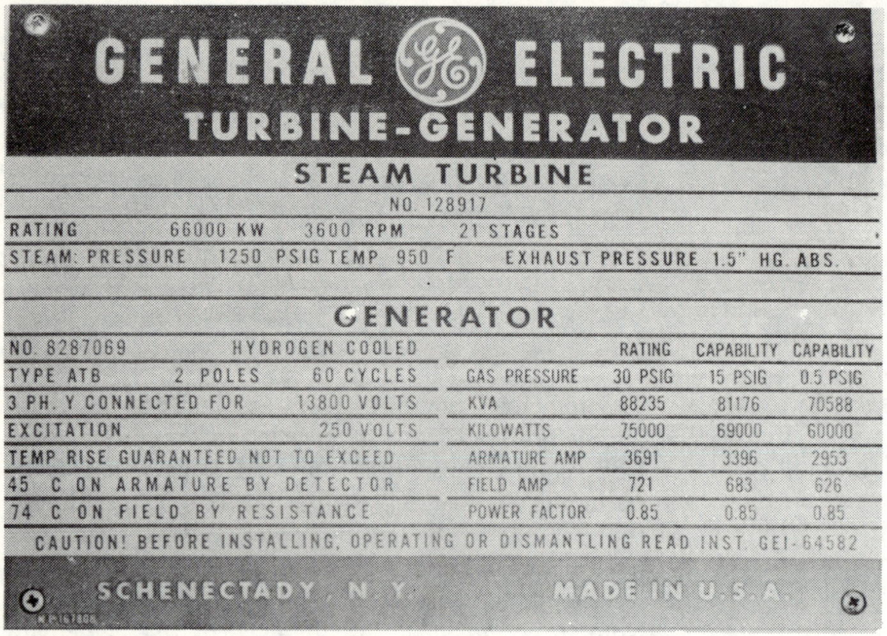

Courtesy General Electric Co.

Fig. 3-18. Nameplate data for a commercial three-phase alternator.

Primary Cells

The operational principle of a primary cell involves the placing of two unlike conductive materials called *electrodes* into a conductive solution called an *electrolyte*. When the chemicals that compose the cell are brought together, their molecular structures are altered. During this alteration, their atoms may either retain additional electrons or leave behind some of their electrons. These atoms then have either a positive or a negative electrical charge and are referred to as *ions*. The ionization process thus develops a chemical solution capable of conducting an electrical current.

When an external load device, such as a lamp, is connected to the cell, electrons will flow from one electrode to the other through the electrolyte material creating an electrical current. Current leaves the cell through its negative electrode, passes through the load device, then reenters the cell through its positive electrode. Thus, a complete circuit is established between the cell (source) and the lamp (load).

The voltage developed by a primary cell depends on the electrode materials used and the type of electrolyte. The familiar carbon-zinc (Leclanche) cell, two types of which are shown in Fig. 3-19, produces approximately 1.5 volts. The negative electrode of this cell is the zinc container and the positive electrode is a carbon rod. A sal-ammoniac paste, which acts as the electrolyte, is placed between the electrodes. This type of cell is usually called a *dry cell.*

An important characteristic of an electrochemical cell is its internal resistance. Since a cell conducts an electrical current, its resistance depends on its cross-sectional area, the length of its current path, the type of materials used, and the operational temperature. The amount of current that a cell will deliver to a load device is expressed as:

$$I = \frac{V}{R_i + R_L}$$

where,

I is the current in amperes,
V is the voltage or electromotive force (emf) of the cell in volts,
R_i is the internal resistance of the cell in ohms,
R_L is the resistance of the load in ohms.

This formula shows that when a cell delivers current to a load, some of the voltage developed must be used to overcome the internal resistance of the cell. As the load current increases, the voltage drop ($I \times R$) within the cell increases. This increased voltage drop causes the terminal voltage of the cell to decrease. Thus,

$$V_T = V - IR_i$$

where,

V_T is the terminal voltage of the cell in volts,
V is the no-load cell voltage in volts,
I is the load current in amperes,
R_i is the internal resistance of the cell in ohms.

Modifying this formula to solve directly for R_i, we obtain:

$$R_i = \frac{V - V_T}{I}.$$

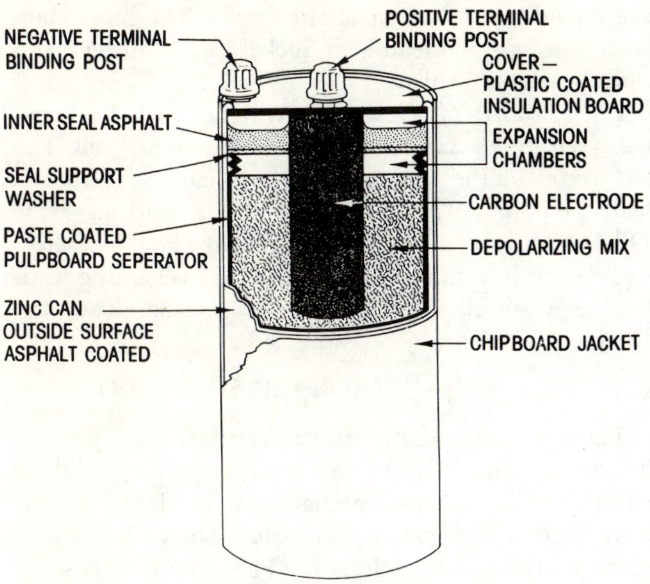

(A) No. 6 dry cell.

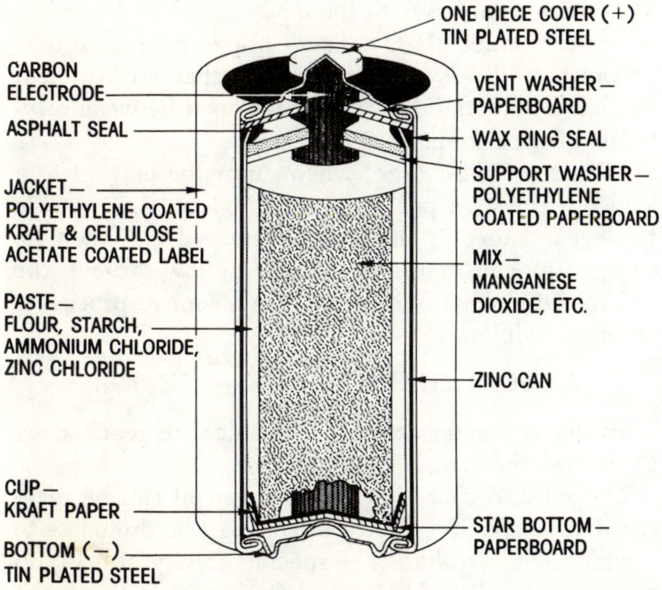

(B) Standard round cell.

Courtesy Union Carbide Corp.

Fig. 3-19. Cutaway drawings of two types of carbon-zinc (Leclanche) primary cells.

Applications of Primary Cells

There are many types of primary cells available today. The carbon-zinc (Leclanche) cell discussed previously is the most often used type. This cell has a low initial cost and is available in a variety of sizes for application mainly with portable equipment or instruments. For uses that require higher voltage or current than one cell can deliver, manufacturers combine several cells in series, parallel, or series-parallel arrangements to form a battery suitable for specialized applications. Carbon-zinc batteries can be obtained in voltage ratings of 1.5 volts up to 3000 volts.

Another type of primary cell is the mercury (zinc-mercuric-oxide) cell shown in Fig. 3-20. This cell was developed to improve upon the carbon-zinc cell. It has a more constant voltage output, a longer active service life, and a smaller physical size. The mercury cell is more expensive and produces a voltage of 1.35 volts, which is lower than the carbon-zinc cell.

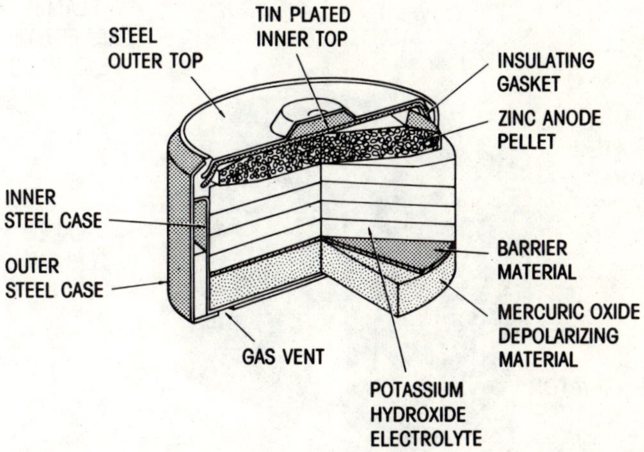

Fig. 3-20. Cutaway drawing of a zinc–mercuric-oxide (mercury) primary cell.

An alkaline (zinc–manganese-dioxide) cell is shown in Fig. 3-21. This cell has a very low internal resistance and a voltage per cell of 1.5 volts. Due to its low internal resistance, the alkaline cell will supply heavy electrical loads. Alkaline cells are capable of a much longer service life than an equivalent carbon-zinc cell.

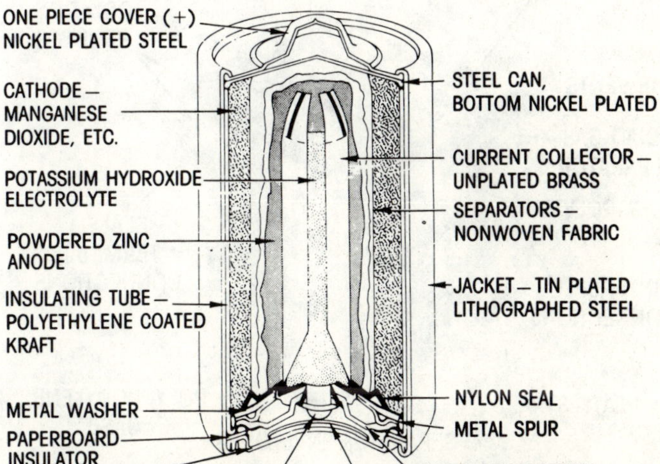

Courtesy Union Carbide Corp.

Fig. 3-21. Cutaway drawing of a zinc–manganese-dioxide (alkaline) primary cell.

Regardless of the type of primary cell, its usable time is limited. When its chemicals are expended, it becomes useless.

Secondary Cells

Direct-current sources of electrical energy whose chemicals may be reactivated by a charging process are called *secondary* or *storage* cells. The most common types are the lead-acid, nickel-iron (Edison), and nickel-cadmium cells.

The Lead-Acid Cell—The widely used lead-acid cell (see Fig. 3-22) is one type of secondary cell. The electrodes of the lead-acid cell are made of lead and lead peroxide. The positive plate is lead peroxide (PbO_2) and the negative plate is lead (Pb). The electrolyte is dilute sulfuric acid (H_2SO_4). When the lead-acid cell supplies current to an external load, the chemical process may be expressed as:

$$Pb + PbO_2 + 2H_2SO_4 \rightarrow 2PbSO_4 + 2H_2O.$$

The sulphuric acid ionizes to produce four positive hydrogen ions (H^+) and two negative sulphate (SO_4^{-}) ions. A negative charge is developed on the lead plate when an SO_4^{-} ion combines with the lead plate to form lead sulphate ($PbSO_4$). The positive hydrogen ions combine with electrons of the lead-peroxide plate and become neutral hydrogen atoms. They next combine with the oxygen (O) of this plate to become water (H_2O). The lead-peroxide plate thus becomes positively charged. A fully charged lead-acid cell has a potential difference developed between electrodes of approximately 2.1 volts.

After a cell discharges when supplying current to a load for a certain period of time, it is no longer able to develop an output voltage. The cell can then be charged by causing direct current to flow through the cell in the opposite direction. The chemical process is reversed as follows:

$$2PbSO_4 + 2H_2O \rightarrow Pb + PbO_2 + 2H_2SO_4.$$

Thus, the original state of the chemicals is reached by the charging action.

The state of charge of a lead-acid cell can be measured by a specific-gravity test using a hydrometer to sample the electrolyte. The specific gravity of a liquid provides an index of how much heavier than water the liquid is. Pure sulphuric acid has a specific gravity of 1.840, while the dilute sulphuric acid of a fully charged lead-acid cell varies from 1.275 to 1.300. During discharge of the cell, water is formed, reducing the specific gravity of the electrolyte. A specific gravity of between 1.120 and 1.150 indicates a fully discharged cell.

The capacity of a battery composed of lead-acid cells is given by an ampere-hour rating. A 60-ampere-hour battery could theoretically deliver 60 amperes for 1 hour, 30 amperes for 2 hours, or 15 amperes for 4 hours. However, this is an approximate rating that depends on the rate of discharge and the operating temperature of the battery. The normal operating temperature is considered to be 80° Fahrenheit.

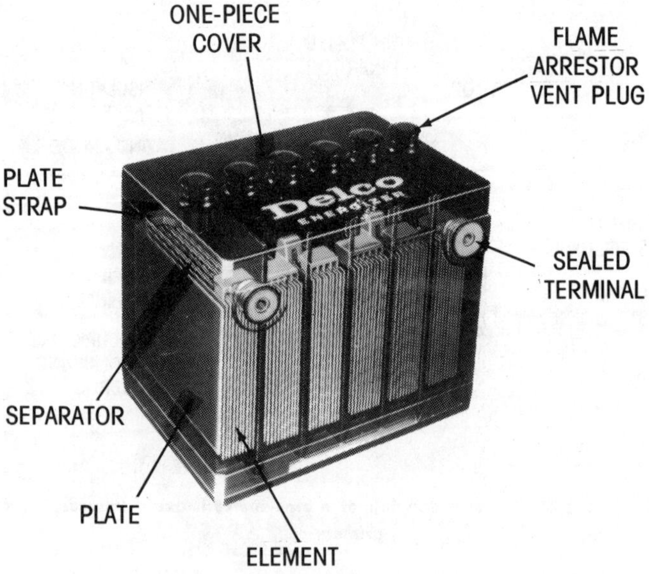

(A) Standard lead-acid battery.

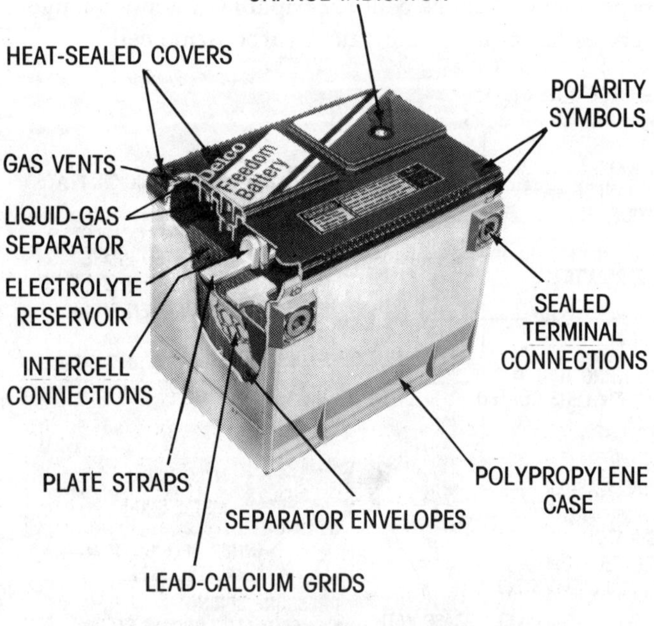

(B) Newer sealed-type lead-acid battery.

Courtesy AC-Delco Div., General Motors Corp.

Fig. 3-22. Cutaway drawings of two types of batteries composed of lead-acid secondary cells.

Nickel-Iron (Edison) Cell—The nickel-iron (Edison) cell, shown in Fig. 3-23, is a secondary cell constructed with an unbreakable case of welded, nickel-plated sheet steel. The positive plate is made of nickel tubes about ¼ inch in diameter and 4¼ inches long. Nickel oxide (NiO_2) is contained in the tubes, which are mounted into steel grids and forced into position under a high pressure. The positive plates are assembled into groups.

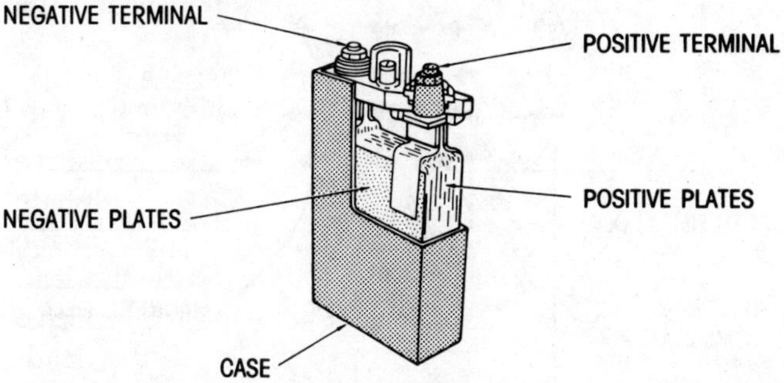

Fig. 3-23. Cutaway drawing of a nickel-iron (Edison) secondary cell.

The negative plates of the Edison cell are made of flat, nickel-plated steel pockets composed of iron oxide. They are built into steel grids in groups in the same way as the positive plates. The electrolyte is a solution of potassium hydroxide (KOH) with lithium hydroxide added.

The chemical action of the nickel-oxide cell is very complex as compared to the lead-acid cell; however, the cells are similar in many respects. The electrolyte, potassium hydroxide (KOH), breaks up into negative and positive ions. The negative ions move toward the iron plate, oxidize the iron, and give up excess electrons to the plate. This plate becomes negatively charged. The positive ions move to the nickel-oxide plate and take electrons from the plate. The deficiency of electrons causes the nickel-oxide plate to become positively charged. The voltage produced by the chemical action is about 1.4 volts per cell.

The internal resistance of the nickel-iron (Edison) cell is higher than the lead-acid cell. This resistance increases quickly as the cell discharges. Unlike the lead-acid cell, the nickel-iron cell cannot be tested using a hydrometer to show its discharge level. A voltmeter is used to indicate the state of charge of the nickel-iron cell. This cell is also charged by applying a direct-current voltage to the positive and negative terminals and causing electron flow within the cells to reverse. The chemicals are thus reactivated so that the cells may be used again.

Other Secondary Cells—Another type of secondary cell, which has gained recent popularity, is the nickel-cadmium cell. These cells are available in large or small sizes with capacities of up to 2000 ampere-hours. Fig. 3-24 shows an exploded view of one of the small, cylindrical-type cells which are used extensively in portable equipment. A larger nickel-cadmium cell is shown in Fig. 3-25. The positive plate of this cell is nickel hydroxide, the negative plate is cadmium hydroxide, and the electrolyte is potassium hydroxide. These cells are extremely reliable, operate over a wide range of temperatures, and have a long life expectancy. A fully charged nickel-cadmium cell has a voltage of approximately 1.25 volts per cell.

There are a few other types of secondary cells in use today, most of which are for specialized applications. Among these other cells are the silver-oxide–zinc cell and the silver-cadmium cell. These cells (see Fig. 3-26) have the highest output per physical size and the longest life, but they are extremely expensive.

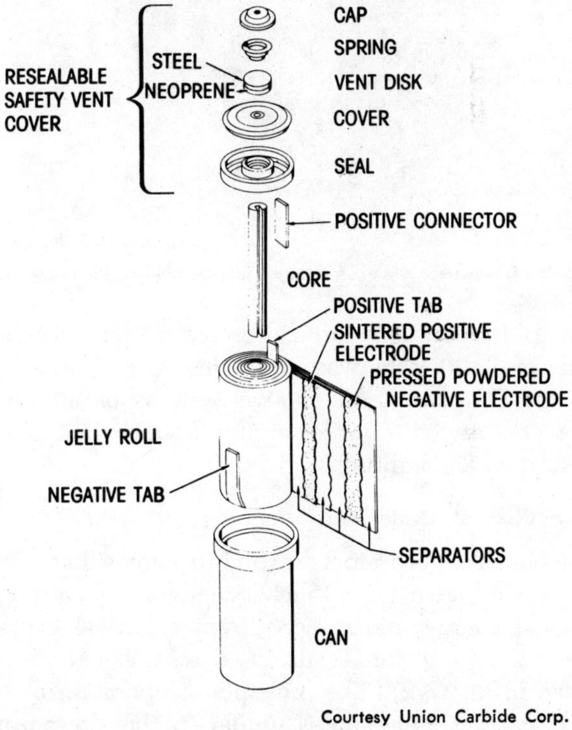

Courtesy Union Carbide Corp.

Fig. 3-24. Exploded view of a small, cylindrical-type, nickel-cadmium secondary cell.

Applications of Secondary Cells

In industry, secondary cells in the form of storage batteries are used to provide emergency power in the event of power failure. Such standby systems are necessary to sustain lighting and some critical operations in industry when power is not available. Industrial trucks and loaders use storage batteries for their everyday operation. Many types of instruments and portable equipment used in industry rely upon batteries for power. Several of these instruments get their power from rechargeable secondary cells rather than primary cells. In addition, many systems that directly affect

Courtesy McGraw-Edison Co.

Fig. 3-25. Transparent view of a larger nickel-cadmium secondary cell.

industrial operations utilize batteries. Batteries are used for lighting railway cars when they are not in motion. Of course, automotive systems of all kinds use secondary batteries to supply direct-current power for starting and lighting.

Direct-Current Generators

Mechanical generators are used in many situations to produce direct current. Such generators convert mechanical energy into direct-current electrical energy. The parts of a simple direct-current generator are shown in Fig. 3-27. The principle of operation of the direct-current generator is similar to the alternating-current generator discussed previously. A rotating

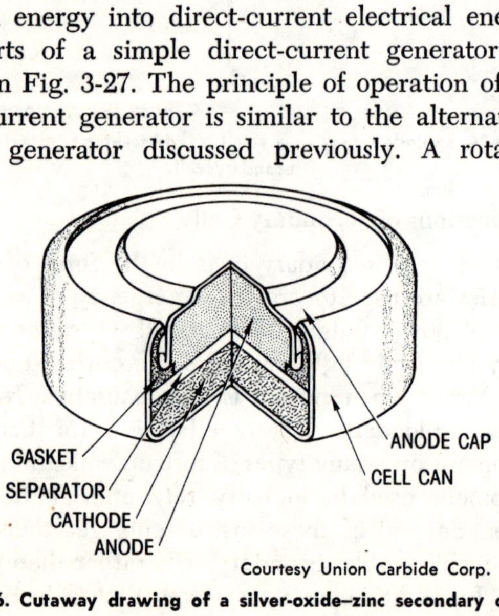

Courtesy Union Carbide Corp.

Fig. 3-26. Cutaway drawing of a silver-oxide–zinc secondary cell.

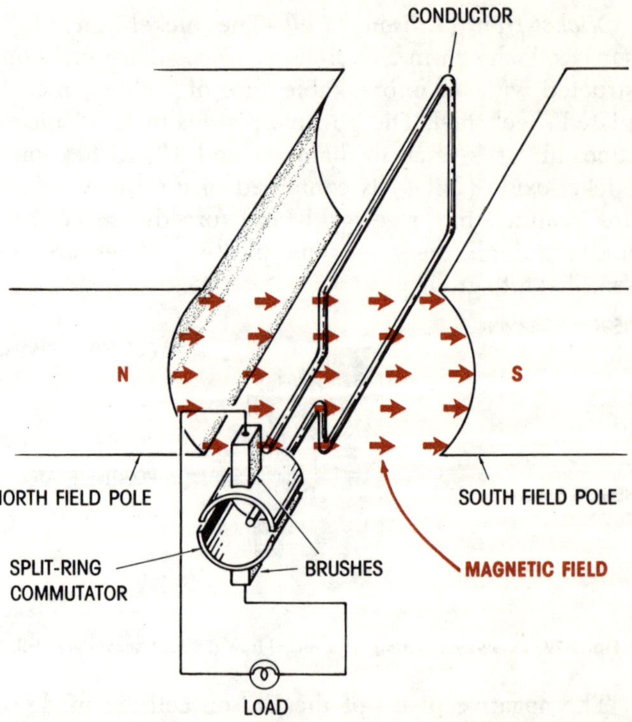

Fig. 3-27. Simplified drawing of the basic parts of a direct-current generator.

armature coil passes through the magnetic field developed between the north and south polarities of permanent magnets or electromagnets. As the coil rotates, electromagnetic induction causes a current to be induced into the coil. The current produced is an alternating current; however, it is possible to convert this alternating current into a form of direct current. The conversion of ac to dc is accomplished through the use of a split-ring commutator. The split-ring commutator, shown in Fig. 3-27, has two segments that are insulated from one another and from the shaft of the machine on which the commutator rotates. An end of each armature conductor is connected to each commutator segment. The purpose of the split-ring commutator is to reverse the armature coil connection to the external load device at the same time that the current induced in the armature coil reverses. This causes direct current of the correct polarity to be applied to the load at all times.

The voltage developed by the single coil example used here would appear as illustrated in Fig. 3-28A. This pulsating direct current is not suitable for most industrial applications. However, by using many turns of wire around the armature, and many split-ring commutator segments, the voltage developed will be a smooth direct current such as that produced by a battery. This type of output is shown in Fig. 3-28B. The voltage developed by a direct-current generator depends on the strength of the magnetic field, the num-

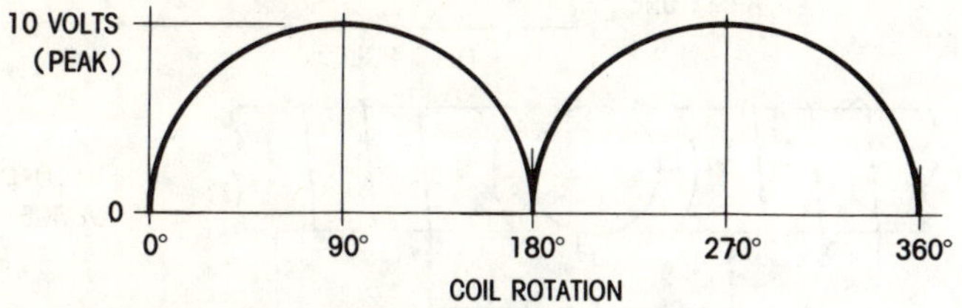

(A) Pulsating dc developed by simple single-coil generator of Fig. 3-27.

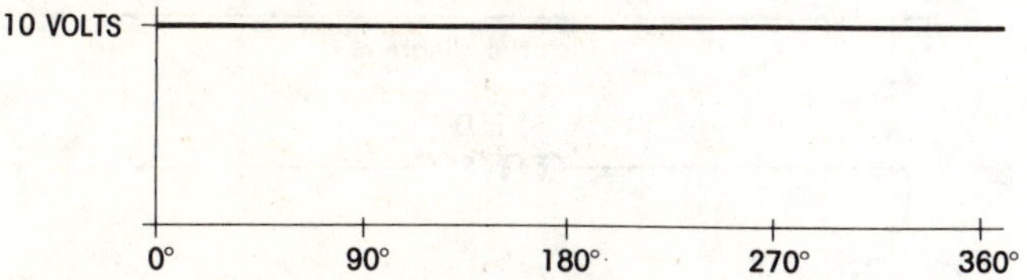

(B) Pure dc developed by more-complex generator using many turns of wire and many commutator segments.

Fig. 3-28. Output waveforms of direct-current generators.

ber of coils in the armature, and the speed of rotation of the armature. By increasing any of these factors, the voltage output can be increased.

Types of Direct-Current Generators

Direct-current generators are classified according to the way in which a magnetic field is developed in the stator of the machine. One method is to use a permanent-magnet field. It is also possible to use electromagnets to develop a magnetic field by applying a separate source of direct current to the electromagnetic coils. However, the most common method of developing a magnetic field is for part of the generator output to be used to supply direct-current energy to the field. Thus, there are three basic classifications of direct-current generators: (1) permanent-magnet field, (2) separately excited field, and (3) self-excited field. The self-excited types are further subdivided according to the method used to connect the armature windings to the field circuit. This may be accomplished by the following connection methods: (1) series, (2) parallel (shunt), or (3) compound.

Permanent-Magnet DC Generator—A simplified diagram of a permanent-magnet dc generator is shown in Fig. 3-29. The conductors shown in this diagram are connected to the split-ring commutator and brush assembly of the machine. The magnetic field is established by using permanent magnets made of alnico (an alloy of aluminum, nickel, cobalt, and iron, or some other natural magnetic material. It is possible to group several permanent magnets together to create a stronger magnetic field.

The armature of the permanent-magnet dc generator consists of many turns of insulated conductors. Therefore, when the armature rotates within the permanent-magnetic field, an induced voltage is developed that

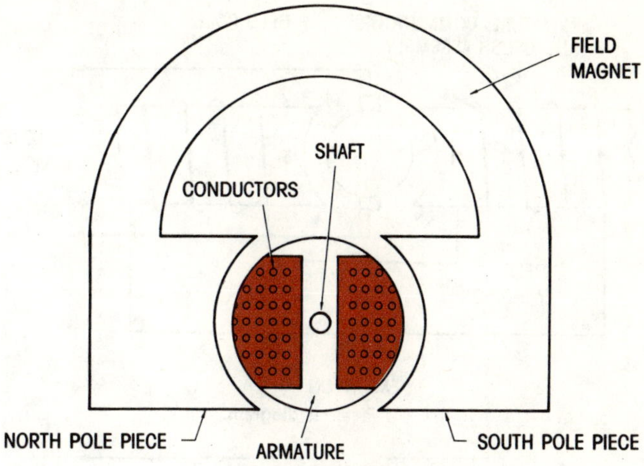

(A) Pictorial diagram.

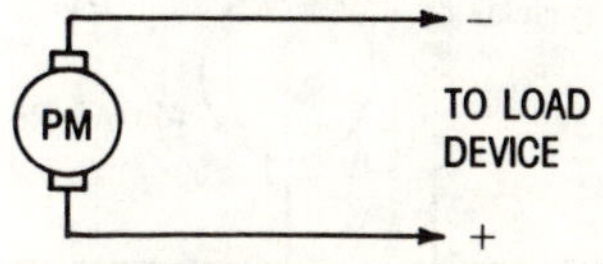

(B) Schematic diagram.

Fig. 3-29. Simplified illustration of a permanent-magnet direct-current generator.

can be applied to a load device. Applications for this type of dc generator are usually confined to those that require low amounts of power. A magneto is an example of a permanent-magnet dc generator.

Separately Excited DC Generator—Where large amounts of direct-current electrical energy are needed, generators with electromagnetic fields are used. Stronger fields can be produced by electromagnets, and it is possible to control the strength of the field by varying the current through the field windings. The output of the generator can thus be controlled with ease.

The direct current used to establish the electromagnetic field is referred to as the *exciting* current. When dc exciting current is obtained from a source separate from the generator, it is called a *separately excited* dc generator. This type of generator is illustrated in Fig. 3-30. Storage batteries are often used to supply dc exciting current to this type of generator. The field current is independent of the armature current, and therefore the separately excited generator maintains a very stable output voltage. Changes in load, which affect the armature current, do not vary the strength of the field. The terminal voltage of a separately excited dc generator can be varied by adjusting the current through the field windings. A high-wattage rheostat in series with the field windings will accomplish field control.

Separately excited dc generators are used only in certain applications where precise voltage control is

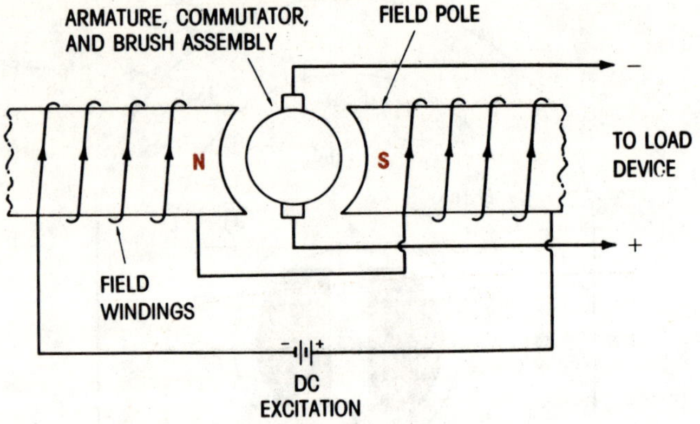

(A) Pictorial diagram.

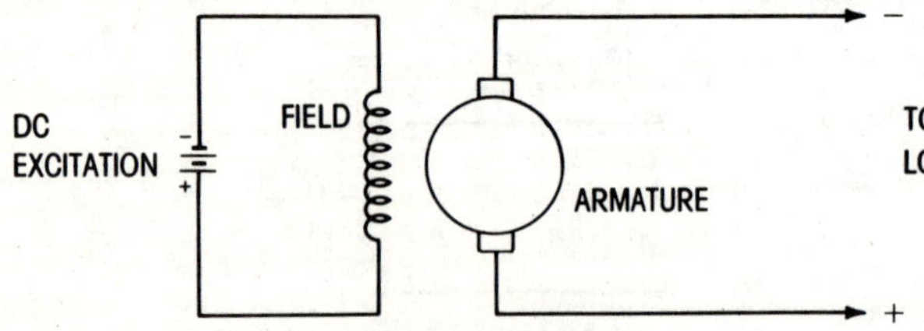

(B) Schematic diagram.

Fig. 3-30. Simplified illustration of a separately excited direct-current generator.

essential. Automatic control processes in industry might require such precision. However, the cost of a separately excited dc generator is usually prohibitive, and other means of obtaining direct-current electrical energy are used.

Self-Excited, Series-Wound DC Generator—Since dc generators produce direct current, it is possible to extract part of a generator's output to obtain exciting current for the field coils. Generators that use part of their own output to supply exciting current for the electromagnetic field are called *self-excited* dc generators. The method used to connect the armature windings to the field windings determines the characteristics of the generator. It is possible to connect armature and field windings in series, parallel (shunt), or series-parallel (compound).

The self-excited, series-wound dc generator has armature windings connected in series with the field windings and the load as shown in Fig. 3-31. In this generator, the total current flowing through the load also flows through the field coils. The field coils are therefore wound with low-resistance wire having only a few turns of large-diameter wire. A sufficient electromagnetic field will then be produced by the large current flowing through the coils.

When the load is disconnected, no current flows through the generator. However, the field coils retain a small amount of magnetism, known as *residual* magnetism. Due to residual magnetism, a current will begin to flow when a load is connected. The current will

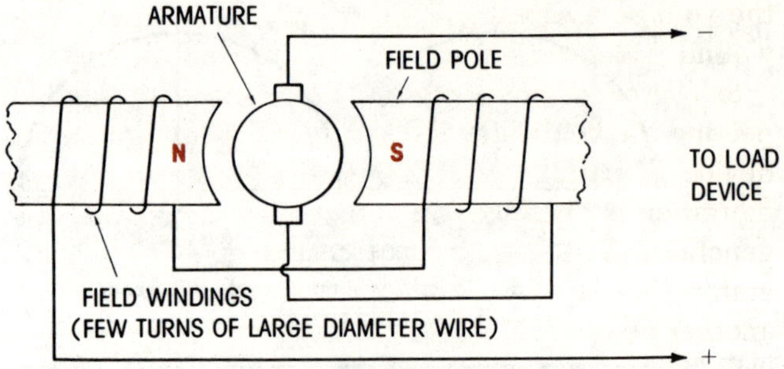

(A) Pictorial diagram.

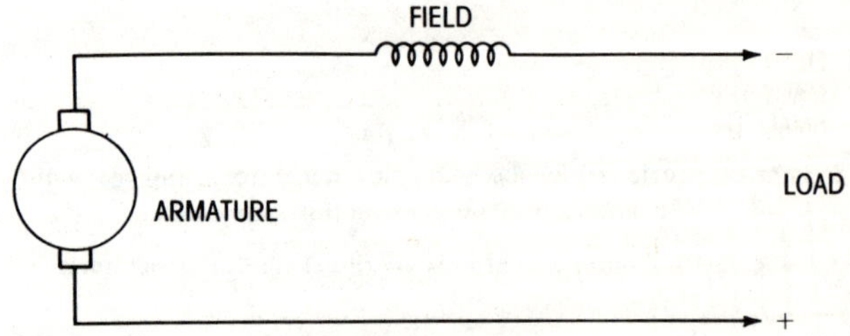

(B) Schematic diagram.

Fig. 3-31. Simplified illustration of a self-excited, series-wound dc generator.

then continue to increase, causing an increase in the magnetic flux of the field. The terminal voltage will rise in proportion to the increases in current. The output curve of this generator is shown in Fig. 3-32. When the peak of the curve is reached, magnetic saturation of the field has occurred, prohibiting an increase in terminal voltage. At this point, an increase in load current will cause a rapid decline in terminal voltage due to circuit losses.

The output of a self-excited, series-wound dc generator varies appreciably with changes in load current. However, beyond the peak of the output curve, the load current remains fairly stable with large variations in voltage. There are specific industrial applications, such as arc welding, that require stable load current as

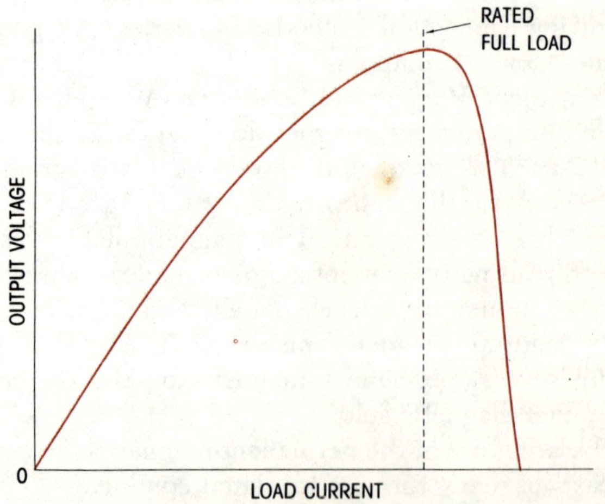

Fig. 3-32. The output curve of a self-excited, series-wound dc generator.

the voltage changes. However, the self-excited, series-wound generator has very few industrial applications.

Self-Excited, Shunt-Wound DC Generator—By connecting the field coils, the armature circuit, and a load device in parallel, a shunt-wound dc generator configuration is obtained. Fig. 3-33 illustrates this type of generator. The output current developed by the generator (I_A) has one path through the load (I_L) and another through the field coils (I_F). The generator is usually designed so that the field current is not more than 5% of the total armature current (I_A).

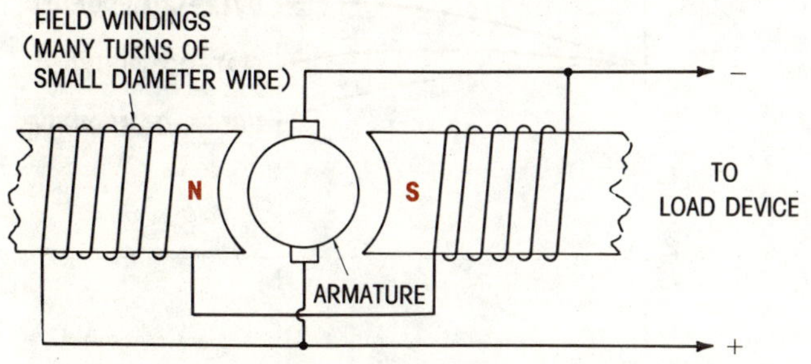

(A) Pictorial diagram.

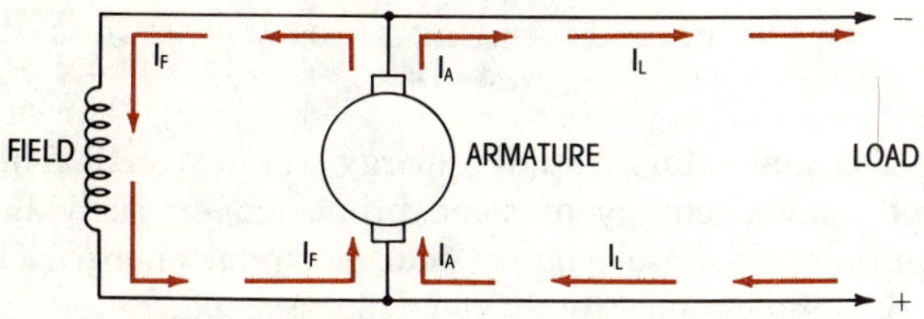

(B) Schematic diagram.

Fig. 3-33. Simplified illustration of a self-excited, shunt-wound dc generator.

In order to establish a strong electromagnetic field and to limit the amount of field current, the field coils are wound with many turns of small diameter wire. These high-resistance coils develop a strong field due to the number of turns, and therefore rely very little on the amount of field current to develop a magnetic flux.

With no load device connected to the shunt-wound dc generator, a voltage is induced as the armature rotates through the electromagnetic field. Again, the presence of residual magnetism in the field coils is critical to the operation of the machine. A current flows in the armature and field circuit as long as there is residual magnetism. As the current increases, the terminal voltage also increases up to a peak level.

When a load device is connected to the generator, the armature current (I_A) increases due to the additional parallel path. This will then increase the I × R drop in the armature, resulting in a smaller terminal voltage. Further increases in load current cause corresponding decreases in terminal voltage as shown in the output curve of Fig. 3-34. With small load currents, the voltage is nearly constant with load current variations. A large load causes the terminal voltage to drop sharply, which is desirable because it provides the generator with a built-in protective feature in case of a short circuit.

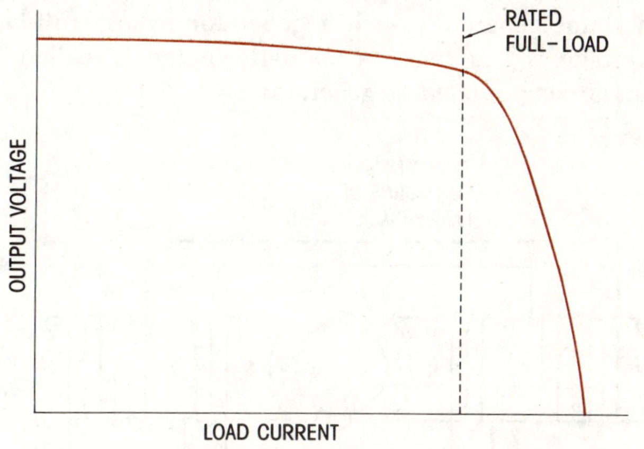

Fig. 3-34. The output curve of a self-excited, shunt-wound dc generator.

The self-excited, shunt-wound dc generator is used where a constant output voltage characteristic is needed. It can be used to supply excitation current to a large ac generator, or to charge storage batteries. However, in applications where initial expense is not critical, the compound-wound generator described next may be more desirable.

Self-Excited, Compound-Wound DC Generator—The compound-wound dc generator has two sets of field windings. One set is made of low-resistance wire and is connected in series with the armature circuit. The other set is made of high-resistance wire and is connected in parallel with the armature circuit. The compound-wound dc generator is illustrated in Fig. 3-35.

As discussed previously, the output voltage of a series-wound dc generator increases with an increase in load current, while the output voltage of a shunt-wound dc generator decreases with an increase in load current. It is possible to produce a dc compound-wound generator, which utilizes both series and shunt windings, that has an almost constant voltage output under changing loads. A constant voltage output may be obtained under varying loads if the series field windings have the proper characteristics to set up a magnetic flux to counterbalance the voltage reduction caused by the I × R drop in the armature circuit.

A constant output voltage is produced by a *flat-compounded* dc generator. The no-load voltage is equal to the rated full-load voltage in this type of machine, as can be seen in the output curves of Fig.

3-36. If the series field has more turns of wire, the generator will possess a series characteristic. The voltage output will increase with an increase in load current. A compound-wound generator whose full-load voltage is greater than its no-load voltage is called an *overcompounded* dc generator. If the series field has fewer turns of wire, the output will be more characteristic of a shunt generator. Such a generator, whose full-load voltage is less than its no-load voltage, is called an *undercompounded* dc generator.

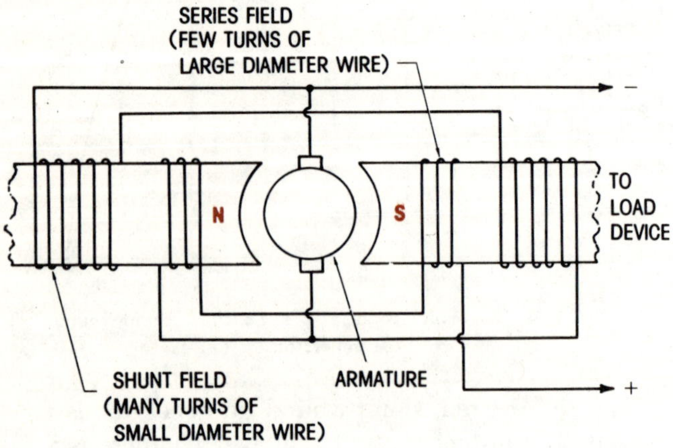

(A) Pictorial diagram.

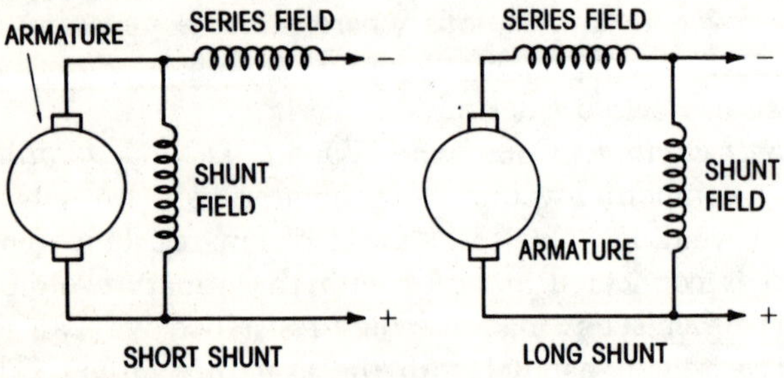

(B) Schematic diagrams.

Fig. 3-35. Simplified illustration of a self-excited, compound-wound dc generator.

Dc compound-wound generators can be constructed so that the series and shunt fields either aid or oppose one another. If the magnetic polarities of adjacent fields are the same, the magnetic fluxes aid each other and are said to be *cumulatively* wound. Opposing polarities of adjacent coils produce a *differentially* wound machine. For almost all applications of compound-wound machines, the cumulative method is used. A generator wound in this way maintains a fairly constant voltage output with variations in load current. The compound-wound dc generator is used more extensively than other dc generators due to its constant voltage output and its design flexibility to obtain various output characteristics.

Common DC Generator Characteristics

Dc generators are used primarily for operation with mobile equipment. In industrial plants, they are used for standby power, battery charging, and for specialized dc operations such as electroplating. In many industries, rectification systems that convert alternating current to direct current have replaced dc generators since they are cheaper to operate and maintain.

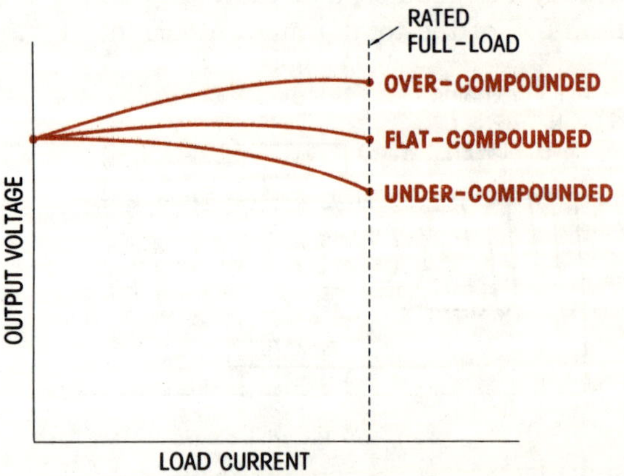

Fig. 3-36. Output curves for self-excited, compound-wound dc generators.

Dc generators supply energy by converting the mechanical energy of some prime mover, such as a gasoline or diesel engine, into electrical energy. The prime mover must rotate at a definite speed.

When dc generators operate, a characteristic known as *armature reaction* takes place. The current through the armature windings produces a magnetic flux, which reacts with the main field flux as shown in Fig. 3-37. The result is that a force is created which tends to rotate the armature in the direction indicated. As the load current increases, the increase in armature current causes a greater amount of armature reaction to take place. This condition can cause a considerable amount of sparking between the brushes and the commutator. However, armature reaction can be reduced by placing windings called *interpoles* between the main field windings of the stator. These windings are connected in series with the armature windings. Thus, an increase in armature current creates a stronger magnetic flux around the interpole, which counteracts the main field distortion created by the armature conductors.

Dc generator output is usually rated in kilowatts, which is the electrical power capacity of a machine. Other ratings, which are specified by the manufacturer on the nameplate of the machine, are current capacity, terminal voltage, speed, and temperature. Dc generators are made in a wide range of physical sizes, with various electrical characteristics.

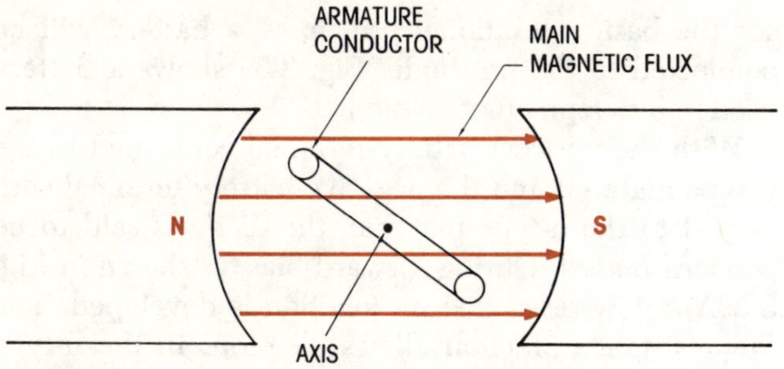

(A) Main magnetic flux with no current in the armature windings.

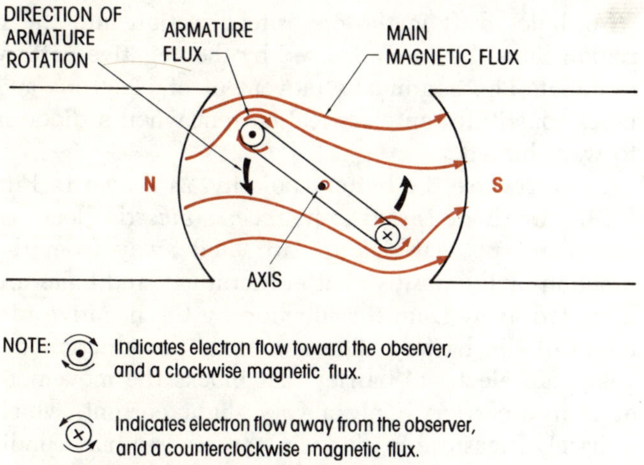

(B) Distortion of main magnetic flux with current in the armature windings.

Fig. 3-37. Illustration of armature reaction.

Direct-Current Conversion Systems

Most of the electrical power supplied to industry is 60-Hz, three-phase alternating current. However, the use of direct current is necessary for many industrial processes. For instance, dc motors have more desirable speed control and torque characteristics than ac motors do. We have already discussed two methods of supplying direct current for industrial use. Batteries and dc generators have been used as direct-current energy sources for many years. In many cases today, it is more economical for industry to use a system for converting alternating current to direct current. Direct-current load devices may be powered by systems, called either *rectifiers* or *converters,* which change alternating current into a form of direct current suitable for use with industrial processes. Some types of ac-dc conversion systems include: (1) semiconductor rectifiers; (2) vacuum-tube rectifiers; (3) gaseous-tube rectifiers; and (4) rotary converters. Semiconductor rectifiers are used much more than the other types.

Single-Phase Rectification

The simplest system for converting alternating current to direct current is single-phase rectification. A single-phase rectifier changes alternating current to pulsating direct current. This process can be accomplished by using either semiconductors, vacuum tubes, gaseous tubes, or rotary converters. The most common and economical method in industry today is the use of low-cost silicon semiconductor rectifiers. Since semiconductors are so prevalent in industry today, it is desirable to review the action of a semiconductor diode before we look at specific rectifier circuits.

The Semiconductor Diode

We know that there are certain materials that conduct electrical current easily (conductors) and other materials that do not conduct current easily (insulators). Between these two extremes, there are other materials that are neither good conductors nor good insulators. These materials, such as silicon and germanium, have characteristics between those of conductors and insulators.

Silicon and germanium are crystalline materials whose electrical characteristics can easily be altered in the manufacturing process. Elements called *impurities* are added to pure silicon or germanium to form either n-type or p-type materials. These materials are used in various combinations to form many types of semiconductors such as diodes, transistors, silicon controlled rectifiers, triacs, and diacs. An n-type material possesses an abundance of electrons or negative charge carriers. A p-type material has an excess of positive charge carriers called *holes.* By controlling the amount of impurities added to the silicon or germanium, the characteristics of the semiconductor material can be determined.

The most basic semiconductor device is the diode, which performs the function of rectification. A diode is merely a pn junction device formed by adjacent p- and n-type materials, as shown in Fig. 3-38A. The junction formed by the adjacent p- and n-type materials will allow current to pass in one direction, but will offer a very high resistance to current in the other direction. This device, called a *pn junction* diode, performs the same electronic function as a vacuum-tube diode or a gaseous-tube diode. Types of pn junction diodes are shown in Fig. 3-38C.

The current carriers of a pn junction diode are in a state of equilibrium when an external potential is not applied. It is possible that, due to temperature, some electrons will cross the junction from the n region to the p region. Likewise, some holes will diffuse from the p region to the n region. After this diffusion has taken place, an electrical barrier is developed between the two regions. Thus, the further movement of charge carriers is prohibited and the diode is in a state of equilibrium.

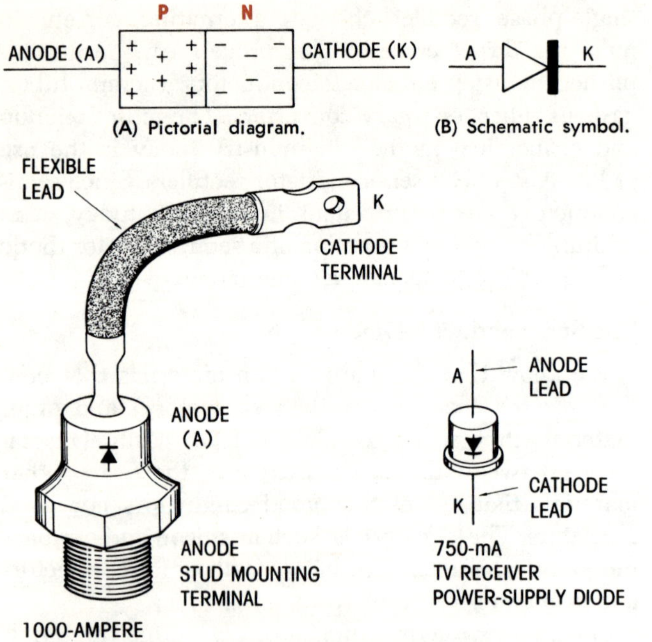

(A) Pictorial diagram. (B) Schematic symbol.

(C) Types of diodes.

Fig. 3-38. Pn junction diodes.

In electrical circuits, diodes are connected to an electrical potential. The instantaneous polarity of the potential connected to each terminal of the diode determines the action performed by the diode. To best see the basic operational principles, a battery will be connected to the pn diode. Fig. 3-39 shows a battery used to bias a pn junction diode.

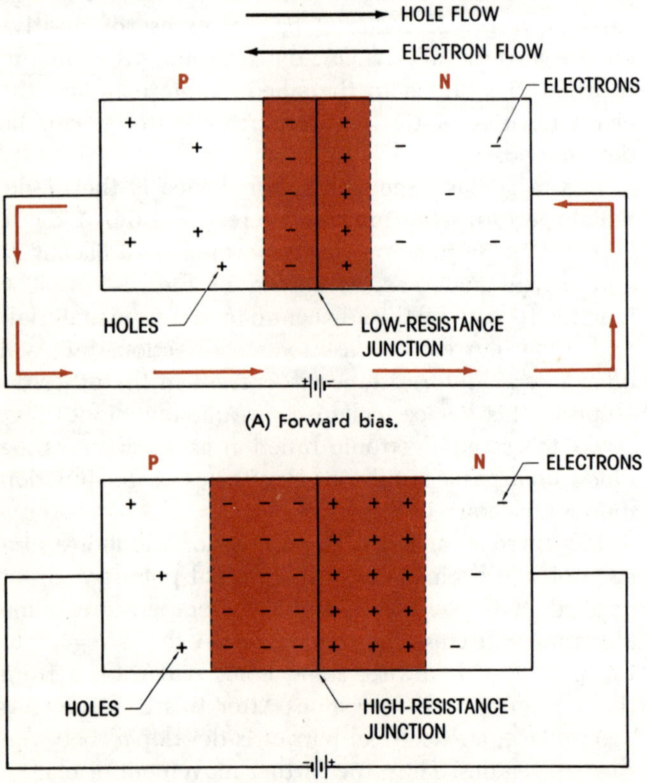

(A) Forward bias.

(B) Reverse bias.

Fig. 3-39. Forward and reverse bias of a pn junction diode.

With the positive battery terminal connected to the p-type material and the negative battery terminal connected to the n-type material, the diode is said to be *forward biased.* During forward bias, as shown in Fig. 3-39A, a low-resistance pn junction is developed. The low-resistance junction allows electrons in the n-type material to drift into the p region since they are attracted by the positive terminal of the battery. Likewise, holes drift in the opposite direction into the n region since they are attracted by the negative battery terminal. The continuous movement of electrons and holes constitutes an electrical current when a diode is forward biased.

If we reverse the battery polarity, as shown in Fig. 3-39B, the diode is said to be *reverse biased.* Electrons in the n-type material are attracted away from the junction by the positive battery terminal, and holes are attracted away from the junction by the negative terminal of the battery. The area near the junction develops an electrical barrier that blocks the movement of charge carriers. Only a very slight current, which is barely measurable, flows in the reverse-bias condition. Since the pn diode conducts while forward biased, and does not conduct while reverse biased, it exhibits the properties of a rectifier.

Single-Phase, Half-Wave Rectification

A single-phase, half-wave rectifier circuit, shown in Fig. 3-40, converts an alternating current into a pulsating direct current. Let us assume that during the positive alternation of the ac cycle, the anode of the diode is positive. The diode will then conduct since it is forward biased and the junction resistance is low. The positive half-cycle of the ac will then appear across the load device (represented by a resistor).

During the negative alternation of the ac cycle, the anode of the diode becomes negative. The diode is now reverse biased and no significant current will flow through the load device. Therefore, there will be no voltage appearing across the load. The input and resulting output waveforms of the half-wave rectifier circuit are shown in Fig. 3-40C. The pulsating direct current of the output has an average dc level. The average value of the pulsating dc produced by single-phase, half-wave rectification is expressed as:

$$V_{dc} = 0.318 \times V_{max}$$

where,

V_{dc} is the average value of rectified voltage,
V_{max} is the peak (maximum) value of applied ac voltage.

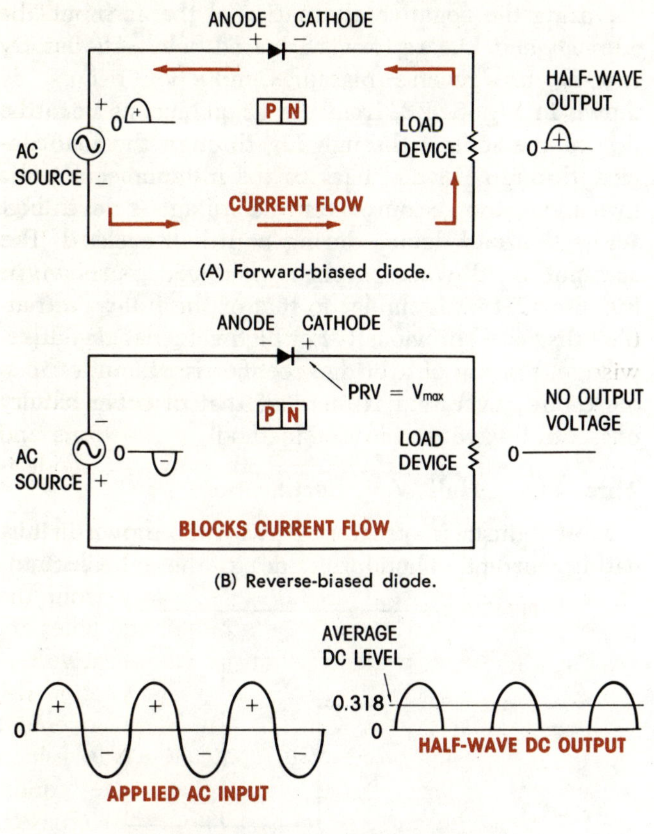

(A) Forward-biased diode.

(B) Reverse-biased diode.

(C) Input/output voltage waveforms.

Fig. 3-40. Single-phase, half-wave rectifier circuit.

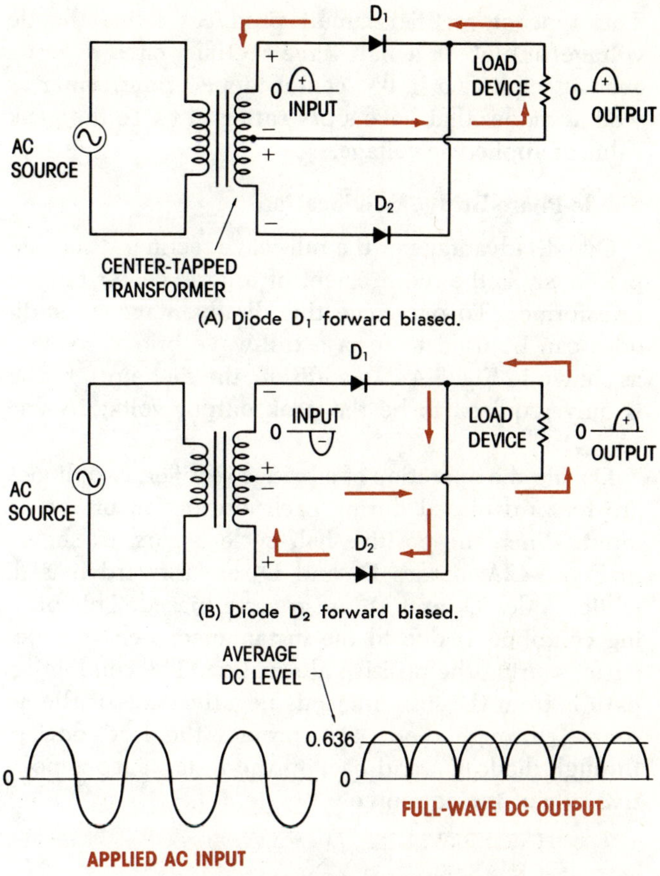

(A) Diode D_1 forward biased.

(B) Diode D_2 forward biased.

(C) Input/output voltage waveforms.

Fig. 3-41. Single-phase, full-wave rectifier circuit.

Certain diode ratings should be considered for half-wave rectifier circuits. The maximum forward current (I_{max}) is the largest current that can flow through the diode while it is forward biased without damaging the device. The peak inverse (reverse) voltage (prv) is the maximum voltage across the diode while it is reverse biased. For the half-wave rectifier, the maximum voltage developed across the diode is V_{max} of the applied ac. As shown in the preceding formula, the prv of a diode used in a half-wave circuit must be much larger than the dc voltage developed.

Single-Phase, Full-Wave Rectification

In order to obtain a purer form of direct-current energy, it is possible to improve upon half-wave rectification systems. Fig. 3-41 shows a single-phase, full-wave rectifier circuit that uses two diodes and produces a dc output voltage during each alternation of the ac input. The rectified output of the full-wave rectifier has twice the dc voltage level of the half-wave rectifier.

The full-wave rectifier utilizes a center-tapped transformer to transfer ac source voltage to the diode rectifier circuit. During the positive half-cycle of ac source voltage, the instantaneous charges on the transformer secondary are as shown in Fig. 3-41A. The peak voltage (V_{max}) is developed across each half of the transformer secondary. At this time, diode D_1 is forward biased and diode D_2 is reverse biased. Therefore, conduction occurs from the center tap, through the load device, through D_1, and back to the outer terminal of the transformer secondary. The positive half-cycle is developed across the load as shown.

During the negative half-cycle of the ac source voltage, diode D_1 is reverse biased and diode D_2 is forward biased by the instantaneous charges shown in Fig. 3-41B. The current path is from the center tap, through the load device, through D_2, and back to the outer terminal of the transformer secondary. The negative half-cycle is also produced across the load, developing a full-wave output as illustrated in Fig. 3-41C.

Each diode in a full-wave rectifier circuit must have a prv rating of twice the value of the peak voltage developed at the output, since twice the peak voltage ($2V_{max}$) is present across the diode when it is reverse biased. The average voltage for a full-wave rectifier circuit is:

$$V_{dc} = 2\,(0.318 \times V_{max}) = 0.636 \times V_{max}.$$

This type of rectifier circuit produces twice the dc voltage output of a half-wave rectifier circuit. However, it requires a bulky, center-tapped transformer as well as diodes that have a prv rating of twice the peak value of applied ac voltage.

Single-Phase Bridge Rectification

One disadvantage of the full-wave rectifier discussed previously is the requirement of a large center-tapped transformer. To overcome this disadvantage, four diodes can be used to form a full-wave bridge rectifier as shown in Fig. 3-42. In addition, the diode prv rating is only required to be the peak output voltage value (V_{max}).

During the operation of a bridge rectifier, two diodes are forward biased during each alternation of the ac input. When the positive half-cycle occurs, as shown in Fig. 3-42A, diodes D_1 and D_3 are forward biased, while diodes D_2 and D_4 are reverse biased. This biasing condition is due to the instantaneous charges occuring during the positive alternation. The conduction path is from the instantaneous negative side of the ac source, through diode D_3, through the load device, through diode D_1, and back to the instantaneous positive side of the ac source.

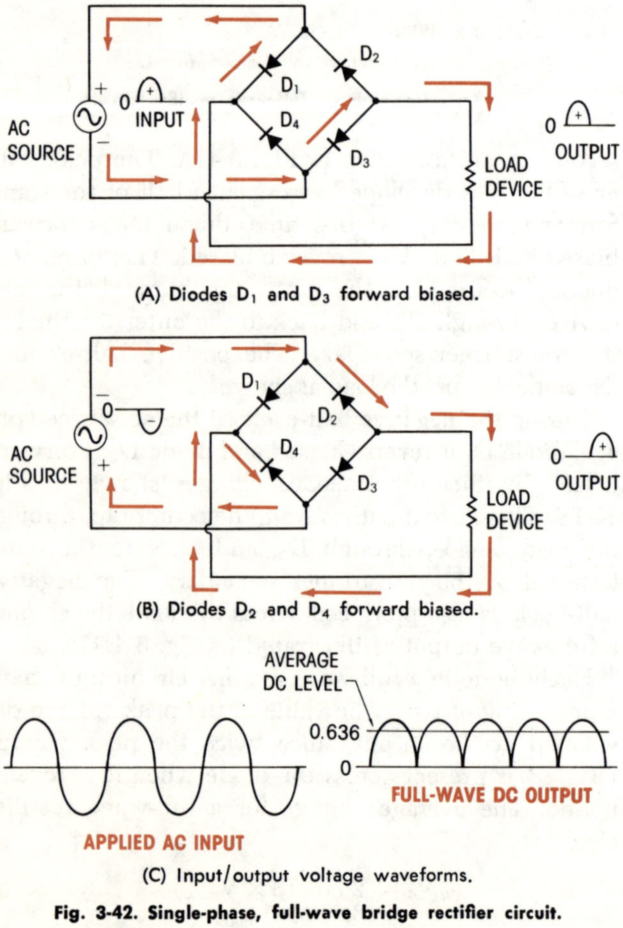

(A) Diodes D_1 and D_3 forward biased.

(B) Diodes D_2 and D_4 forward biased.

(C) Input/output voltage waveforms.

Fig. 3-42. Single-phase, full-wave bridge rectifier circuit.

During the negative alternation of the ac input, diodes D_2 and D_4 are forward biased, while diodes D_1 and D_3 are reverse biased. Conduction occurs, as shown in Fig. 3-42B, from the instantaneous negative side of the source, through D_2, through the load device, through D_4, and back to the instantaneous positive side of the ac source. Since a voltage is developed across the load device during both half-cycles of the ac input, a full-wave output is produced, as shown in Fig. 3-42C, that is similar to that of the full-wave rectifier discussed previously. For high values of dc output voltage, the use of a bridge rectifier is desirable since the diode prv rating is one-half that of other single-phase, full-wave rectification methods.

Three-Phase, Half-Wave Rectification

Most industries are supplied with three-phase alternating current. Therefore, due to the inherent ad-

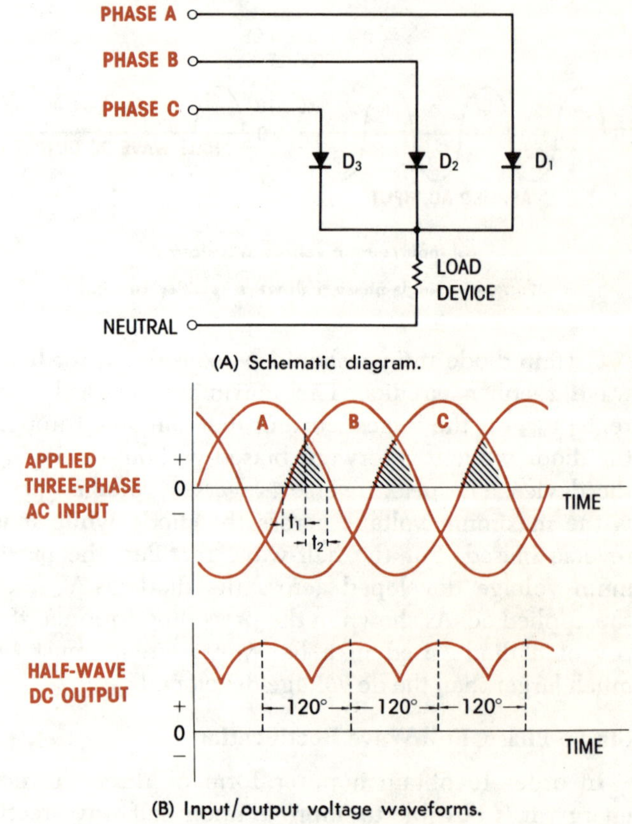

(A) Schematic diagram.

(B) Input/output voltage waveforms.

Fig. 3-43. Three-phase, half-wave rectifier circuit.

vantages of three-phase power, it is beneficial to use three-phase rectifier circuits to supply dc voltage for industrial use. Single-phase rectifier circuits are ordinarily used where low amounts of dc power are required. To supply large amounts of dc power for industrial requirements, a three-phase rectifier circuit, such as the one shown in Fig. 3-43, could be employed. Three-phase rectifier circuits produce a purer dc voltage output than single-phase rectifier circuits do, thus wasting less ac power.

Fig. 3-43 shows a three-phase, half-wave rectifier circuit that does not use a transformer. Phases A, B, and C of the wye-connected, three-phase source supply voltage to the anodes of diodes D_1, D_2, and D_3. The load device is connected between the cathodes of the diodes and the neutral of the wye-connected source. When phase A is at its peak positive value, maximum conduction occurs through diode D_1, since it is forward biased. No conduction occurs through D_1 during the negative alternation of phase A. The other diodes operate in a similar manner, conducting during the positive ac input alternation and not conducting during the associated negative ac alternation. In a sense, this circuit combines three single-phase, half-wave rectifiers to produce a half-wave dc output, as shown in Fig. 3-43B. Of course, the voltages appearing across the diodes are 120° out of phase. There is a period of time during each ac cycle when the positive alternations overlap one another, as shown in the shaded areas of the diagram. During overlap time period t_1, the phase A voltage is more positive than the phase B voltage, while during the t_2 interval, phase B is more positive. Diode D_1 will conduct until the time period t_1 ends, then D_2 will conduct beginning at the end of t_1 until the next area of overlapping is reached.

Note that the voltage across the load device rises to a peak value twice during each phase alternation of the ac input voltage. These peaks are 120° apart. Since the dc output voltage never falls to zero, less ac ripple is present, which results in a purer form of dc than single-phase rectifiers produce. The average dc output voltage (V_{dc}) is expressed as:

$$V_{dc} = 0.831 \times V_{max},$$

which compares very favorably with single-phase, full-wave rectifier circuits.

A disadvantage of this type of three-phase rectifier circuit is that the ac lines are not isolated. This lack of isolation, which is a direct connection to the ac lines, could be a hazardous safety factor. To overcome this disadvantage, a transformer can be used, as shown in Fig. 3-44, to form a similar three-phase, half-wave rectifier. The secondary voltage can either be increased or decreased by the proper selection of the transformer, providing a variable dc voltage capability. The circuit illustrated uses a delta-to-wye connected transformer. The operation of this circuit is identical to that of the three-phase, half-wave circuit previously discussed; however, line isolation has been accomplished.

Three-Phase, Full-Wave Rectification

The full-wave counterpart of the three-phase, half-wave rectifier circuit is shown in Fig. 3-45A. This type of circuit is popular for many industrial applications. Six rectifiers are required for operation of the circuit. The anodes of D_4, D_5, and D_6 are connected together at point A, while the cathodes of D_1, D_2, and D_3 are connected together at point B. The load device is connected across these two points. The three-phase ac lines are connected to the anode-cathode junctions of D_1 and D_4, D_2 and D_5, and D_3 and D_6. This circuit does not require the neutral line of the three-phase source; therefore, a delta-connected source could be used.

The resulting dc output voltage of the three-phase, full-wave rectifier circuit is shown in Fig. 3-45B. The operation of the circuit is similar to a single-phase bridge rectifier in many respects. At any single instant of time along the three-phase ac input cycle, the anode voltage of one of the diodes is more positive than all others, while the cathode voltage of another diode is more negative than all others. These two diodes will then form the conduction path for that time period. This conducting action is similar to a bridge rectifier, since two diodes conduct during a time interval. Each rectifier in this circuit conducts during one-third of an ac cycle (120°). Peak positive dc output voltage occurs during every 60° of the three-phase ac input.

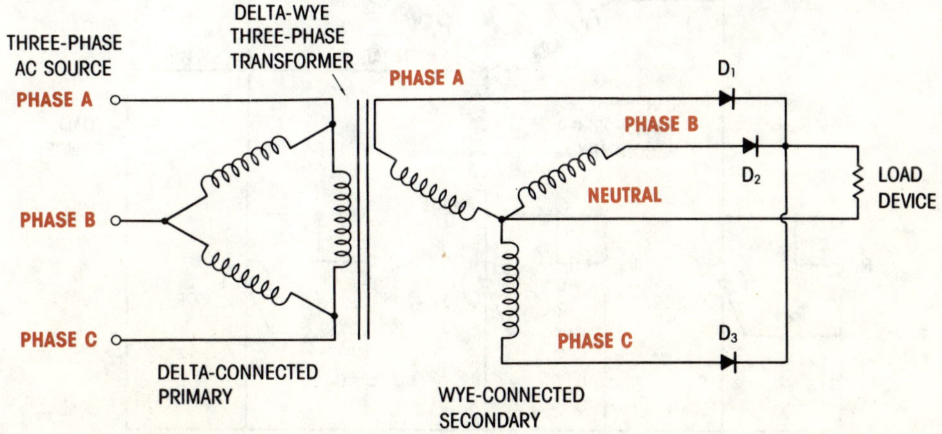

Fig. 3-44. Three-phase, half-wave rectifier circuit using a transformer.

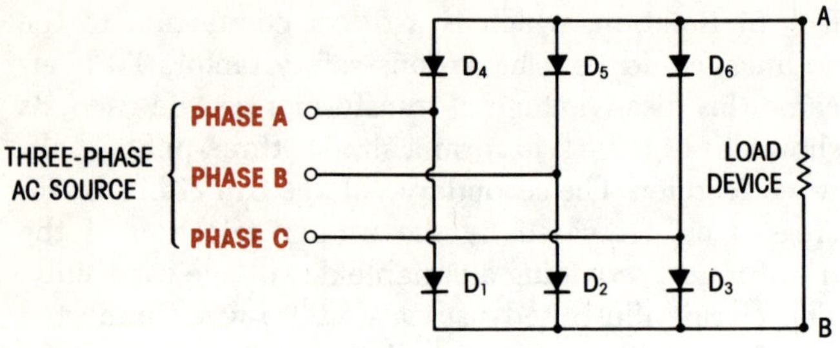

(A) Schematic diagram.

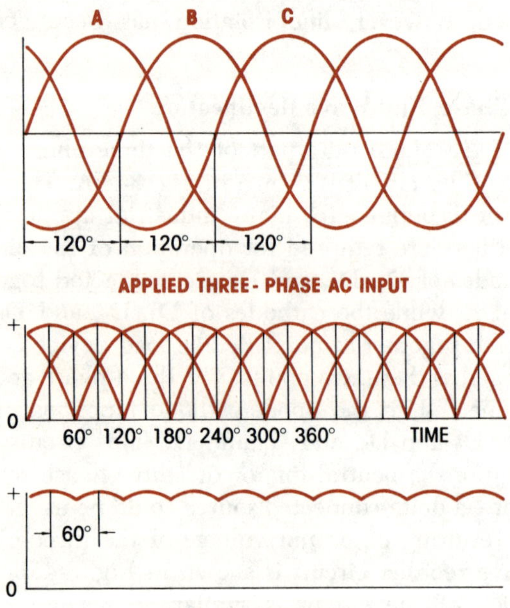

(B) Input/output voltage waveforms.

Fig. 3-45. Three-phase, full-wave rectifier circuit.

Three-Phase Rectification Using SCRs

The rectification circuits previously discussed have no means of controlling the output voltage and current levels. In the circuit shown in Fig. 3-46, silicon controlled rectifiers (SCRs) are used in place of semiconductor diode rectifiers. The conduction time of a silicon controlled rectifier can be varied to control dc output voltage and current levels.

The use of an SCR as a controlled rectifying device is illustrated in Fig. 3-47. The SCR conducts when its forward breakover voltage ($V_{(BO)}$) is reached. The $V_{(BO)}$ is determined by the gate control circuit. When the SCR conducts, current flows through the load device as shown in Fig. 3-47D. When the ac source voltage level drops to that shown at point 2 (holding current level), the SCR turns off and remains off during the negative half-cycle of the ac input.

Voltage is developed across the load only from point 1 to point 2 in Fig. 3-47D. When the gate control circuit is used to change the $V_{(BO)}$ value, the result is to change the conduction time of the SCR, as shown by the dotted lines in Fig. 3-47D. The dotted lines illustrate the effect of a reduced $V_{(BO)}$ on the anode-cathode voltage of the SCR and the corresponding voltage developed across the load. Similar rectifying action can be accomplished by using gaseous thyratron tubes in place of SCRs.

In order to control the conduction time of the SCRs of the three-phase rectifier, a phase-shift network is used. In Fig. 3-46, R_1C_1, R_2C_2, and R_3C_3 are identical RC phase-shifting networks. The three rheostats are ganged together so that control is accomplished by varying only one knob. The conduction time of the circuit is increased by increasing the resistance of the rheostats. Operationally, this circuit is the same as the three-phase, half-wave rectifier discussed previously. However, the dc output can be easily controlled by varying the rheostat setting.

Vacuum-Tube and Gaseous-Tube Rectifiers

For many years, industries have been concerned with the electronic control of industrial equipment and processes. There are still methods other than semiconductor rectifier circuits that are used to supply

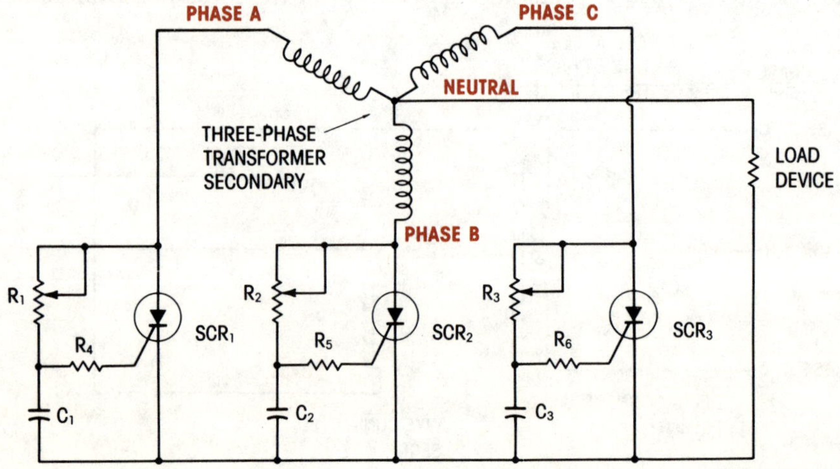

Fig. 3-46. Three-phase, half-wave rectifier circuit using SCRs.

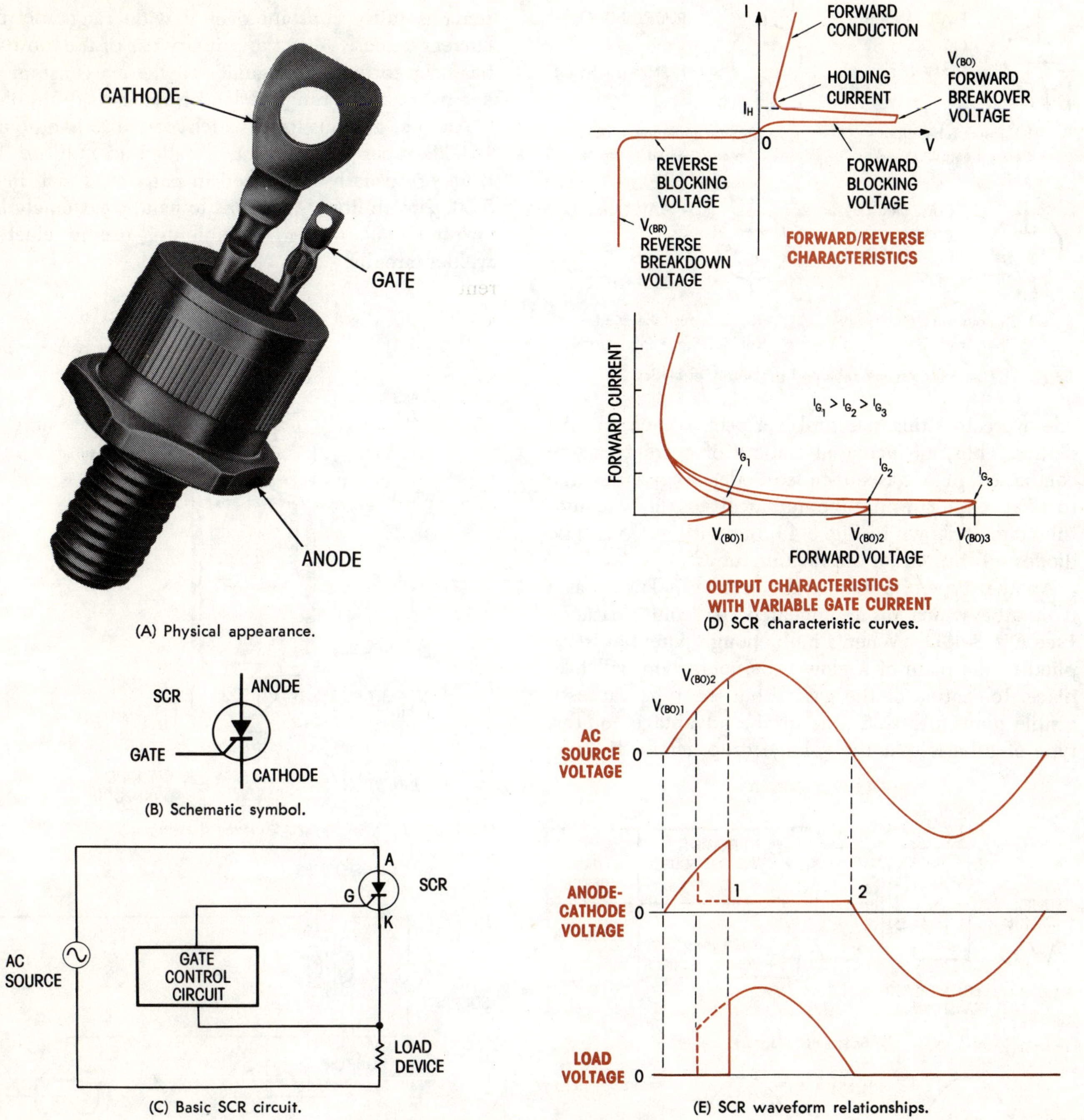

Fig. 3-47. The silicon controlled rectifier.

direct-current power. In some cases, semiconductors are limited by the amount of current they are capable of handling. Where control of extremely large amounts of current is required, vacuum-tube or gaseous-tube rectifiers may be used.

Vacuum-tube diodes, whose symbols are shown in Figs. 3-48A and B, consist of a filament, cathode, and plate (or anode). The filament must be heated by the application of an electrical potential, usually alternating current. A vacuum-tube diode, full-wave rectifier circuit is shown in Fig. 3-49. Compare this circuit with its semiconductor counterpart of Fig. 3-41. A vacuum tube will conduct only when its plate (anode) is made positive relative to its cathode. Tube V_1 will conduct when point A is positive, and tube V_2 will conduct when point B becomes positive. A full-wave output will appear across the load device.

A problem of the vacuum-tube diode is that not all of the electrons emitted by the cathode reach the plate. The ionization that takes place in a gaseous-tube di-

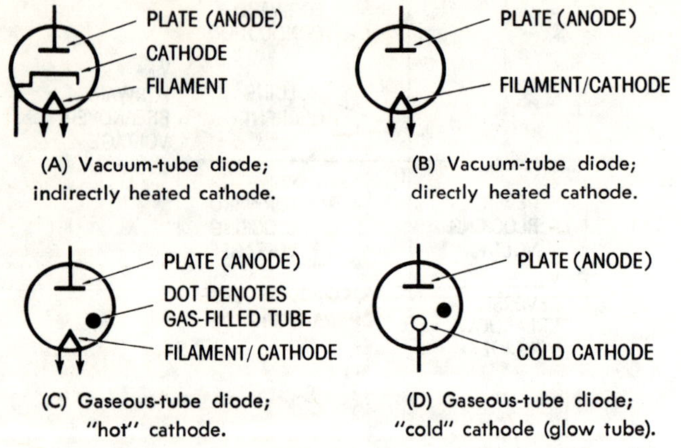

(A) Vacuum-tube diode; indirectly heated cathode.

(B) Vacuum-tube diode; directly heated cathode.

(C) Gaseous-tube diode; "hot" cathode.

(D) Gaseous-tube diode; "cold" cathode (glow tube).

Fig. 3-48. Vacuum-tube and gaseous-tube diodes.

ode overcomes this undesirable aspect of vacuum-tube diodes. Thus, an increased amount of current can be conducted in a gaseous-tube diode. A gaseous-tube rectifier circuit would be identical to the vacuum-tube circuit shown in Fig. 3-49, but with gaseous-tube diodes substituted for the vacuum tubes.

Another type of gaseous-tube rectifier is known as a *glow* tube, which has an unheated (or "cold") cathode (see Fig. 3-48D). When a high enough potential is applied to the plate of a glow tube, ionization will take place. Ionization of the gas within the tube causes a visible glow to result. One of the advantages of this type of tube is that the voltage drop across the tube

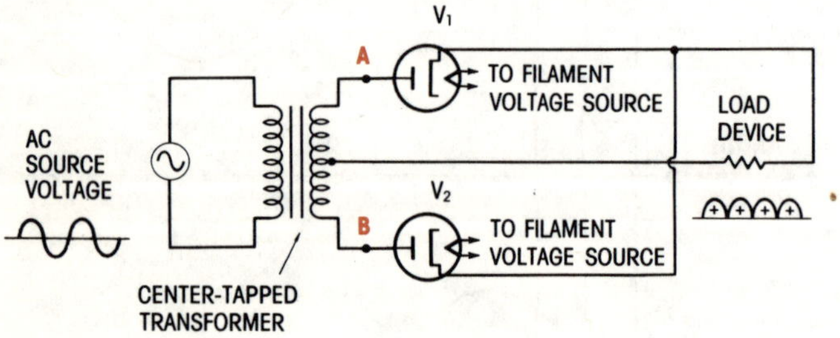

(A) Schematic diagram.

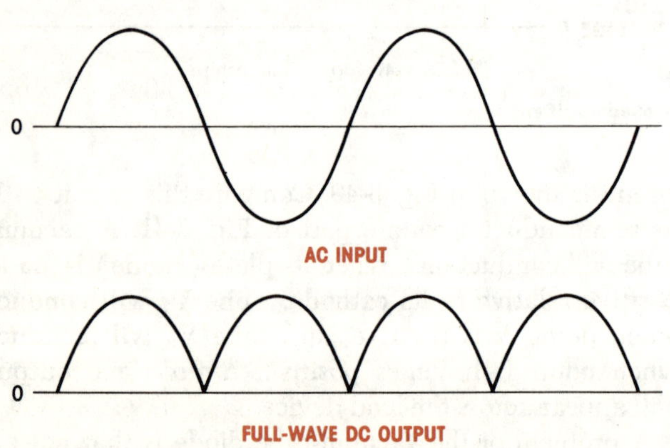

(B) Input/output voltage waveforms.

Fig. 3-49. Vacuum-tube diode, full-wave rectifier circuit.

remains fairly constant over a wide range of plate-current values. Thus, the primary use of the glow tube has been for voltage regulation, where a constant voltage over a wide range of load currents is desirable.

Another gaseous tube, which is used to handle up to 10,000 amperes of current, is called an *ignitron.* Ignitrons are usually connected in pairs as shown in Fig. 3-50. The ability of ignitrons to handle extremely large currents makes them suitable for use in electronic welder circuits.

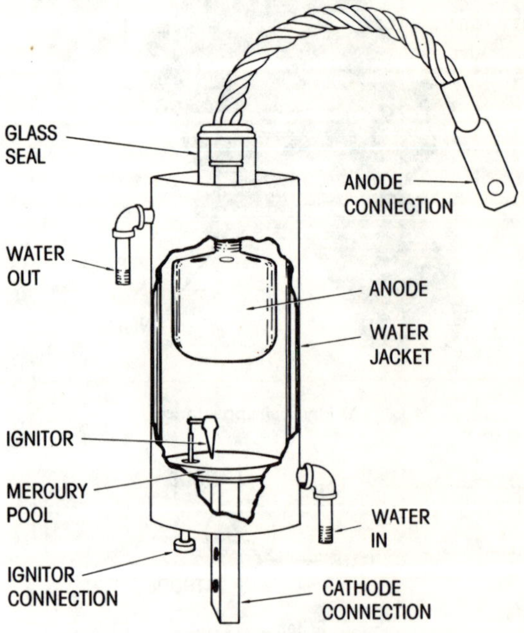

(A) Cutaway drawing.

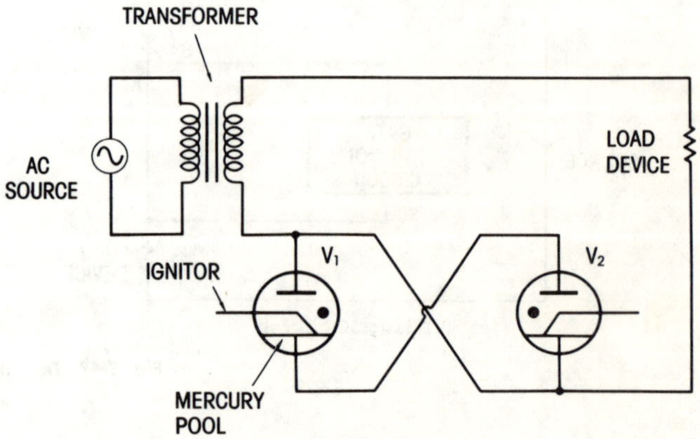

(B) Circuit diagram.

Fig. 3-50. The ignitron tube.

The ignitron is a gaseous tube that is not heated, but uses a pool of liquid mercury as its cathode. If some means is used to establish an arc within the ignitron, a large quantity of electrons will be attracted to the anode when a positive charge is placed on it. The ignitor of the tube extends to near the mercury pool (see Fig. 3-50A). When a positive potential of high enough value is applied to the ignitor, an arc is started between it and the mercury pool. Electrons are then passed to the anode due to the ionization that takes place. When current flows through the tube, the ignitor arc is no longer needed, so it is stopped. A thyratron or saturable reactor circuit could be used to control ignitor firing. The high current flowing in an ignitron tube causes a great deal of heat to be produced. Ordinarily, water is circulated through the tube to remove this heat.

Rotary Converters

An additional method that has been used to convert alternating current to direct current is by means of a rotary converter. Rotary ac to dc converters are seldom used today. However, motor-driven generator units, such as the one illustrated in Fig. 3-51, can also

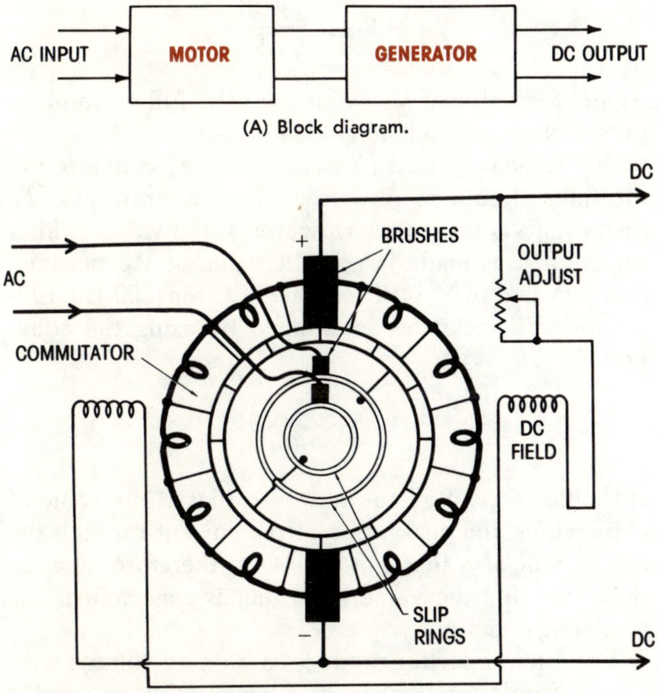

(B) Schematic diagram.

Fig. 3-51. Rotary ac to dc converter.

be used to convert direct current to alternating current (called an *inverter*). When operated as a converter to produce dc, the machine is run off an ac line. The ac is transferred to the machine windings through slip rings, and converted to dc by a split-ring commutator located on the same shaft. The amount of dc voltage output is determined by the magnitude of the ac voltage applied to the machine. Converters may be designed as two units, with motor and generator shafts coupled together, or as one unit housing both the motor and generator.

Direct-Current Filtering Methods

The pulsating direct current produced by both single-phase and three-phase rectifier circuits is not pure direct current. A certain amount of alternating-current ripple is evident in each type of rectifier. For many applications, a smooth dc output voltage with the ac ripple removed is required. Circuits used to remove ac variations of rectified direct current are called *filter* circuits.

The output of a rectifier has a dc value and an ac ripple value as shown in Fig. 3-52. To gain a relative

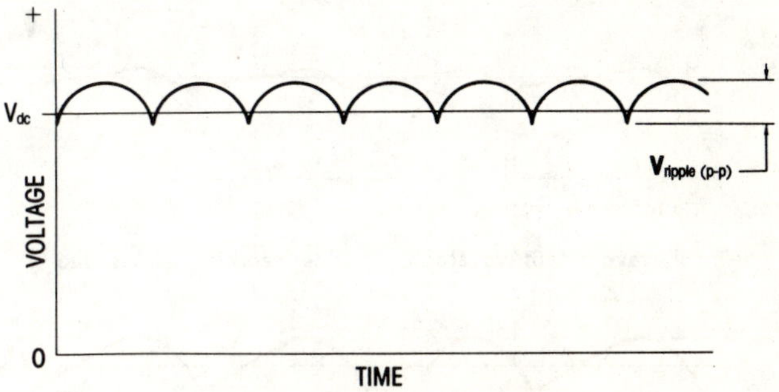

Fig. 3-52. Rectifier output voltage.

index of the amount of ac variation, the ripple factor of the output waveform of a rectifier can be determined by:

$$r = \frac{V_{r(rms)}}{V_{dc}}$$

where,

r is the ripple factor,
$V_{r(rms)}$ is the rms value of the ac component,
V_{dc} is the average value of the rectified dc voltage.

Another index used to express the amount of ac variation in the output of a rectifier is the percentage of ripple. Ripple percentage can be determined by:

$$\%\ \text{ripple} = \frac{V_{r(rms)}}{V_{dc}} \times 100.$$

A full-wave rectified voltage has less percentage of ripple than does a half-wave rectified voltage. When a dc supply must have a low amount of ripple, a full-wave rectifier circuit should be used.

Capacitor Filter

A simple capacitor filter can be used to smooth the ac ripple of a rectifier output. Fig. 3-53 shows the result of adding a capacitor across the output of a 60-Hz, single-phase, full-wave bridge rectifier. The output waveform after the capacitor has been added is shown in Fig. 3-53C.

The ideal filtered dc voltage would be one with no ac ripple and a value equal to the peak voltage (V_{max})

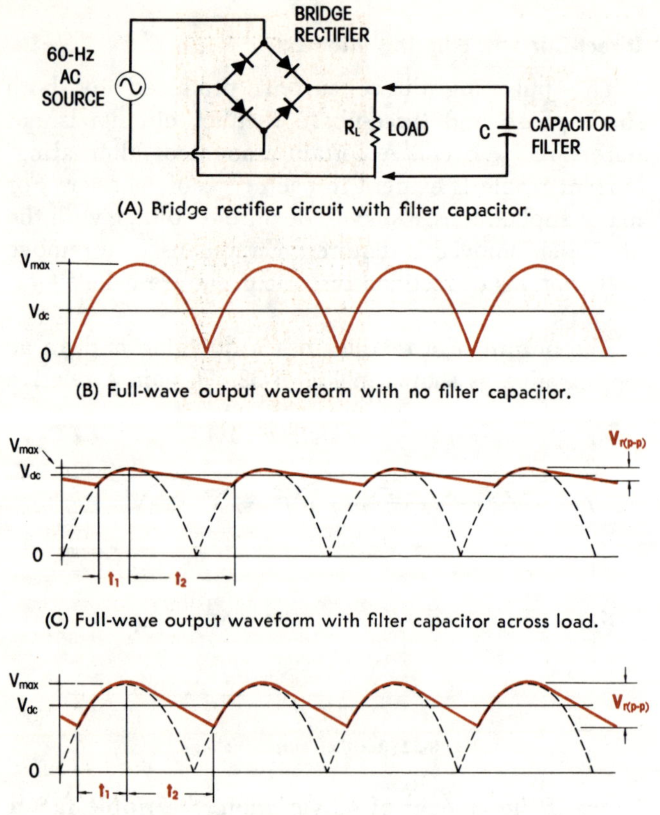

(A) Bridge rectifier circuit with filter capacitor.

(B) Full-wave output waveform with no filter capacitor.

(C) Full-wave output waveform with filter capacitor across load.

(D) Full-wave output waveform with filter capacitor across heavier (lower resistance) load.

Fig. 3-53. Filter capacitor operation.

from the rectifier output. Note that in Fig. 3-53C the value of V_{dc} is approaching that of V_{max}, and compare this to the full-wave rectified voltage of Fig. 3-53B. There are two time intervals shown in Fig. 3-53C. Time period t_1 represents diode conduction, which charges the filter capacitor (C) to the peak rectified voltage (V_{max}). Time period t_2 is the time required for the capacitor to discharge through the load (R_L).

A different value of filter capacitor would result in a change in the rate of discharge. If C discharged a very small amount, the value of V_{dc} would be closer to the value of V_{max}. With light loads (high resistance), the capacitor filter will supply a high dc voltage with little ripple. However, with a heavy (low resistance) load connected, the dc voltage level will drop due to a greater ripple. The increased ripple is caused by the lower resistance discharge path for the filter capacitor. The effect of a heavier load on the filter capacitor is shown in Fig. 3-53D.

By utilizing the values indicated on the waveforms of Fig. 3-53, it is possible to express V_{dc} and $V_{r(rms)}$ as:

$$V_{dc} = V_{max} - \frac{V_{r(p\text{-}p)}}{2},$$

and

$$V_{r(rms)} = \frac{V_{r(p\text{-}p)}}{2\sqrt{3}}.$$

We can also express the amount of ripple of a 60-Hz, full-wave filter-capacitor circuit with a light load as:

$$V_{r(rms)} = \frac{2.4 I_{dc}}{C},$$

or

$$r = \frac{2.4}{R_L C},$$

where,

I_{dc} is the load current in milliamperes,
C is the filter capacitor value in microfarads,
R_L is the load resistance in kilohms.

The average dc voltage output can be expressed as:

$$V_{dc} = V_{max} - \frac{4.16 I_{dc}}{C}.$$

Again, the value of V_{dc} is for a 60-Hz, full-wave filter-capacitor circuit with a light load.

With a heavier load (lower resistance) connected to the filter circuit, more current (I_{dc}) is drawn. As I_{dc} increases, V_{dc} decreases. However, if the value of filter capacitor C is made larger, the value of V_{dc} becomes closer to that of V_{max}. The value of C for a 60-Hz, full-wave rectifier can be determined by using the equation:

$$C = \frac{2.4 I_{dc}}{V_{r(rms)}}.$$

It should be pointed out, however, that as the value of C increases, the peak value of the current through the diodes will also increase. There is, therefore, a practical limit for the value of C that is reflected in the foregoing equation.

The filter capacitor produces a high dc voltage with low ac ripple for light loads. However, its major disadvantages are higher ripple and lower V_{dc} at heavier loads, poor voltage regulation, and high peak current through the diodes.

RC Filter

It is possible to improve upon the previous filter circuit by using an RC filter. Fig. 3-54 shows an RC filter stage. This filter has lower ripple than a capacitor filter does, but has a lower average dc voltage due to the voltage dropped across R_1. The purpose of R_1 and C_2 is to add another filter network, which further reduces the ripple. This circuit also operates best with light loads connected. It is possible to have many stages of RC filters to further reduce ac ripple.

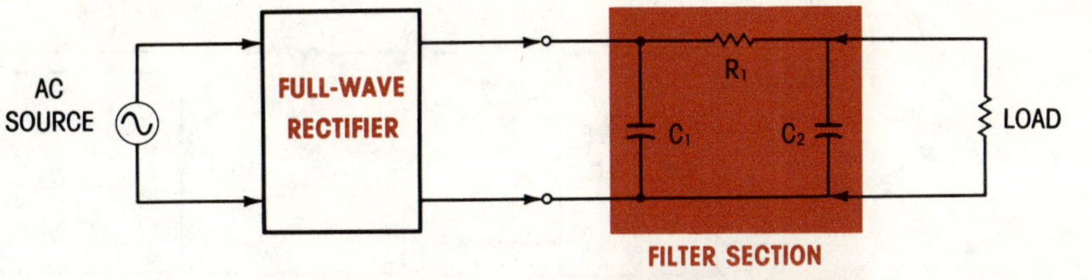

Fig. 3-54. An RC filter circuit.

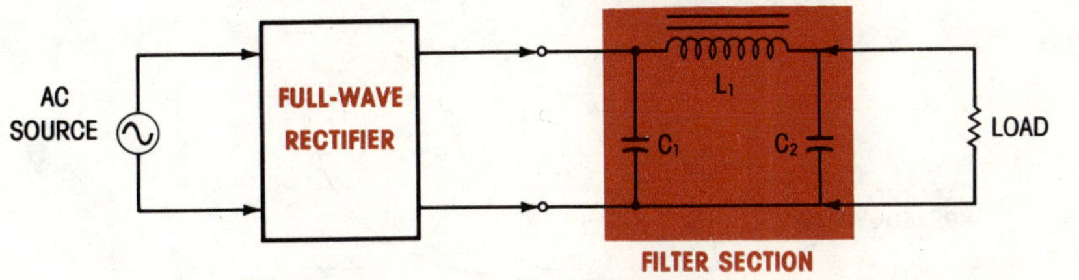

Fig. 3-55. A pi (π)-type filter circuit.

Pi (π)-Type Filter

The use of a resistor (R_1) in the RC filter is not desirable in many cases, since it reduces the average dc output of the circuit. To compensate for the dc voltage reduction, a pi (π)-type filter can be used. Fig. 3-55 shows this type of filter. The advantage of the choke coil (L_1) over the resistor (R_1) of the RC filter is that it offers only a small dc series resistance, but its ac impedance is much larger. Therefore, it passes dc and blocks the ac component of the rectified voltage. Several sections of this type of filter can also be used to further reduce ac ripple.

Voltage Multiplier Circuits

The purpose of a voltage multiplier circuit is to increase the value of rectified dc voltage output applied to a load. Voltage multiplier circuits are modifications of filter circuits. The applications of these circuits are limited in industry; however, there are specific types of equipment that use voltage multiplier circuits.

The basic voltage multiplier circuits are the voltage doublers shown in Fig. 3-56. In the half-wave doubler circuit, shown in Fig. 3-56A, when point A has an instantaneous positive charge, diode D_1 will conduct and diode D_2 will be reverse biased. Capacitor C_1 will charge to the peak rectified voltage (V_{max}). The right side of capacitor C_1 will obtain a positive charge. During the next half-cycle of ac input, the instantaneous polarities will be reversed, placing point B at a positive potential. Diode D_2 will now conduct while D_1 will be reverse biased, placing a charge on capacitor C_2 equal to V_{max}. During the next half-cycle of ac input, the charge on capacitor C_1 will accumulate on capacitor C_2 by discharging through D_2. Also during this half-cycle, capacitor C_2 will discharge through the load. The sum of the voltages accumulated on C_2 will be equal to $2V_{max}$. The output is similar to a half-wave rectified waveform that has a simple capacitor filter. The diode prv rating must be $2V_{max}$ in value.

The operation of the full-wave voltage doubler of Fig. 3-56B is similar to the half-wave doubler. When point A has a positive potential, diode D_1 will conduct, allowing C_1 to charge to the value of V_{max}. The next half-cycle of ac input will place an instantaneous positive charge on point B. Diode D_2 will then conduct, charging capacitor C_2. With no load connected, the dc voltage at the output terminals is $2V_{max}$. The prv rating of the diodes must also be $2V_{max}$. More sections of these diode/capacitor circuits can be connected together to form voltage tripler or voltage quadrupler circuits. Extremely high dc voltages can be produced in this way.

Direct-Current Regulation Methods

The concept of voltage regulation can best be understood by referring to the formula:

$$VR = \frac{V_{NL} - V_{FL}}{V_{FL}} \times 100$$

where,

VR is the voltage regulation in percent,
V_{NL} is the no-load terminal voltage,
V_{FL} is the rated full-load terminal voltage.

In an ideally regulated circuit, V_{NL} would equal V_{FL}.

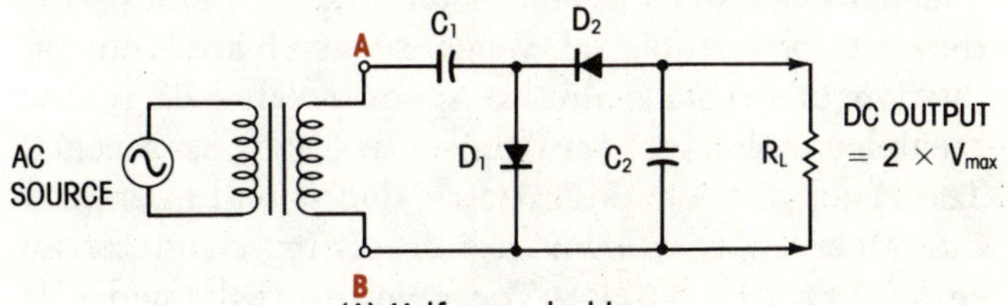

(A) Half-wave doubler.

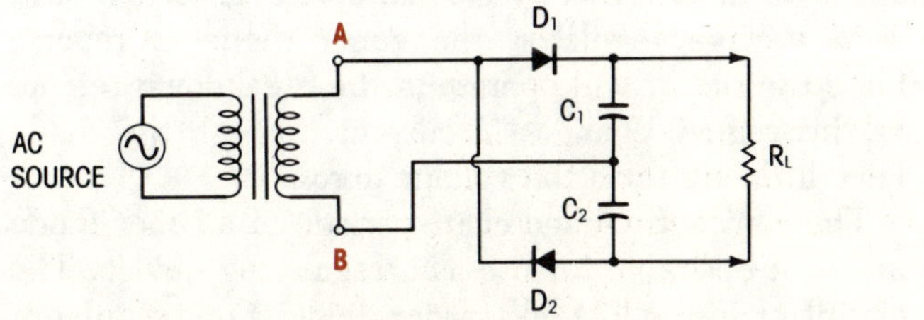

(B) Full-wave doubler.

Fig. 3-56. Voltage doubler circuits.

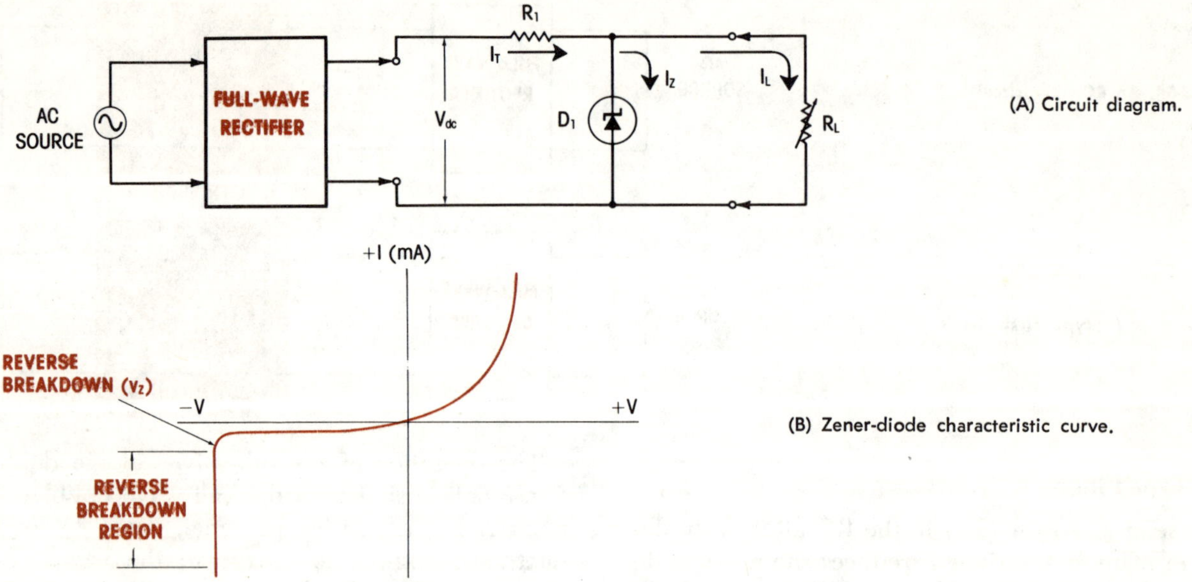

Fig. 3-57. Zener-diode voltage regulator.

In all types of power supplies that convert ac to dc, the dc output levels are affected by variations in the load. The lower the percentage of regulation (approaching 0%), the better regulated the circuit is. For instance, power supplies are capable of having a voltage regulation of less than 0.01%, which means that the value of the load has little effect on the dc output voltage produced. A well-regulated dc power supply is necessary for many industrial applications.

Zener Diode Regulators

A simple voltage regulator circuit uses a zener diode as shown in Fig. 3-57. This circuit consists of a series resistor (R_1) and a zener diode (D_1) connected to the output of a rectifier circuit. We should review zener diode operation before discussing the circuit operation.

Zener diodes are similar to pn junction diodes when they are forward biased. When reverse biased, no conduction takes place until a specific value of reverse breakdown voltage (or "zener" voltage) is reached. The zener diode is designed so that it will operate in the reverse breakdown region of its characteristic curve (see Fig. 3-57B). The reverse breakdown voltage is predetermined by the manufacturer. When used as a voltage regulator, the zener diode is reverse biased so that it will operate in the breakdown region. In this region, changes in current through the diode have little effect on the voltage across it.

The constant-voltage characteristic of a zener diode makes it desirable for use as a regulating device. The circuit of Fig. 3-57A is a zener-diode shunt regulator. The zener establishes a constant voltage across the load resistance within a range of rectified dc voltages and output load currents. Over this range, the voltage drop across the zener remains constant. The current flow through the zener (I_Z) will vary to compensate for changes in load resistance, since $I_Z = I_T - I_L$. Thus, the output voltage will remain constant.

Transistor Voltage Regulators

An improvement over the zener-diode voltage regulator is a transistorized regulator such as the circuit of Fig. 3-58. This regulator has a transistor (Q_1) placed in series with the load device (R_L). Transistor Q_1 acts as a variable resistance to compensate for changes in input voltage. The collector-emitter resistance of Q_1 varies automatically with changes in the circuit conditions. The zener diode establishes the dc bias placed on the base of Q_1. When this circuit is operating properly, if the voltage across the load (R_L) increases, the rise in emitter voltage makes the base less positive. The current through Q_1 will then be reduced, which results in an increase in the collector-emitter resistance of Q_1. The increase in resistance will cause a larger voltage drop across Q_1, which will now compensate for the change in voltage across the load. Opposite conditions

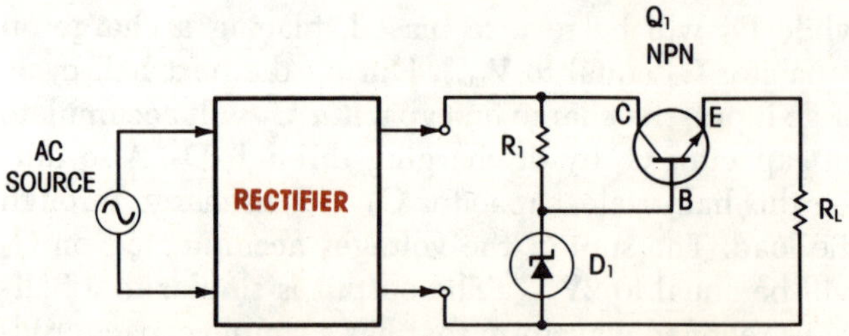

Fig. 3-58. Series transistor voltage regulator circuit.

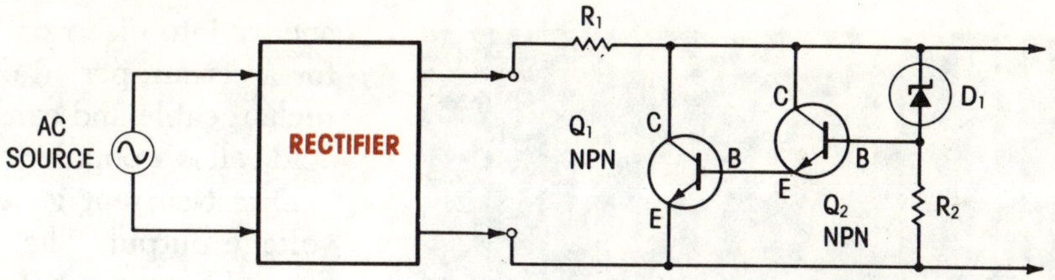

Fig. 3-59. Shunt transistor voltage regulator circuit.

would occur if the load voltage were to decrease. Many variations of this circuit are used in regulated power supplies today.

Shunt transistor regulators are also used in dc power supplies. The circuit of Fig. 3-59 is a shunt voltage regulator. Again, the zener diode (D_1) is used to establish a constant dc bias level. Therefore, voltage variations across the dc output will be sensed only by resistor R_2. If the dc output voltage rises, an increased positive voltage will be present at the base of Q_1. The increased forward bias on Q_1 will cause it to conduct more, which makes the base of Q_2 more positive. Transistor Q_2 will then conduct more heavily. Increased current flow through both transistors causes an increase in the voltage drop across R_1, which will then counterbalance the rise in output voltage. Thus, the dc output voltage will remain stabilized. Decreases in dc output voltage will cause the circuit action to reverse. Shunt transistor voltage regulator circuits of many types are also used extensively in power supplies. Some examples of commercial regulated dc power supplies are shown in Figs. 3-60 through 3-63.

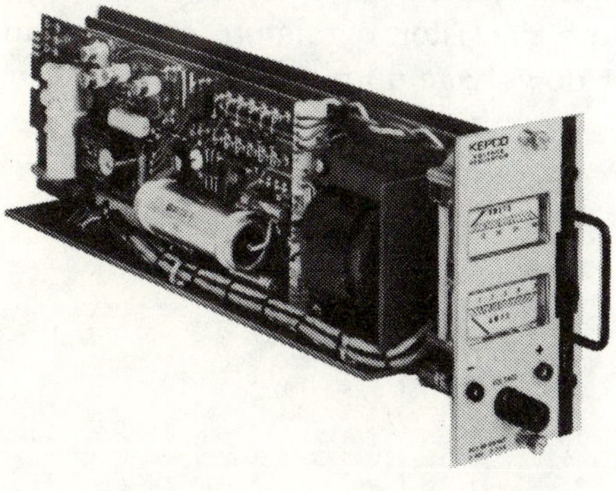

Courtesy Kepco, Inc.

Fig. 3-60. A regulated dc power supply.

SUMMARY

In this chapter, we have discussed electrical energy sources for industry. Traditional energy sources such as fossil-fuel systems, hydroelectric systems, and nuclear-fission systems continue to be our largest electrical power sources. Recent energy shortages and an increasing awareness of our shrinking supply of natural resources have brought about experimentation with new energy sources such as solar cells, nuclear fusion,

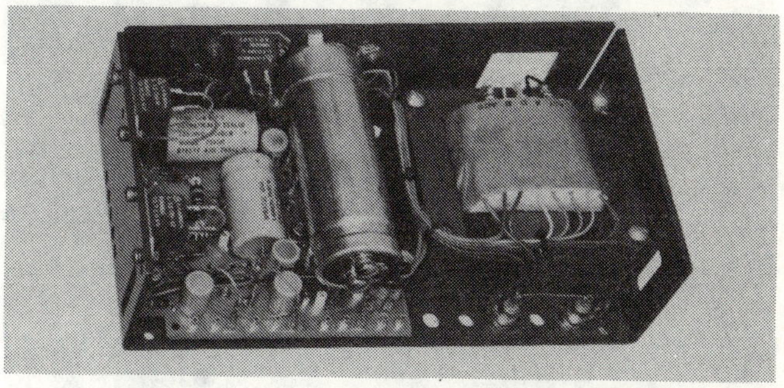

Courtesy Lambda Electronics

Fig. 3-61. A microprocessor power supply.

geothermal systems, wind-powered systems, MHD systems, and fuel cells.

Most of the electrical power produced in the United States is three-phase alternating current. However, both three-phase and single-phase generators have specific applications. Two basic methods of producing power are the rotating armature and rotating field methods. Three-phase generators ordinarily have large power outputs and, therefore, use the rotating field method. The windings of a three-phase generator can be connected in either a wye or delta configuration. In the wye system, $V_L = V_P \times 1.73$ and $I_L = I_P$, while in the delta system, $V_L = V_P$ and $I_L = I_P \times 1.73$.

Courtesy Honeywell, Inc.

Fig. 3-62. An ac to dc power controller.

Courtesy Westinghouse Electric Corp.

Fig. 3-63. A heavy-duty industrial dc power supply.

Direct-current sources are used in industry for many specialized applications. Several methods may be used to produce direct-current energy. Among these methods are primary and secondary chemical cells, direct-current generators, and single-phase and three-phase rectification systems. Chemical cells convert chemical energy into electrical energy. Primary cells are usable for a certain period of time, while secondary cells are rechargeable and can be used over and over. The lead-acid cell is a popular type of secondary cell.

Direct-current generators are used to produce a dc voltage output. The basic types of dc generators are the: (1) permanent-magnet field; (2) separately excited field; (3) self-excited, series-wound field; (4) self-excited, shunt-wound field; and (5) self-excited, compound-wound field. Each of these types of dc generators has a different set of characteristics.

Rectification systems are used in industry to convert alternating current to direct current. Single-phase and three-phase conversion systems utilize the action of semiconductor diodes, vacuum-tube diodes, gaseous-tube diodes, or rotary converters. Semiconductors are used much more frequently than other methods of rectification. The types of rectification systems used in industry today include the single-phase, half-wave; single-phase, full-wave; single-phase bridge; three-phase, half-wave; and three-phase, full-wave.

In order to produce a purer form of direct current, a rectifier output must be filtered. Several filtering methods are commonly used in power supplies that convert ac to dc. Capacitor filters, RC filters, and pi (π)-type filters are among the most used types. In some cases, power supplies contain voltage multiplier or voltage regulation circuits. Voltage multiplier circuits are used to increase the dc voltage output. Voltage regulators, such as zener-diode circuits and series and shunt transistor regulators are used to assure a constant dc voltage output under changes in load conditions.

Study Questions

1. *What are the three most used electrical energy sources?*
2. *What are some alternative electrical energy sources?*
3. *Briefly describe the following power systems:*
   a. *Fossil-fuel system.*
   b. *Hydroelectric system.*
   c. *Nuclear-fission system.*
   d. *Solar energy system.*
   e. *Geothermal system.*
   f. *Wind-power system.*
   g. *Magnetohydrodynamic system.*
   h. *Nuclear-fusion system.*
   i. *Fuel-cell system.*
4. *What three conditions must be met in order to convert mechanical energy into electrical energy?*
5. *Describe the rotating armature and rotating field methods of producing electrical energy.*
6. *How is the frequency of an alternator determined?*
7. *What effect does load have on an alternator?*
8. *Define voltage regulation of an alternator.*
9. *Define efficiency regulation of an alternator.*
10. *Make a sketch of the wye and delta systems for connecting three-phase alternator windings.*
11. *What are the voltage and current relationships in wye and delta systems?*
12. *What are some methods used to produce direct-current energy?*
13. *Describe primary cells and secondary cells.*
14. *What is meant by the internal resistance of a chemical cell?*
15. *What are some types of primary cells? Secondary cells?*
16. *Discuss the operation of a lead-acid cell.*
17. *What is meant by specific gravity? Ampere-hour rating?*
18. *What are the types of direct-current generators?*
19. *What is the difference between a separately excited and a self-excited dc generator?*
20. *What methods may be employed to convert alternating current to direct current?*
21. *Discuss the operation of a semiconductor diode.*
22. *Describe the following types of rectification systems:*
    a. *Single-phase, half-wave.*
    b. *Single-phase, full-wave.*
    c. *Single-phase bridge.*
    d. *Three-phase, half-wave.*
    e. *Three-phase, full-wave.*
23. *Discuss vacuum-tube and gaseous-tube rectifiers.*
24. *How does a rotary converter operate?*
25. *What methods may be used to filter direct current?*
26. *What is a voltage multiplier?*
27. *What are some methods used for voltage regulation?*

Problems

1. If the peak voltage of a 60-Hz ac generator is 170 volts, what is the instantaneous voltage after a conductor has rotated 70°?
2. What is the instantaneous voltage after the conductor in Problem 1 has rotated 140°? 220°? 310°?
3. At what speed must a two-pole ac alternator rotate in order to produce a frequency of 50 Hz?
4. What is the voltage regulation of a generator that has a full-load output of 240 volts and a no-load output of 260 volts?
5. What is the efficiency of a generator whose rated output is 60 kW when the input is 80 hp?
6. The phase voltage of a three-phase, wye-connected generator is 277 volts. What is the line voltage?
7. If a wye-connected generator has a phase voltage of 120 volts, what is the line voltage?
8. The phase current of a delta-connected generator is 80 amperes. What is the line current?
9. Three loads, each having a resistance of 10 ohms, are placed across the output of a wye-connected generator. The generator output is 480 volts. Find the line current and total three-phase power.
10. Calculate the line current and total three-phase power for the generator of Problem 10 if it is delta connected.
11. A shunt-wound dc generator delivers 50 amperes to a load. If the field current is 2 amperes, what is the armature current?
12. A shunt-wound dc generator is rated at 300 kW and 240 volts. Its field resistance is 50 ohms. Find: (a) line current, (b) field current, (c) armature current.
13. What is the average value of dc output voltage of a half-wave rectified voltage of 200 volts peak value?
14. If a 120-volt (rms) voltage is rectified by a half-wave circuit, what is the dc output voltage?
15. What is the dc voltage obtained from a full-wave rectifier if the peak ac input is 150 volts?
16. Calculate the ripple factor and percentage ripple of a filtered rectifier circuit whose dc voltage is 20 volts and ac ripple voltage is 1.5 volts (rms).
17. The no-load voltage of a dc power supply is 160 volts. The full-load voltage is 150 volts. What is the voltage regulation?
18. What is the ripple voltage of a full-wave rectifier with a capacitor filter of 25 μF and a load current of 50 mA?
19. In the circuit of Problem 18, if the peak ac voltage input is 150 volts, what is the filtered dc voltage?
20. A three-phase, half-wave rectifier has an ac voltage input of 240 volts (rms). What is the average dc voltage output?

4

Power Distribution Systems in Industry

Although the field of industrial electronics has become very sophisticated in recent years, there are still numerous electrical devices associated with industrial systems. In order to transfer electrical power from an alternating-current or direct-current source to a load device, some type of distribution network must be utilized. The method used to distribute power from source to load may be quite simple. For example, a battery (source) could be connected directly to a motor (load), with only a set of wires to connect them together.

More-complex power distribution systems are commonly used for industrial equipment control. Distribution systems might employ such equipment as transformers, circuit breakers, protective devices, printed-circuit boards, and metallic terminal connectors. At this time, we should distinguish between three types of power distribution. The first type is distribution *to* the industry. The second type is distribution *inside* the industrial plant. The third type is the distribution of power *within* industrial equipment and its associated control system.

DISTRIBUTION OF POWER TO INDUSTRY

The distribution of power in the United States is normally three-phase, 60-Hz alternating current. This power, of course, can be manipulated or changed in many ways by the use of electrical circuitry. For instance, a rectification system is capable of converting 60-Hz alternating current to a form of direct current, as discussed in Chapter 3. Single-phase power is generally suitable for lighting and small appliances such as those used in the home. However, three-phase power is more economical where a large amount of electrical energy is required. Since industries use well over 40% of all electrical power produced, three-phase power is distributed directly to most industries. Electrical substations located near industries, such as the one shown in Fig. 4-1, employ massive transformers and associated equipment such as oil-filled circuit breakers, high-voltage conductors, and huge strings of insulators to distribute power to the industry. From these substations, power is brought into the industrial site so that the many kinds of equipment used for mass production and specialized industrial processes can be energized.

TRANSFORMER PRINCIPLES

The heart of power distribution systems is an electrical device known as a *transformer*. This device is capable of controlling massive amounts of power for efficient distribution. Industries also use transformers for many other applications. Before we continue, we should review the theory of transformer operation.

Transformers provide a means of converting an alternating-current voltage from one value to another. The basic construction of a transformer is illustrated in Fig. 4-2. Notice that the transformer consists of two sets of windings that are not physically connected. The only connection between the primary and secondary windings is the magnetic coupling effect, known as *mutual induction,* which takes place when the circuit is energized by an alternating-current source. The metallic core plays an important role in transferring magnetic flux from the primary winding to the secondary winding. A single-phase, 25-kVA distribution transformer is shown in Fig. 4-3.

If an alternating current, which is constantly changing in value, flows in the primary winding, the magnetic flux produced around the primary will be transferred to the secondary winding. Thus, an induced voltage is developed across the secondary winding. In this way, electrical energy is transferred from the source to a load.

The efficient transfer of energy from primary to secondary windings depends on the flux linkage of the magnetic field between these sets of windings. Ideally,

Fig. 4-1. An electrical substation.

all magnetic flux lines developed around the primary would be transferred by flux linkage to the secondary. However, a certain amount of flux leakage takes place as some lines of force escape to the surrounding air.

In order to decrease the amount of flux leakage, transformer windings are wound around iron cores. Iron cores concentrate the magnetic flux so that better flux linkage between primary and secondary is accomplished. Two types of cores are illustrated in Fig. 4-4. These cores are made of laminated iron to reduce undesirable eddy currents, which are induced into the core material. The diagram of Fig. 4-4A shows closed-core transformer construction where the windings are

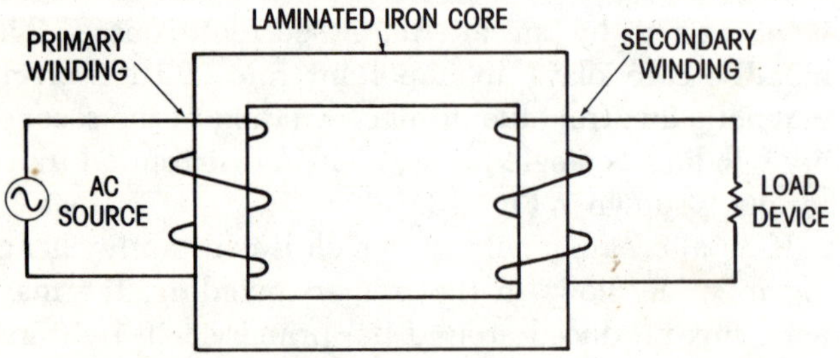

(A) Pictorial diagram.

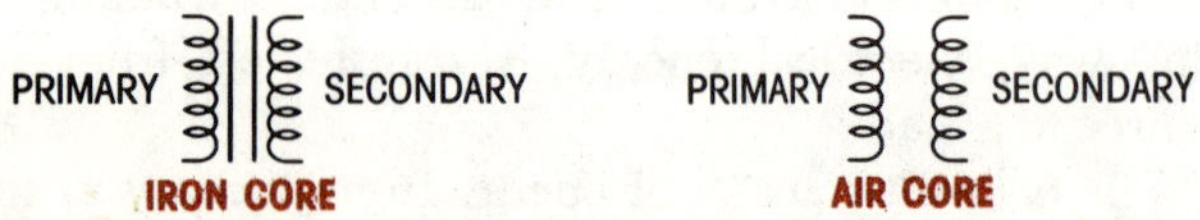

(B) Schematic diagrams.

Fig. 4-2. Basic transformer construction.

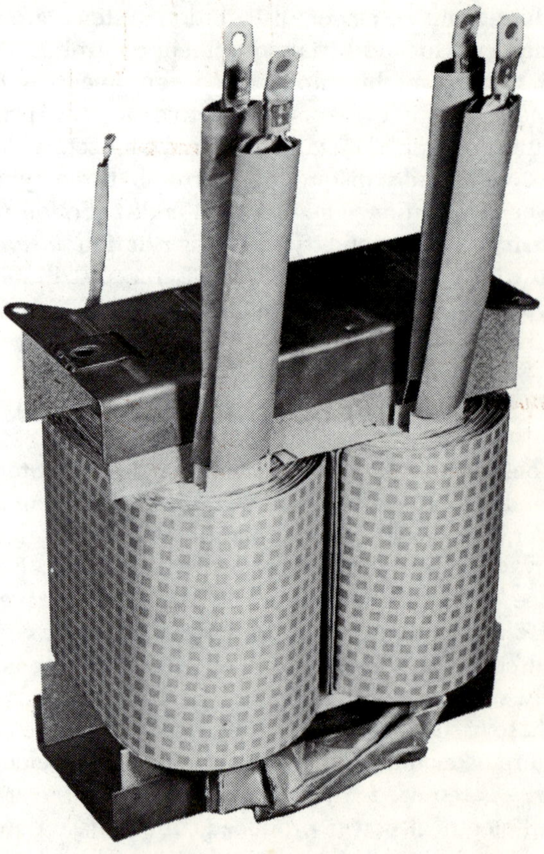

Courtesy General Electric Co.

Fig. 4-3. A single-phase, 25-kVA distribution transformer.

placed along the outside of the metal core. Fig. 4-4B shows the shell-core construction, which produces better magnetic coupling since the windings are surrounded by metal on both sides. Note that the primary and secondary windings are placed adjacent to one another for better coupling.

Transformers are very efficient electrical devices. A typical efficiency rating would be around 98%. Recall that efficiency is expressed as:

$$\text{Efficiency } (\%) = \frac{P_{out}}{P_{in}} \times 100$$

where,

P_{out} is the power output in watts,
P_{in} is the power input in watts.

The losses that reduce efficiency, in addition to flux leakage, are copper and iron losses. Copper loss is the I^2R loss of the windings, while iron losses are those caused by the metallic core material. The insulated laminations of the iron core help to reduce iron losses.

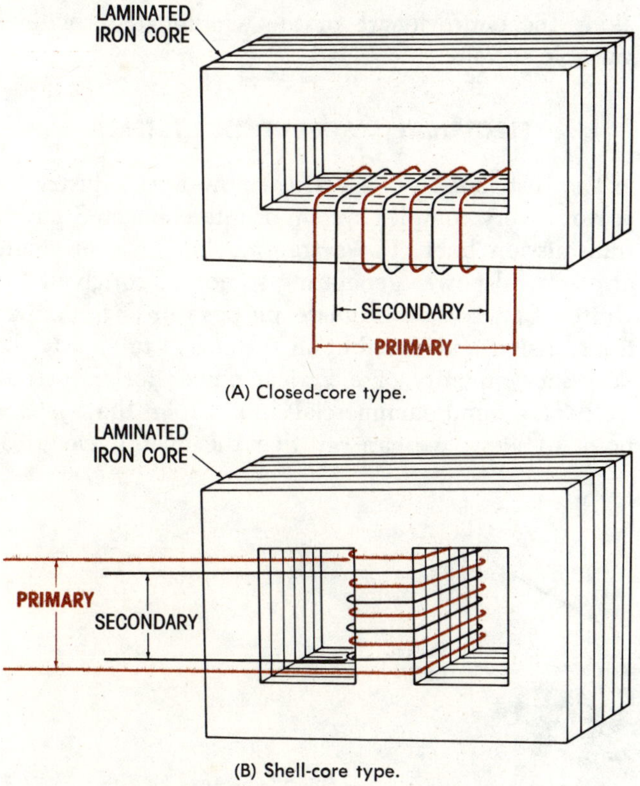

(A) Closed-core type.

(B) Shell-core type.

Fig. 4-4. Types of iron-core transformers.

Transformers are functionally classified as step-up or step-down types. These types are illustrated in Fig. 4-5. The step-up transformer of Fig. 4-5A has fewer turns of wire on the primary than on the secondary. If the primary winding has 50 turns of wire and the secondary has 500 turns, a turns ratio of 1:10 is developed. Therefore, if 12 volts ac is applied to the primary from the source, 10 times that voltage, or 120 volts ac, will be transferred to the secondary load (assuming no losses).

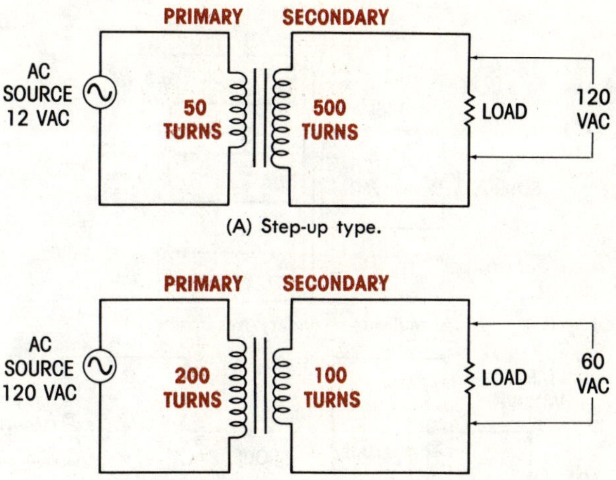

(A) Step-up type.

(B) Step-down type.

Fig. 4-5. Step-up and step-down transformers.

The example of Fig. 4-5B is a step-down transformer. The step-down transformer has more turns of wire on the primary than on the secondary. The primary winding of the example has 200 turns while the secondary winding has 100 turns, or a 2:1 ratio. If 120 volts ac is applied to the primary from the source, then one-half that amount, or 60 volts ac, will be transferred to the secondary load.

From the preceding examples, we can see that there is a direct relationship between the primary and secondary turns and the voltages across each winding. This relationship can be expressed as:

$$\frac{V_p}{V_s} = \frac{N_p}{N_s}$$

where,

V_p is the voltage across the primary winding,
V_s is the voltage across the secondary winding,
N_p is the number of turns in the primary winding,
N_s is the number of turns in the secondary winding.

The transformer is a power-control device; therefore, the following relationship can be expressed:

$$P_p = P_s + \text{losses}$$

where,

P_p is the primary power,
P_s is the secondary power,

and the losses are those that ordinarily occur in a transformer. In transformer theory, an ideal device is usually assumed, and losses are not considered. Thus, since $P_p = P_s$ and $P = V \times I$, then

$$V_p \times I_p = V_s \times I_s$$

where,

I_p is the primary current,
I_s is the secondary current.

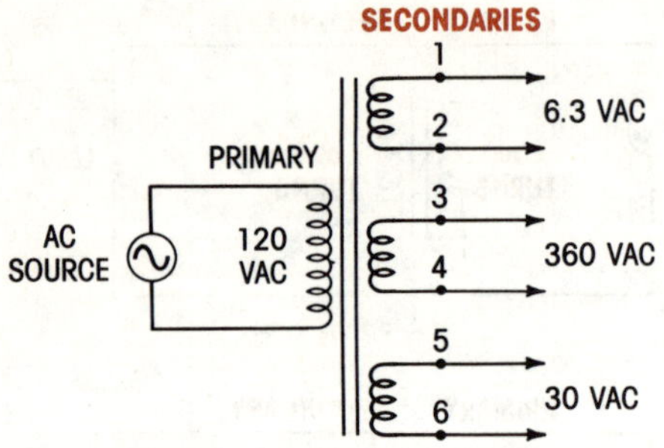

(A) Multiple-secondary transformer.

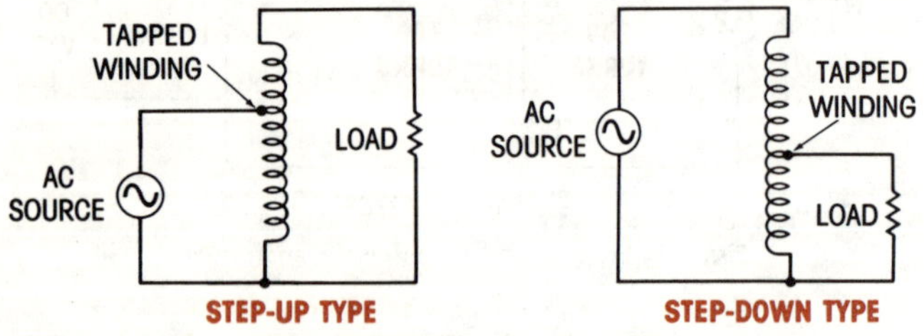

(B) Autotransformers.

Fig. 4-6. Specialized types of transformers.

Therefore, if the voltage across the secondary is stepped up to twice the voltage across the primary, then the secondary current will be stepped down to one-half the primary current. The current relationship of a transformer is thus expressed as:

$$\frac{I_p}{I_s} = \frac{N_s}{N_p}.$$

Note that whereas the voltage-turns ratio is a *direct* relationship, the current-turns ratio is an *inverse* relationship.

It is also possible to construct a transformer that has multiple secondary windings, as shown in Fig. 4-6A. This transformer is connected to a source of 120 volts ac, which produces the primary magnetic flux. The secondary has two step-down windings and one step-up winding. Between points 1 and 2, a voltage of 6.3 volts ac could be supplied. This voltage is one standard voltage for vacuum-tube filament supplies. Between points 5 and 6, a voltage of 30 volts ac can be obtained, and between points 3 and 4, a voltage of 360 volts ac can be supplied to a load, such as the plate of a vacuum tube or gaseous tube. This type of transformer is often used for the power supply of various types of electronic equipment and instruments.

Another specialized type of transformer is the autotransformer, illustrated in Fig. 4-6B. The autotransformer has only one winding, with a common connection between the primary and secondary. The principle of operation of the autotransformer is similar to other transformers. Both the step-up and step-down types are shown in Fig. 4-6B. Fig. 4-7 shows a variable autotransformer in which the winding tap can be adjusted along the entire length of the winding to provide a variable ac voltage to a load.

ELECTRICAL DISTRIBUTION SYSTEMS

The distribution of electrical power to industry involves a very complex system of interconnected power transmission lines. These transmission lines originate at electrical power generating stations throughout the United States. The ultimate purpose of these power transmission and distribution systems is to supply the necessary quantity of electrical power for industrial, residential, and commercial use. From the systems point of view, we can say that the overall electrical

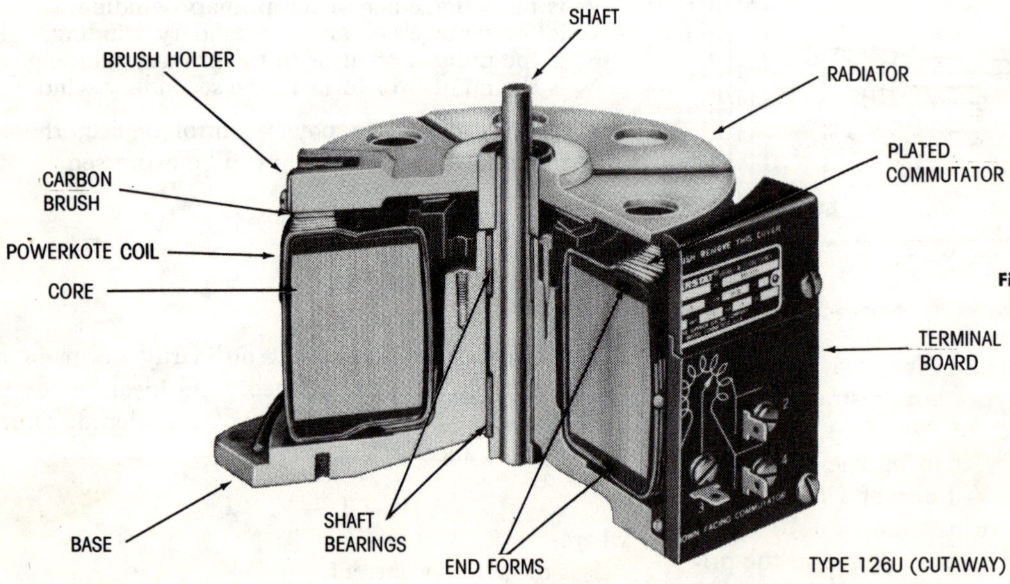

Fig. 4-7. Cutaway view of a variable autotransformer.

Courtesy The Superior Electric Co.

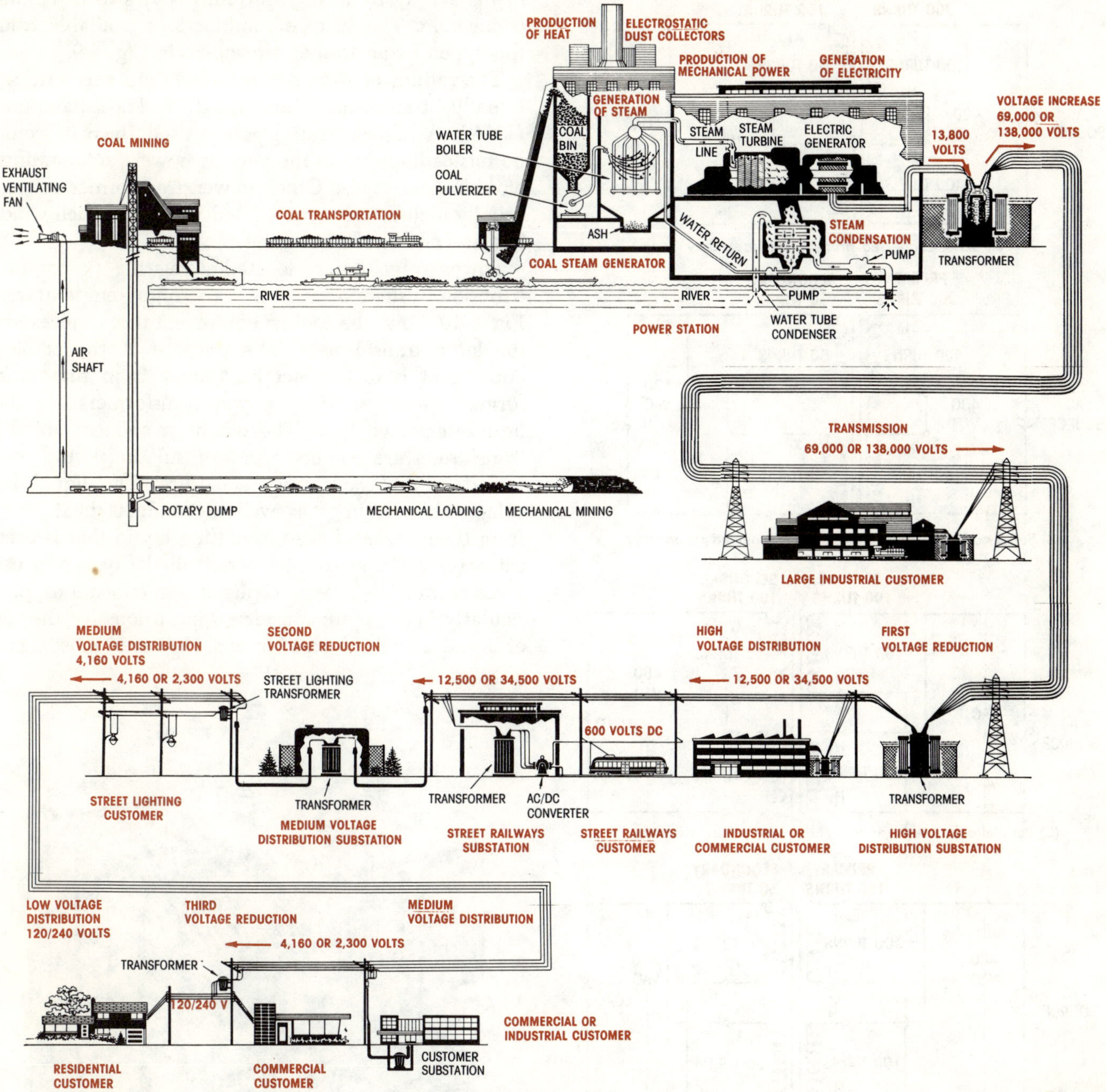

Courtesy Kentucky Utilities Co.

Fig. 4-8. A typical electrical power distribution system.

power system delivers power from the source to the load that is connected to it. A typical electrical power distribution system is illustrated in Fig. 4-8.

The distribution of alternating-current power is dependent upon the use of transformers at many points along the power distribution system. To transmit electrical power over long distances, less current is required at high voltages, since line loss (I^2R) is reduced significantly. A typical high-voltage transmission line may extend a distance of 50–100 miles from the generating station to the first substation. These high-voltage power transmission lines may operate at 300,000 to 500,000 volts by using step-up transformers to increase the voltage produced by the ac generators at the power station. Various substations are encountered along the power distribution system, where transformers are used to reduce the high transmission voltages to a voltage level such as 480 volts, which is suitable for industrial motor loads, or 120/240 volts for residential use.

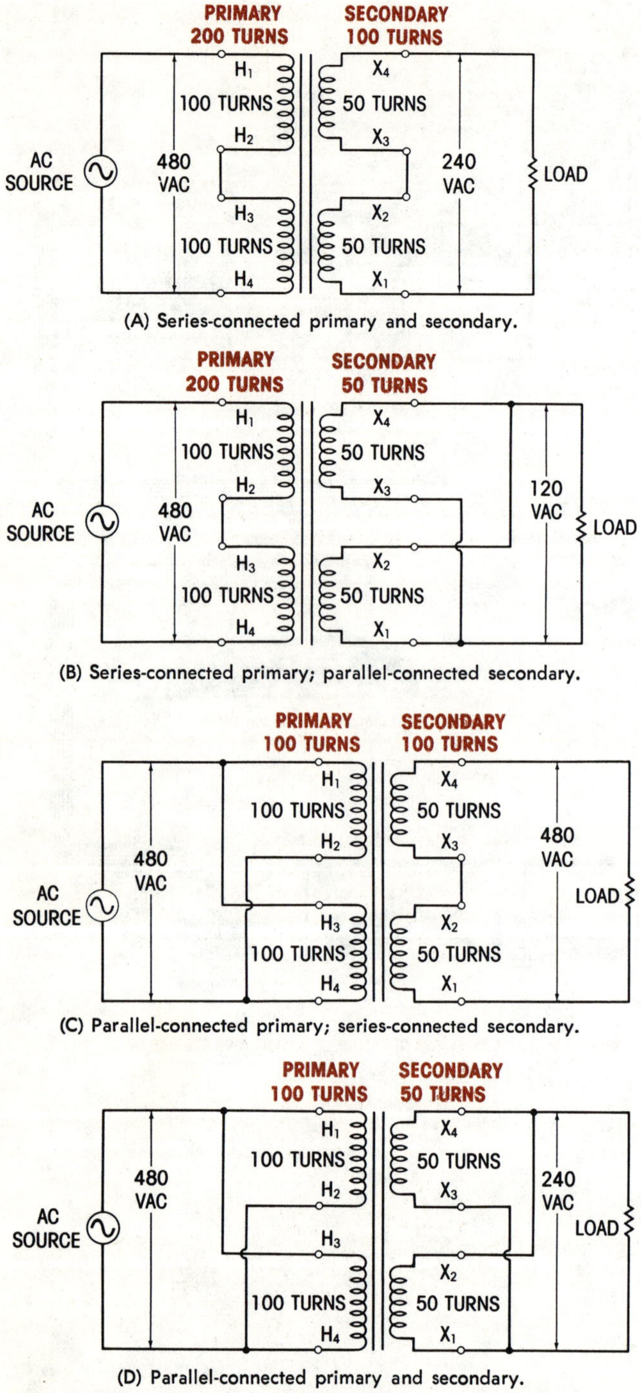

Fig. 4-9. Transformer connection methods for various voltage combinations using a distribution transformer with a 2:1 turns ratio.

Power distribution transformers usually have polarity markings so that their windings can be connected in parallel to increase their current capacity. The standard markings are H_1, H_2, H_3, etc., for the high-voltage windings, and X_1, X_2, X_3, etc., for the low-voltage windings. Many power transformers have two similar primary windings and two similar secondary windings to make them adaptable to different voltage requirements simply by changing from a series to a parallel connection. The voltage combinations available from this type of transformer are shown in Fig. 4-9.

The ratings of power transformers are very critical. Usually, transformers are rated in kilovoltamperes (kVA). A kilowatt rating is not used since it would be misleading due to the various power factor ratings of industrial loads. Other power transformer ratings usually include frequency, rated voltage of each winding, and an impedance rating.

Power transformers located along the power distribution system operate at very high temperatures. Fig. 4-10 shows the cooling equipment that is necessary for large transformers. The purpose of the cooling equipment is to conduct heat away from the transformer windings. Many power transformers are the liquid-immersed type. The windings and core of the transformer are immersed in an insulating liquid contained in the transformer enclosure. The liquid insulates the windings as well as conducts heat away from them. *Askarel* is an insulating liquid that is used extensively. Some transformers, called dry types, use forced air or inert gas as coolants. Some locations, particularly indoors, are considered hazardous for the use of liquid-immersed transformers. However, most trans-

Courtesy Kuhlman Electric Co.

Fig. 4-10. A large three-phase power transformer.

formers rated at over 500 kVA are liquid filled. Fig. 4-11 shows the core and coil assembly for a 500-kVA, 60-Hz, three-phase transformer.

Three-Phase Power Distribution Systems

Since industries predominately use three-phase power, they rely upon three-phase distribution systems to supply this power. Usually, when single-phase voltages are required, they are taken from three-phase lines. Large three-phase distribution transformers are usually located at substations adjoining industrial plants. Their purpose is to supply the proper ac voltages to meet industrial load requirements. The ac voltages transmitted to the industrial substations are high voltages that must be stepped down by three-phase transformers.

There are five methods that can be used to connect the primary and secondary windings of three-phase transformers. These methods are the: (1) delta-delta, (2) delta-wye, (3) wye-wye, (4) wye-delta, and (5) open delta. These basic methods are illustrated in Fig. 4-12. The delta-delta connection is used for some lower-voltage applications. The delta-wye method is commonly used for stepping voltages up, since an inherent step-up factor of 1.73 times is obtained due to the voltage characteristic of the wye-connected secondary. The wye-wye connection is ordinarily not used, while the wye-delta method may be advantageously used to step voltages down. The open-delta connection is used if one transformer winding becomes damaged or is taken out of service. The transformer will still deliver three-phase power, but at a lower current and power capacity. This connection may also be desirable where the full capacity of three transformers is not needed until a later time. Two identical single-phase transformers could be used to supply power to the load until, at a later time, the third transformer is needed to meet increased load requirements.

Three-phase power distribution systems that supply industries may be classified according to the number of phases and number of wires required. These systems, shown in Fig. 4-13, are the: (1) three-phase, three-wire system; (2) three-phase, three-wire system with neutral; and (3) three-phase, four-wire system. The primary winding connection method is not considered here. The three-phase, three-wire system can be used to supply motor loads of 240 volts or 480 volts, for instance. Its major disadvantage is that it only supplies one voltage. Only three hot lines are supplied to the load. The usual insulation color code on these hot lines is black, red, or blue, as specified in the *National Electrical Code.*

The disadvantage of the three-phase, three-wire system can be partially overcome by adding one center-tapped phase winding as shown in the three-phase, three-wire with neutral system of Fig. 4-13B. This system could be used as a supply for 120/240 volts or 240/480 volts. Assume that it is used to supply 120/240 volts. The voltage from the hot line at point 1 and the hot line at point 2 to neutral would be 120 volts,

Fig. 4-11. Core and coil assembly for a 500-kVA, 60-Hz, three-phase transformer.

Courtesy General Electric Co.

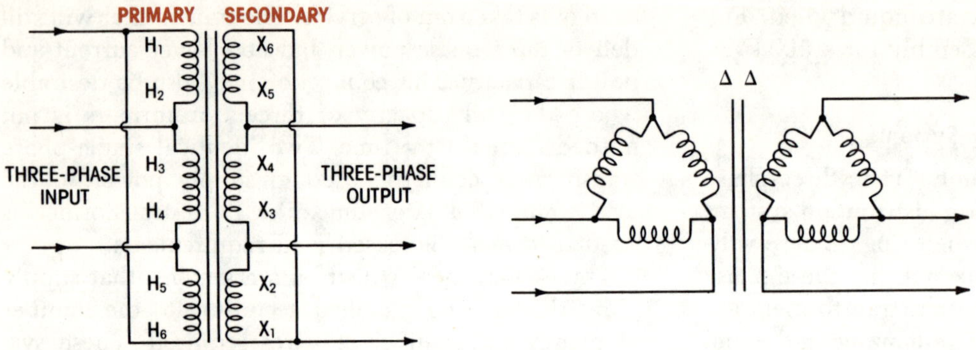

(A) The delta-delta connection.

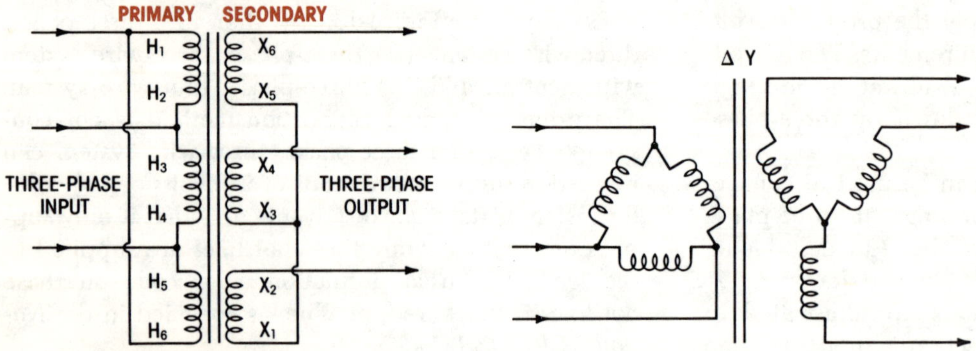

(B) The delta-wye connection.

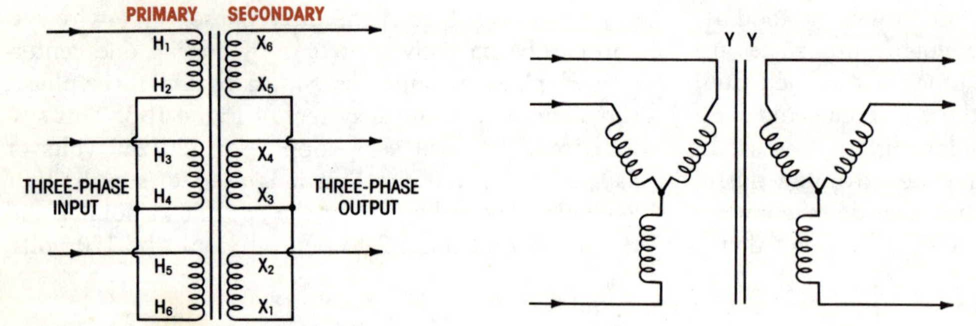

(C) The wye-wye connection.

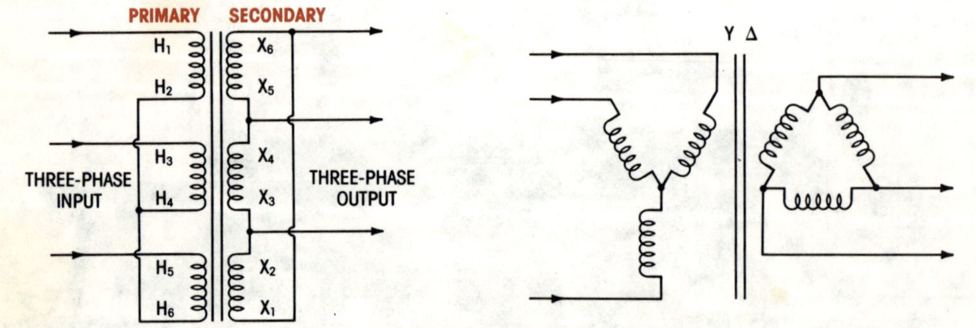

(D) The wye-delta connection.

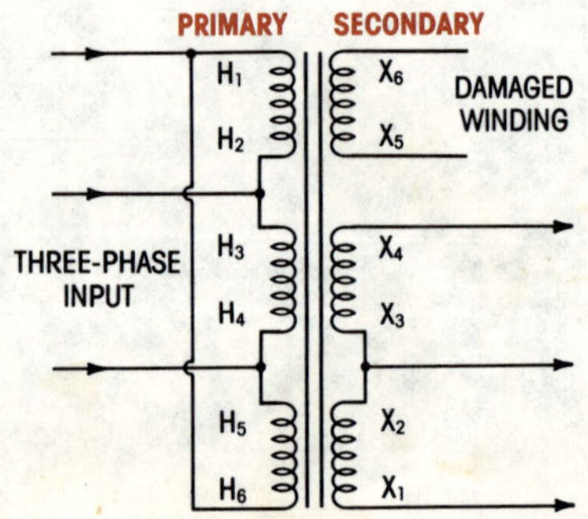

(E) The open-delta connection.

Fig. 4-12. Basic three-phase transformer connection methods.

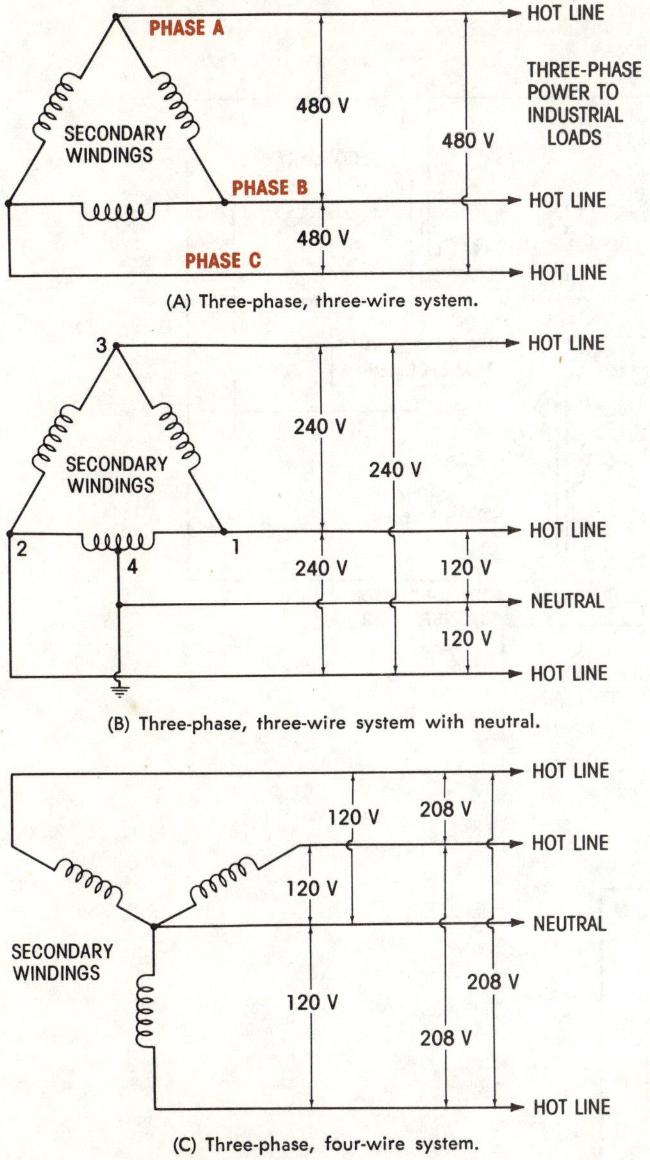

Fig. 4-13. Industrial power distribution systems.

due to the center-tapped winding. Across any two hot lines, 240 volts would still be available. The neutral wire is color coded with white or gray insulation. The disadvantage of this system is that, when making wiring changes, it would be possible to connect a 120-volt load between the neutral and point 3 (sometimes called the "wild" phase). The voltage present here would be the combination of three-phase voltages between points 1 and 4 and points 1 and 3. This voltage would be in excess of 300 volts! Although the "wild" phase situation exists, this system is capable of supplying both high-power loads and low-voltage loads such as lighting and small equipment.

The most widely used industrial distribution system is the three-phase, four-wire system. This system, shown in Fig. 4-13C, commonly supplies 120/208 volts and 277/480 volts for industrial load requirements. The 120/208-volt system is illustrated here. From neutral to any hot line, 120 volts for lighting and low-power loads may be obtained. Across any two hot lines, 208 volts is present for supplying motors or other high-power loads. The most popular system for industrial and commercial power distribution is the 277/408-volt system that is capable of supplying both three-phase and single-phase loads. A 240/416-volt system is also used for industrial loads, while the 120/208-volt system is often used for underground distribution in urban areas. Note that this system is based on the voltage characteristics of the three-phase wye connection and that the relationship $V_L = V_P \times 1.73$ exists for each application of this system.

Single-Phase Distribution Systems

Single-phase distribution systems can be considered in the same way as three-phase systems. Although single-phase systems are used mainly for residential power distribution systems, there are some industrial applications of single-phase systems. Single-phase power distribution is usually originated from three-phase power lines. It is possible to obtain three separate single-phase voltages from a three-phase line. Thus, our power systems are capable of supplying both three-phase and single-phase loads from the same power lines. Fig. 4-14 shows a typical power distribution system from the power station (source) to the various loads that are connected to the system.

Single-phase systems could be of two major types: (1) single-phase, two-wire systems and (2) single-phase, three-wire systems. A single-phase, two-wire system is shown in Fig. 4-15A. This system uses a transformer whose secondary produces one single-phase voltage, such as 120 or 240 volts. In early residential distribution systems, this type was the most used to provide 120-volt service. However, as appliance power requirements increased, the need for a dual voltage system was evident.

To meet the demand for more residential power, the single-phase, three-wire system is now used. A home service entrance could be supplied with 120/240-volt electrical energy by the system of Figs. 4-15B or C. Each of these systems is derived from a three-phase power line. The single-phase, three-wire system has two hot lines and a neutral. The hot lines (black and red insulated wire) are connected to the ends of the transformer secondary windings. The neutral (white insulated wire) is connected to the center tap of the distribution transformer. Thus, from neutral to either hot line, 120 volts for lighting and low-power requirements is available. Across the hot lines, 240 volts is supplied for high-power requirements. Therefore, the

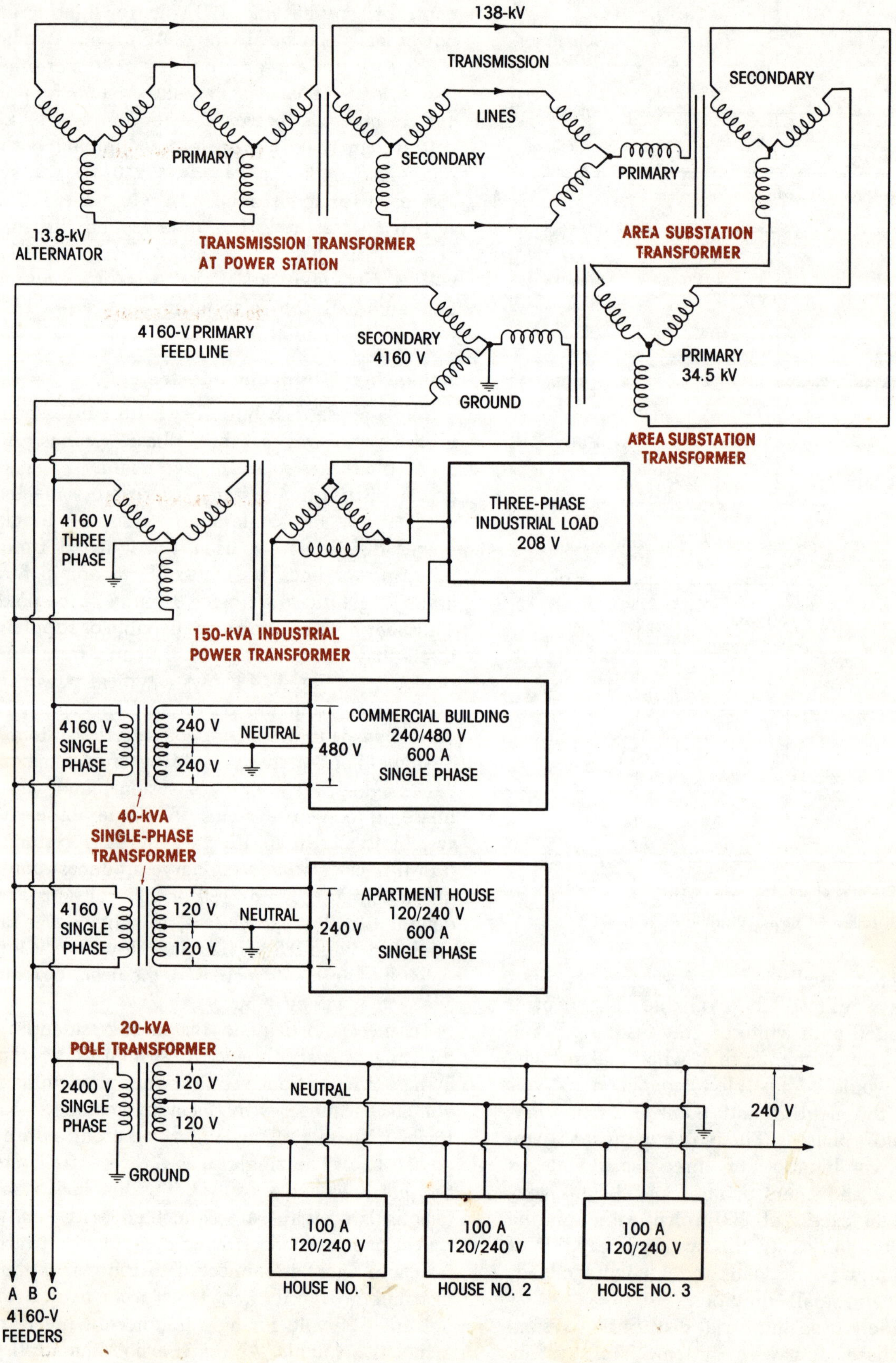

Fig. 4-14. A typical power distribution system.

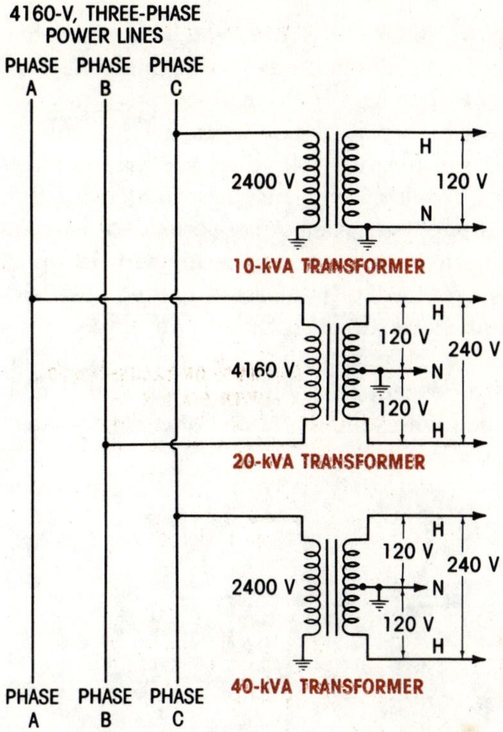

(A) Single-phase, two-wire system.

(B) Single-phase, three-wire system.

(C) Single-phase, three-wire system.

Fig. 4-15. Single-phase power systems.

current requirement for large power consuming equipment is cut in half, since 240 volts rather than 120 volts is used. Either the single-phase, two-wire system or the single-phase, three-wire system could be used to supply single-phase power to industry. However, these single-phase systems are mainly for residential power requirements.

Specialized Distribution Equipment

In order to distribute electrical power to industry, there are many types of specialized equipment required. The electrical power system consists of such specialized equipment as overhead transmission lines, underground cables, transformers, fuses, circuit breakers, lightning arresters, power-factor correcting capacitors, and power metering systems. Some of this specialized equipment will be discussed in this section.

Overhead Transmission Lines

Large overhead transmission lines extend for many miles between power plants and substations. The major design problem of transmission lines is the high voltages at which they operate. Usually, transmission lines are uninsulated conductors supported by large steel towers. Many transmission lines now operate at voltages as high as 500 kV and 765 kV, with voltages of 69 kV and 138 kV being fairly typical. The developmental trend is toward even higher voltages of above 1000 kV. When high-voltage lines are insulated, massive amounts of insulation must be placed around the conductors.

Underground Distribution

The use of underground cable is ordinarily confined to short lengths required in congested urban areas. The cost of underground cable is much more than aerial cable. To improve underground cable power-handling capability, research is being done in forced-cooling techniques such as circulating oil, with compressed gas insulation. Another possible method is the use of cryogenic cables (or "superconductors"), which operate at extremely low temperatures and have a large power-handling capability.

Substations

Substations are an important part of the electrical power system. The link between the high-voltage transmission lines and low-voltage distribution is the substation. The function of a substation is to receive power from a high-voltage system and convert it to voltage levels suitable for industrial, commercial, or residential distribution. The major functional component of a substation is the transformer, which was discussed previously. However, many other specialized types of equipment are required at the substation.

High-Voltage Fuses

Since short-circuited lines are frequently encountered, various protective devices are used to prevent damage to power lines and equipment. This protective equipment must be designed to handle high voltages and currents. Either fuses or circuit breakers may be used to protect high-voltage power lines. High-voltage fuses (those used for over 600 volts) are made in several ways. An expulsion-type fuse has an element that will melt and vaporize when it is overloaded, causing the power line it is connected in series with to open. Liquid fuses are used to immerse a melted fuse element in a liquid to extinguish the arc developed by a faulty current in the power line. A solid fuse is similar to a liquid fuse except that the arc is extinguished in a chamber filled with solid material. Some high-voltage fuses are shown in Fig. 4-16.

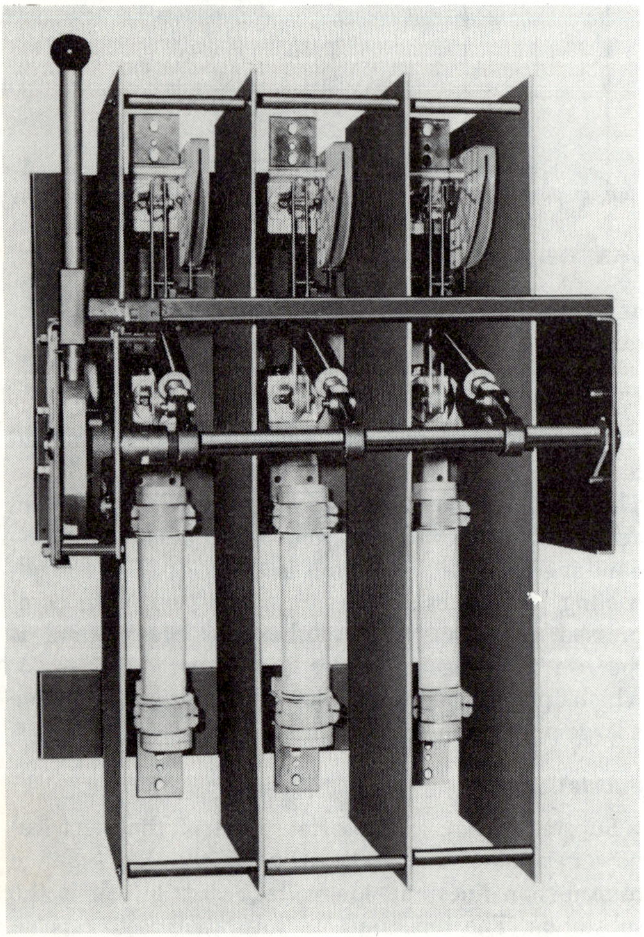

Courtesy Square D Co.

Fig. 4-16. High-voltage fuses.

Ordinarily, high-voltage fuses at substations are mounted adjacent to air-break disconnect switches. These switches provide a means of switching lines and disconnecting lines for repair. The fuse and switch enclosure is usually mounted near the overhead power lines at a substation.

High-Voltage Circuit Breakers

Circuit breakers are also located at electrical substations. Many outdoor substations use oil-filled circuit breakers. In this type of circuit breaker, the contacts are immersed in an insulating oil contained in the metal enclosure. Another type of high-voltage circuit breaker is the magnetic air breaker in which the contacts separate in air when the power line is overloaded. The arc that results when the contacts open is extinguished by magnetic blowout coils, which develop a magnetic field that concentrates the arc into arc chutes. A modification of this type is the compressed-air circuit breaker. In this type, a stream of

Fig. 4-17. An outdoor substation showing a high-voltage circuit breaker.

compressed air is concentrated on the contacts when the power line is opened. The compressed air aids in extinguishing the arc that is developed when the contacts open. It should be pointed out that large arcs are present whenever a high-voltage circuit is interrupted. This problem is not encountered to any great extent in low-voltage protective equipment. A substation circuit breaker is shown in Fig. 4-17.

Lightning Arresters

Lightning arresters (see Fig. 4-18) are also part of outdoor substations. Lightning is a major cause of short-circuit situations on overhead power lines. The purpose of lightning arresters is to provide a path to ground for flashes of lightning. This eliminates flashover between the power lines, which causes short circuits. Valve-type lightning arresters are used frequently. They are two-terminal devices in which one terminal is connected to the power line and the other is connected to ground. The path from line to ground is of such high resistance that it is normally open. However, when lightning, which is a very high voltage, strikes a power line, it causes conduction from line to ground before flashover between lines occurs. After the lightning surge has been conducted to ground, the valve assembly then causes the arrester to be nonconductive once more.

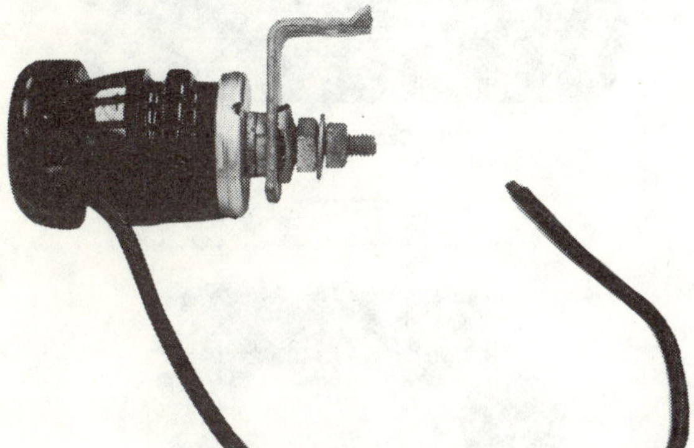

Fig. 4-18. A lightning arrester.

Insulators and Conductors

All power lines must be isolated so as not to be safety hazards. Large strings of insulators are used at substations and at other points along the power distribution system to isolate the current-carrying conductors from their steel supports or any other ground-mounted equipment. Insulators may be made of porcelain, rubber, or thermoplastic material. Some high-voltage insulators are shown in Fig. 4-19.

Fig. 4-19. High-voltage insulators.

The conductors used for power distribution are ordinarily uninsulated aluminum or aluminum-conductor steel-reinforced (ACSR) for long-distance transmission, and insulated copper for shorter distances. Some typical types of conductors for high voltages are shown in Fig. 4-20.

There are yet more types of specialized equipment used in power distribution systems. However, the major types have been discussed in this section. Later sections will discuss equipment associated with power distribution systems within the industrial plant and within the equipment itself.

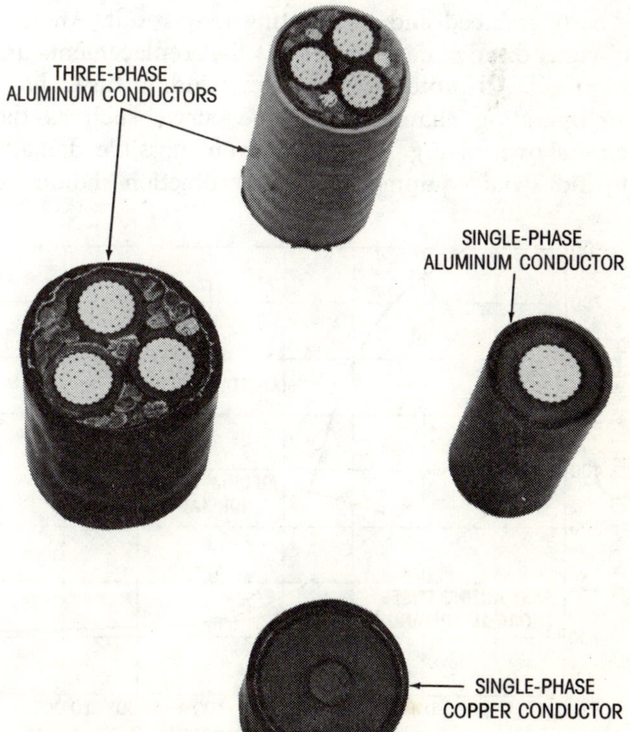

Fig. 4-20. High-voltage conductors.

ELECTRICAL DISTRIBUTION SYSTEMS IN INDUSTRIAL PLANTS

Another category of power distribution is the power distribution network within an industrial plant. After electrical power has been delivered to an industry, it must be distributed to the specialized machinery and equipment that the industry uses. In this phase of power distribution, the industrial technician is concerned with power switchgear, indoor conductors, grounding equipment, protective equipment, and other associated equipment located in the industrial plant.

System Protective Equipment

Several types of protective devices are used in industrial installations. The purpose of any type of protective device is to cause a current-carrying conductor to become inoperative when an excessive amount of current flows.

Fuses

The simplest type of protective device is the fuse. Fuses are low-cost items and have a fast operating speed. However, in three-phase systems, if one fuse burns out, two lines are still operative since each hot line must be fused. Three-phase motors will continue to run when one phase is removed. This condition is undesirable in most instances since motor torque is greatly reduced and overheating may result. Another obvious disadvantage of fuses is that replacements are required. All protective devices, including fuses, have an operating characteristic time curve, such as the ones shown in Fig. 4-21. To prevent possible damage to industrial equipment, circuit protection should be planned by utilizing these curves. They show the response time required for a protective device to interrupt a circuit when an overload condition occurs.

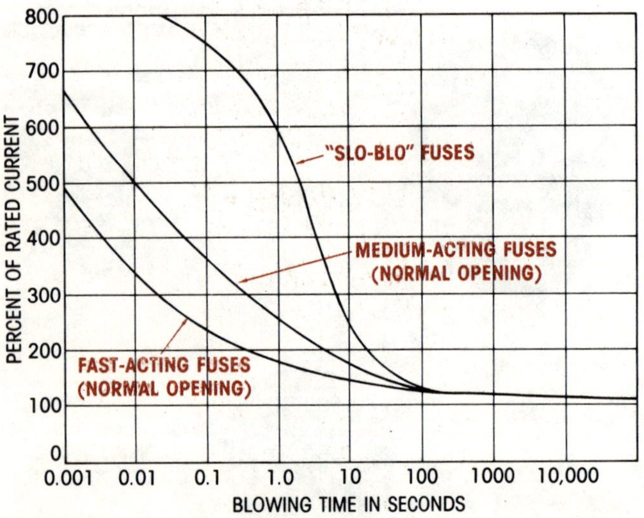

Courtesy Littlefuse, Inc.

Fig. 4-21. Typical operating characteristic time curves for three types of fuses.

Fuses are used in safety switches and power distribution panels. The plug fuse, shown in Fig. 4-22A, is a common type of fuse. Standard sizes of this fuse are 10, 15, 20, 25, and 30 amperes at voltages of 125 volts or less. These fuses have a zinc or alloy fusible element enclosed in a case made of an insulating material.

(A) Plug fuses.

(B) Cartridge fuse.

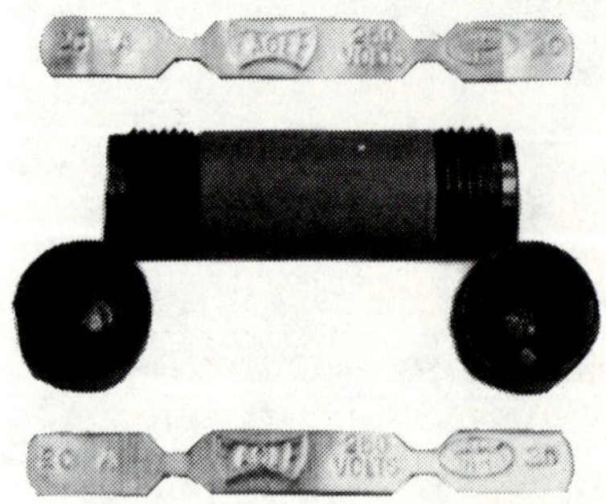

(C) Renewable-element fuse.

Fig. 4-22. Common types of fuses.

Cartridge fuses, such as the one shown in Fig. 4-22B, are commonly used in industrial power distribution systems at voltages up to 600 volts. They have a zinc or alloy fusible element housed in a round fiber enclosure. One type has a nonrenewable element, while another type has a renewable element, such as the one shown in Fig. 4-22C. Standard sizes for cartridge fuses are 60, 100, 200, 400, 600, and 1000 amperes.

Another modification of either the plug or cartridge fuse is called a *time-lag* (or *slo-blo*) fuse. This type of fuse is used to delay the circuit interrupting action. It is useful where momentary high currents exist periodically, such as motor-starting currents. The fuse element melts only when an excessive current is sustained over the time-lag period, thus sufficient circuit protection is still provided.

Circuit Breakers

Although their function is the same as fuses, circuit breakers are somewhat more sophisticated and are much more versatile. In three-phase systems, circuit breakers can open all three hot lines when an overload occurs. They can also be activated by remote-control relays. Relay systems can cause circuit breakers to open due to changes in frequency, voltage, current, or other circuit variables. The internal construction of a typical circuit breaker is shown in Fig. 4-23. The cir-

Courtesy Airpax Electronics, Inc.

Fig. 4-23. A typical circuit breaker, partially disassembled to show its internal construction.

cuit breakers used in industrial plants are usually of the low-voltage variety (less than 600 volts) and are not nearly as complex as their high-voltage counterparts discussed previously. Most low-voltage circuit breakers are housed in molded plastic cases that mount in metal power distribution panels. Breakers are designed so that they will automatically open when a current that exceeds their rating occurs. They must ordinarily be reset manually. Most circuit breakers employ either a thermal tripping element or a magnetic tripping element. Ratings of circuit breakers extend into current ranges of 800 to 2000 amperes.

Distribution Switching Equipment

In addition to circuit protection, power distribution systems must have equipment for connecting or disconnecting the entire system or parts of the system. Various types of switching devices are used to perform this function. A simple type of switch is the safety

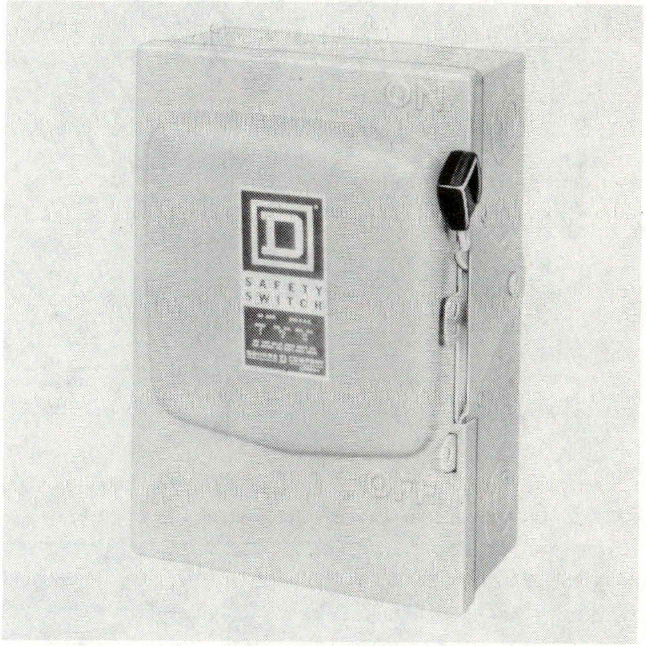

(A) Exterior view.

(B) Interior view.

Courtesy Square D Co.

Fig. 4-24. Exterior and interior views of a single-phase safety switch.

switch that is mounted in a metal enclosure and operated by means of an external handle. Single-phase and three-phase safety switches are shown in Figs. 4-24 and 4-25. Safety switches are used only to turn a circuit off or on; however, fuses are often mounted in the same enclosure.

Another type of switch is the kind used in conjunction with a circuit-breaker panelboard. Panelboards are metal cabinets enclosing a main disconnect switch and branch-circuit protective equipment. Distribution panelboards are usually located between the power feed lines within the industrial plant, and branch circuits that supply the equipment in a section of the plant. Circuit breakers are placed inside the panelboard to protect each of the branch circuits connected. A typical panelboard is shown in Fig. 4-26.

Metal-enclosed, low-voltage switchgear is used in many industrial plants. These are used as a distribution control center to house the circuit breakers, bus bars, and terminal connections that are part of the industrial distribution system. Ordinarily, a combination of switchgear and distribution transformers are placed in adjacent metal enclosures, such as that shown in Fig. 4-27. This combination is referred to as a *load center*

Courtesy Square D Co.

Fig. 4-25. Interior view of a three-phase safety switch.

Courtesy Square D Co.

Fig. 4-26. A typical power distribution panelboard.

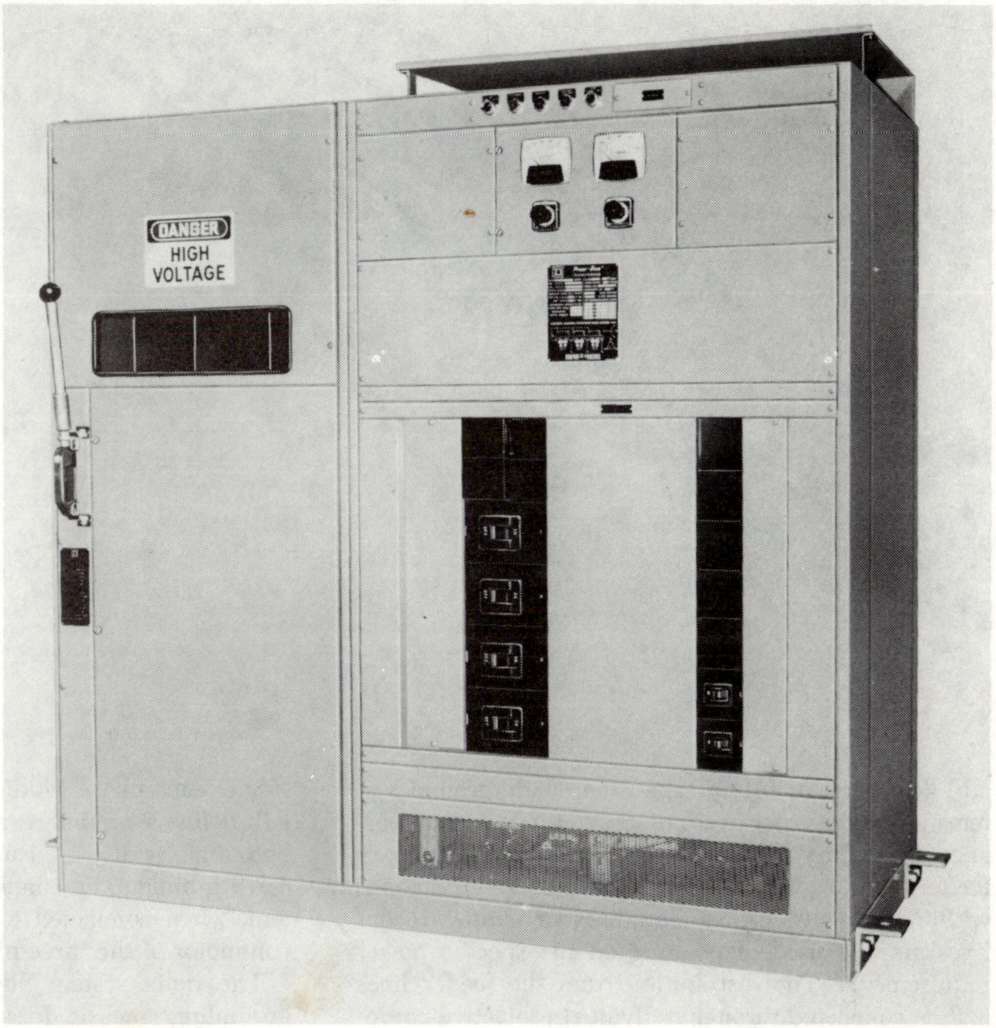

Fig. 4-27. A typical load center unit substation.

Courtesy Square D Co.

unit substation since it is the central control for several industrial loads. The rating of these load centers is usually 15,000 volts or less for the high-voltage section, and 600 volts or less for the low-voltage section. Load centers provide flexibility in the electrical power distribution design of industrial plants.

DISTRIBUTION TO INDUSTRIAL LOADS

Electrical power is delivered to industry, then distributed by the power system within the plant. Various types of circuit breakers and switchgear are employed for in-plant distribution. Another factor involved in industrial power distribution is the distribution of electrical energy to the many types of loads connected to the system. This part of the distribution system is concerned with the conductors, feeder systems, branch circuits, grounding methods, and machine protective and control devices used.

Raceways

Most electrical distribution to industrial loads is contained in raceways. Raceways may be large metal ducts, such as the one shown in Fig. 4-28, or rigid metal conduit. Raceways contain the conductors that distribute power to the various equipment within the industrial plant. Copper conductors are ordinarily used for in-plant power distribution. The physical size of each conductor required is dependent upon the current rating of the branch circuit.

Feeders and Branch Circuits

The conductors that carry current to the electrical load devices in industry are called *feeders* and *branch circuits.* Feeder lines supply power to branches that are connected to them. Primary feed lines in plants may be either overhead or underground. In several situations, overhead lines are preferred because they

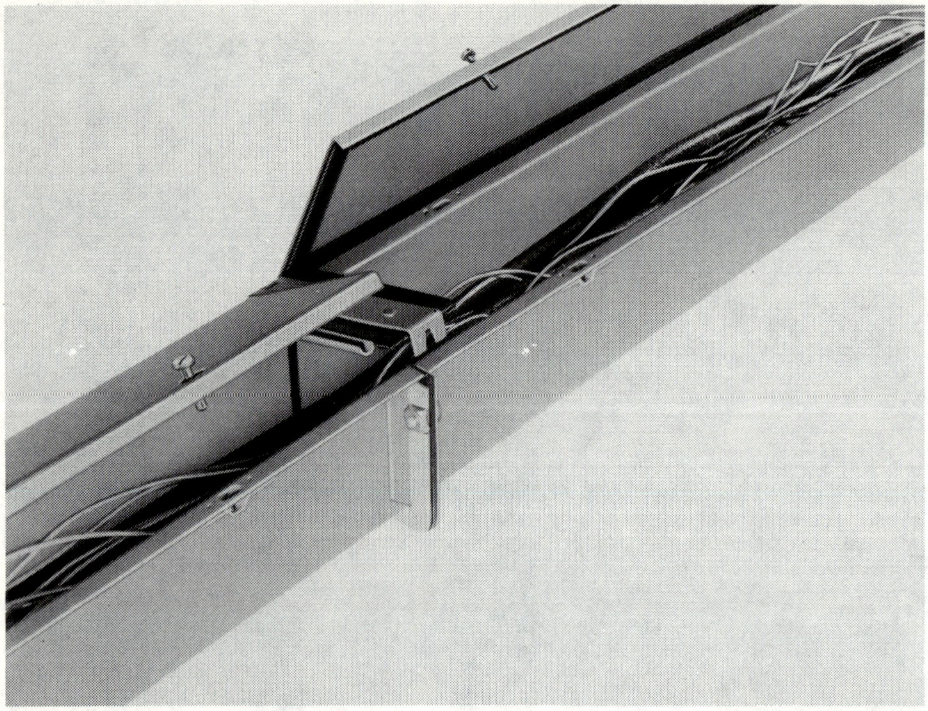

Courtesy Square D Co.

Fig. 4-28. A typical metallic raceway.

add flexibility for future expansion. Underground systems are more expensive, but present a better appearance. Secondary feeders are connected to the primary feeder lines to supply power to individual sections within the plant. Either aluminum or copper feeder lines may be used, depending on the specific power requirements. The distribution from the feeder lines is then connected through individual protective equipment to branch circuits that supply the various industrial loads. Each branch circuit has protective devices according to the needs of that particular branch. The overall feeder-branch system in industry may be a very complex network of switching equipment, transformers, conductors, and protective equipment.

Distribution System Grounding

An important factor in the operation of any electrical distribution system is grounding. Proper grounding techniques are required for safety as well as circuit performance. There are two types of grounding: (1) system grounding and (2) equipment grounding. Another important grounding factor is ground-fault protection.

System Grounding

System grounding involves the actual grounding of a current-carrying conductor (usually called the neutral) of a power system. Three-phase systems may be either the wye or delta type. The wye system has an obvious advantage over the delta system since one side of each phase winding is connected to ground. We will define a ground as a reference point of zero volts potential, which is usually an actual connection to earth ground. The common terminals of the wye system, when connected to ground, become the neutral conductor of the three-phase, four-wire system.

The delta system does not readily lend itself to grounding since it does not have a common neutral. The problem of ground faults (line-to-ground shorts) occurring in ungrounded delta systems is much greater than in wye systems. A common method of grounding a delta system is to use a wye-delta transformer connection and ground the common terminals of the wye-connected primary. However, the wye system is now used more often for industrial distribution since the secondary is easily grounded and it provides overvoltage protection from lightning or line-to-ground shorts.

Single-phase 120/240- or 240/480-volt systems are grounded in a manner similar to a three-phase ground. The neutral of the single-phase, three-wire system is grounded by a metal rod driven into the earth at the transformer location. System grounding conductors are insulated with white material for easy identification.

Equipment Grounding

The second type of ground is the equipment ground which, as the term implies, places operating equipment at ground potential. The conductor used for this purpose is either bare wire or a green insulated wire. The *National Electrical Code* describes conditions that re-

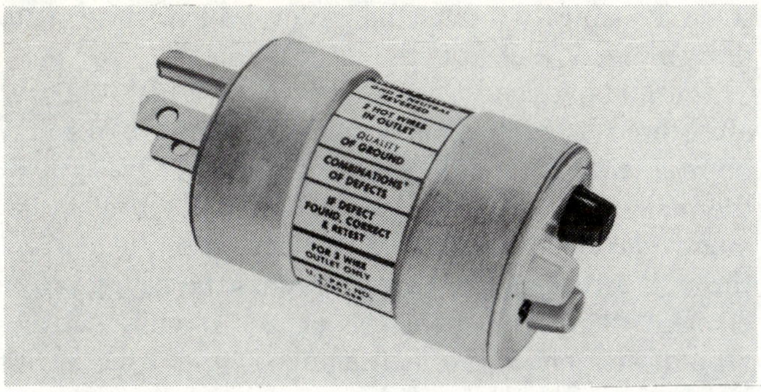

Courtesy Circle F Industries

Fig. 4-29. A plug-in ground tester.

quire fixed electrical equipment to be grounded. Usually all fixed electrical equipment located in industrial plants should be grounded. Types of equipment that should be grounded include enclosures for switching and protective equipment for load control, transformer enclosures, electric motor frames, and fixed electronic test equipment. Industrial plants should use 120-volt, single-phase duplex receptacles of the grounded type for all portable tools. The grounding of these receptacles can be checked by using a plug-in tester such as the one shown in Fig. 4-29.

Ground Fault Protection

Ground fault interrupters (GFIs), such as the one shown in Fig. 4-30, are now used extensively in industrial and residential power distribution systems. It is required by the *National Electrical Code* that all 120-volt, single-phase, 15- or 20-ampere receptacle outlets that are outdoors or in bathrooms have ground fault interrupters installed. These devices are designed to eliminate electrical shock hazards resulting from individuals coming in contact with a hot ac line (line-to-ground short). The circuit interrupter is designed to sense any change in circuit conditions, such as would occur when a line-to-ground short exists.

Courtesy Leviton Manufacturing Co., Inc.

Fig. 4-30. A ground fault interrupter (GFI).

One type of GFI has control wires that extend through a magnetic toroidal loop (see Fig. 3-31). Ordinarily, the ac current flowing through the conductors inside the loop is equal in magnitude and opposite in direction. Any change in this equal and opposite condition is sensed by the magnetic toroidal loop. When a line-to-ground short occurs, an instantaneous change in circuit conditions occurs. The change causes a magnetic field to be induced into the toroidal loop. The induced current is amplified to a level sufficient to cause the circuit breaker to open. Thus, any line-to-ground short will cause the ground fault interrupter to open. The operating speed is so fast (only a very small current opens the circuit) that the shock hazard to individuals is greatly reduced.

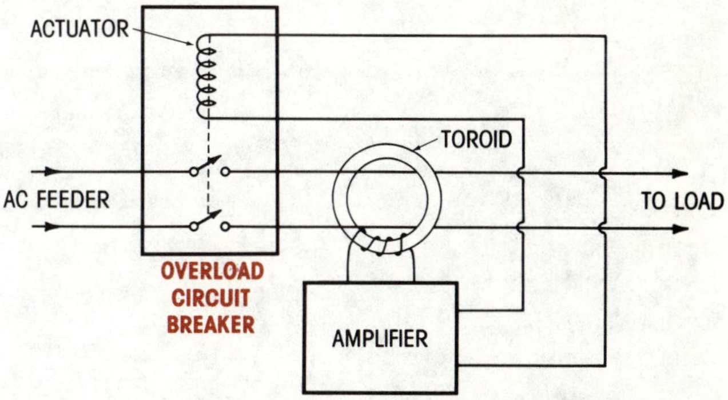

Fig. 4-31. Simplified schematic of a ground fault interrupter.

Construction sites, where temporary wiring is set up, are also required to use GFIs for protection of workers using electrical equipment. Ground fault protection of industrial equipment is provided for wye-connected systems of 150 to 600 volts for each distribution panelboard rated at over 1000 amperes. In this situation, the GFI will open all ungrounded conductors at the panelboard when a line-to-ground short occurs.

Motor Protective Devices

The distribution of power to electrical motors is an important function of industrial distribution networks. Since motors are used in some way with most industrial processes, the efficient distribution of energy to electric motors must be considered. Basic functions that motors are expected to perform in conjunction with industrial processes are starting, stopping, reversing, and speed variation. These functions may be manually or automatically controlled. Various types of protective devices are used for efficient distribution of power to electric motors.

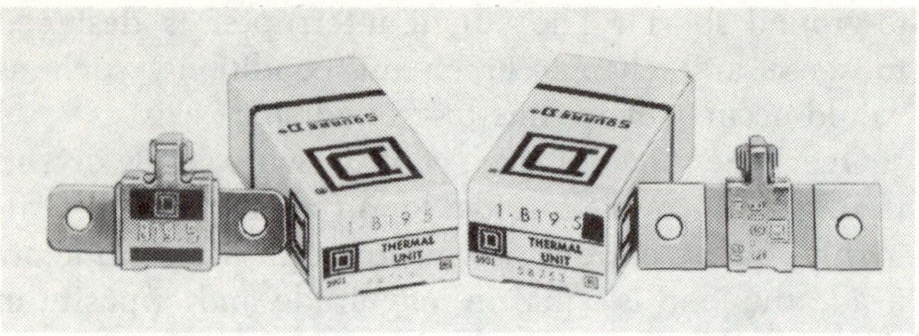

Courtesy Square D Co.

Fig. 4-32. Thermal-overload relays (melting-alloy type).

Overload protection is the most important motor protective function. Such protection should serve the motor, its branch circuit, and associated control equipment. The major cause of motor overload is an excessive mechanical load on the motor, which causes it to draw more current from the source.

Thermal-overload relays, shown in Fig. 4-32, are often used as protective devices. Thermal relays may be reset either manually or automatically. One type of thermal overload relay uses a bimetallic heater element. The element bends as it is heated by the current through it. When the current reaches the rating of the element, the relay opens the branch circuit. Another type of element is the melting-alloy type. This device has contacts held closed by a ratchet wheel. At the rated current capacity of the device, the fusible alloy melts causing the ratchet wheel to turn. A spring then causes the device to open the circuit.

Fig. 4-33. Electrical distribution within industrial equipment.

Courtesy Bridgeport-Tektron Co.

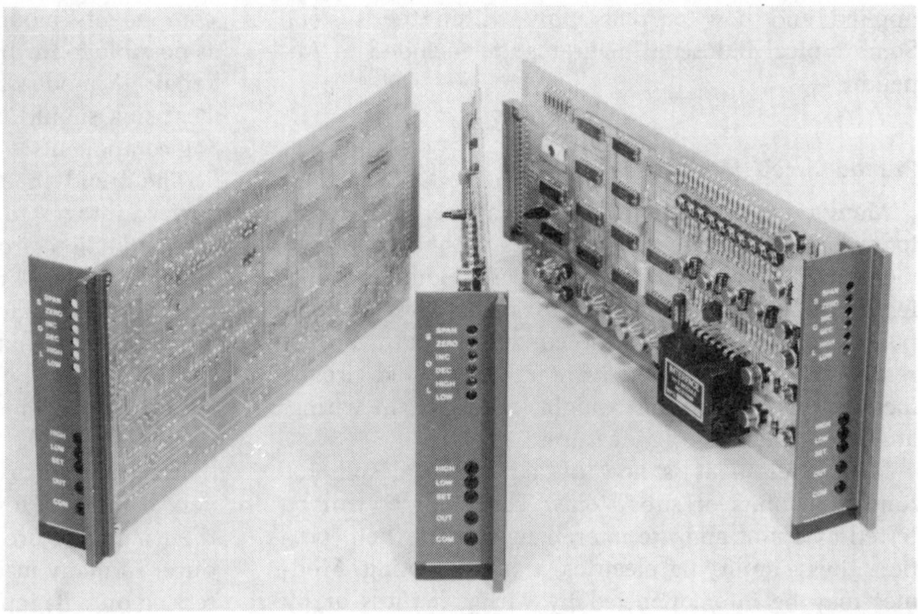

Courtesy The Foxboro Co.

Fig. 4-34. A printed-circuit module—front and rear views.

ELECTRICAL DISTRIBUTION IN INDUSTRIAL EQUIPMENT

After electrical power has been delivered to an industry and it has been distributed to the various types of equipment located in the industrial plant, it is necessary to distribute this electrical power within the equipment. Some equipment distribution networks are fairly simple. An electric motor, for instance, requires only copper conductors (windings), a switching (on and off) system, and some protective devices. Other types of equipment, such as testing equipment, may require several printed-circuit modules, interconnection terminals for wire harnesses, and other complex distribution methods (see Fig. 4-33). In this section, we will consider some of the methods used for power distribution within more complex industrial equipment.

Schematics and Interconnection Diagrams

In Chapter 2, we discussed the use of schematic diagrams in industry. Circuit diagrams for power distribution that may be of interest to an industrial technician include conduit and wire routing diagrams, terminal connection diagrams, and the schematic diagram that shows how components within the equipment are actually interconnected.

Wiring diagrams that show how various parts of a system are joined together may be shown in several ways. Connection diagrams should illustrate the location of parts and all identifying numbers and letters of conductors within the equipment. This facilitates testing and troubleshooting procedure. Sometimes single line diagrams are used to simplify complex electrical networks. The parts of the system are shown connected together by single lines, which show how voltages are

Courtesy Brand-Rex Co.

Fig. 4-35. Flexible cable interconnectors.

applied and how currents flow within the system. Some typical industrial diagrams are included in Appendix 4.

Printed-Circuit Modules

Many types of electrical equipment used in industry today have printed-circuit modules within the equipment. An example of a printed-circuit module is shown in Fig. 4-34. Note the neatly arranged lines of components on the front and the conductive paths on the rear of the module. The advantages of printed circuits include smaller size and weight, reduction in wiring, and more reliable design features.

Most equipment is now of a modular design for functional units or subsystems. The modules can be tested as a unit and are interchangeable at their location, thus simplifying electrical troubleshooting. Modules may be interconnected by wiring harness or by flexible cable such as shown in Fig. 4-35.

When printed-circuit modules were first introduced, they were difficult to troubleshoot and service. However, circuit boards are now designed to facilitate repair. Care must be used in desoldering and soldering components mounted on printed-circuit modules. It is possible to further damage a circuit while attempting repair. A moderate temperature soldering iron and a heat sink should be used when unsoldering or soldering components.

The trend in electronic equipment design now is toward integrated circuitry. Terms such as *micromodule* and *microcircuit* are now used to describe electronic circuits. Miniaturization has become widespread in most electronic equipment design. By using integrated circuits, many functional electronic components can be placed in one small package. The electrical distribution to components within a microcircuit can be seen in Fig. 4-36.

Electrical distribution within equipment may be very complex. However, advanced techniques used in their manufacture have simplified the servicing procedure. In many instances, a module can be checked by testing only its input and output terminals. A defective module can be replaced in a short period of time. Thus, printed-circuit modules and microcircuits have had a significant impact on the servicing of industrial equipment as well as on its size and operational characteristics.

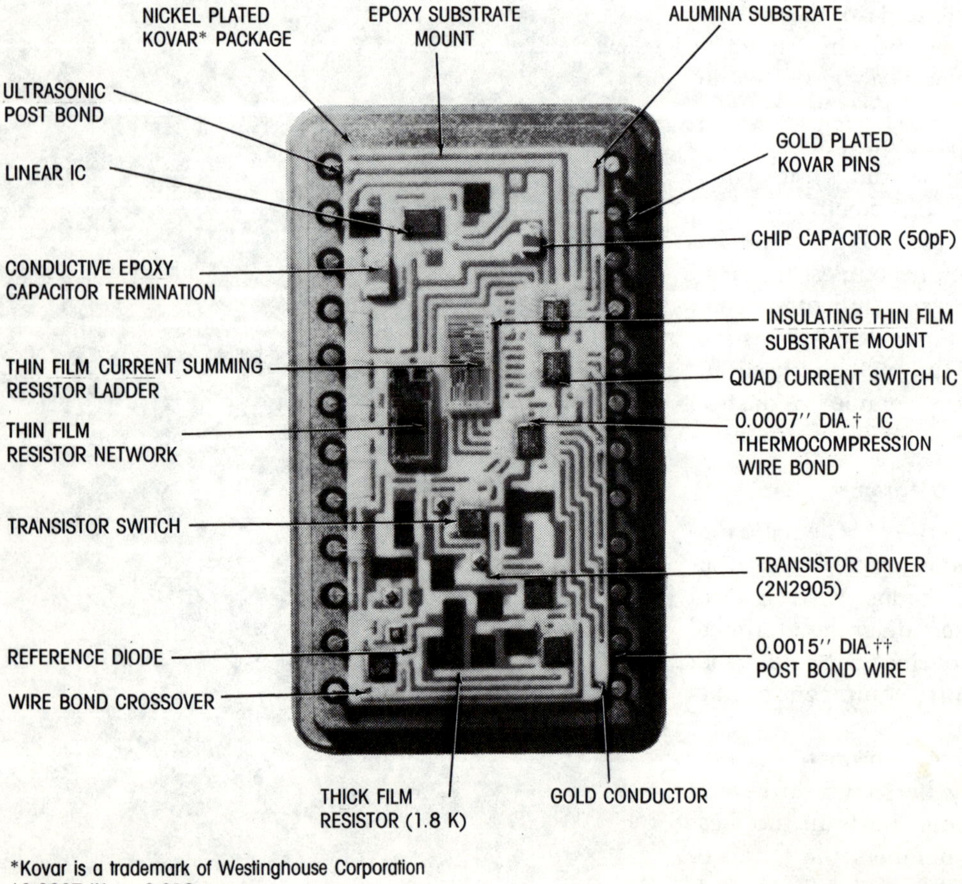

Fig. 4-36. Layout of a typical microcircuit.

*Kovar is a trademark of Westinghouse Corporation
†0.0007 IN. ≅ 0.018 mm.
††0.0015 IN. ≅ 0.038 mm.

Courtesy Beckman Instruments, Inc.

SUMMARY

We have traced the distribution of electrical power from its origin at the power station to the distribution within the industrial plant and on to the distribution networks within equipment. Distribution of power to industry is accomplished by a system of power plants that are interconnected in order to supply the load demand of industry at any given time. Three-phase alternators and other high-voltage equipment, such as circuit breakers, conductors, protective devices, and transformers, play important roles in the power distribution system. Transformers are vital to the overall operation of power systems. Without them, power could not be transmitted over long distances.

Industries primarily use three-phase power. Single-phase voltages are derived from three-phase distribution systems. The types of three-phase distribution systems used in industry are the: (1) three-phase, three-wire system, (2) three-phase, three-wire with neutral system, and (3) three-phase, four-wire system. The latter system is used most commonly for industrial distribution since it provides two voltages—one for power equipment and a lower voltage for lighting. Single-phase systems can either be two-wire or three-wire systems. Ordinarily, three-wire systems are used for residential distribution.

Specialized equipment such as overhead transmission lines, underground cable, high-voltage fuses, high-voltage circuit breakers, and lightning arresters are used at various points along the power distribution system. This equipment must be designed to handle the high voltages used along the system.

Distribution in industrial plants involves the equipment needed to supply electrical power to the various load devices. Protective equipment, such as fuses and circuit breakers, are essential to the safe operation of industrial plants. Switching equipment is needed to supply the loads with the proper branch circuits. Another important aspect of industrial distribution is grounding. Both system grounding and equipment grounding are important for proper operation of industrial electrical systems. Ground fault protection is now required for added safety in electrical systems. Such protection reduces the possibility of electrical shock due to line-to-ground shorts.

Power distribution must continue beyond the switching equipment and branch circuits in an industrial plant. Power must be distributed within the equipment that it causes to operate. This type of distribution is most complex in many respects. It could involve printed-circuit modules, interconnected components, or microcircuits.

Study Questions

1. Describe the power distribution route from the power plant to industry.
2. What are some of the types of equipment used in electrical power distribution systems?
3. Discuss the operation of a single-phase transformer.
4. How do step-up and step-down transformers differ?
5. What is a multiple-secondary transformer? An autotransformer?
6. What is meant by transformer polarity?
7. Why is power distribution equipment rated in VA or kVA rather than watts or kW?
8. What are the methods that may be used to connect the primary and secondary windings of three-phase transformers?
9. What are the three methods used to distribute three-phase power to industry? Make a symbolic sketch of each system.
10. Describe the following:
a. Overhead transmission lines.
b. Underground distribution.
c. Substations.
d. High-voltage fuses.
e. High-voltage circuit breakers.
f. Lightning arresters.
g. High-voltage insulators.
h. High-voltage conductors.
11. Describe the electrical distribution in an industrial plant.
12. What are three types of fuses and how do they differ?
13. How do circuit breakers operate?
14. What are some types of electrical switching equipment used in industry?
15. What is the difference between system grounding and equipment grounding?
16. What is the purpose of a ground fault interrupter?
17. How does a thermal overload relay operate?
18. How have printed-circuit modules and microcircuits simplified troubleshooting and equipment maintenance?

Problems

1. A 500-turn primary winding of a transformer has 240 volts applied. The secondary winding has an output of 60 volts. How many turns does the secondary have?
2. When a 20-ohm resistance is placed across a 100-volt secondary of a transformer, the primary current is 15 amperes. What is the primary voltage?
3. A transformer has a 20:1 turns ratio. The primary voltage is 1000 volts. What is the secondary voltage?
4. A transformer has a primary winding rated at 120 volts and a secondary rated at 360 volts. The primary has 500 turns. How many turns are on the secondary winding?
5. A 4800/480-volt transformer has a primary current of 18 amperes and a secondary current of 170 amperes. What is its efficiency?
6. A 40-kVA transformer has a primary voltage of 480 volts and a secondary voltage of 120 volts. What is the value of primary current and secondary current?

5

Industrial Electrical Loads

The electrical load devices used in industry are very important parts of industrial systems. The load of any system performs a function that involves power conversion. A load converts one form of energy to another form. An electrical load converts electrical energy to some other form of energy, such as heat, light, or mechanical energy. Industrial loads may be classified according to the function they perform (lighting, heating, mechanical) or by the electrical characteristics they exhibit (resistive, inductive, capacitive). Fluorescent lights, electric motors, and welders are common examples of industrial electrical loads.

CHARACTERISTICS OF ELECTRICAL LOADS

In order to plan for industrial load requirements, it is necessary to understand the electrical characteristics of all loads connected to the power source. The types of power supplies and distribution systems that a plant uses are determined by the load characteristics. All loads can be considered as either resistive, inductive, capacitive, or a combination of these. Since industries primarily use three-phase alternating-current sources, we should be aware of the effects that various types of loads will have on the power source. The nature of alternating current causes certain electrical circuit properties to exist.

Resistive Loads

The simplest type of electrical load is a resistive load, illustrated in Fig. 5-1. A resistive load device offers the same type of opposition to ac as it does to pure dc. We know that in dc circuits,

$$V = I \times R,$$

$$I = \frac{V}{R},$$

and

$$R = \frac{V}{I}.$$

These basic electrical relationships show that when the voltage applied to a circuit is increased, the current in the circuit increases proportionally. Also, as the resistance of a circuit is increased, the current in the circuit decreases. By looking at the waveforms of Fig. 5-1B, we can see that the voltage and current in a

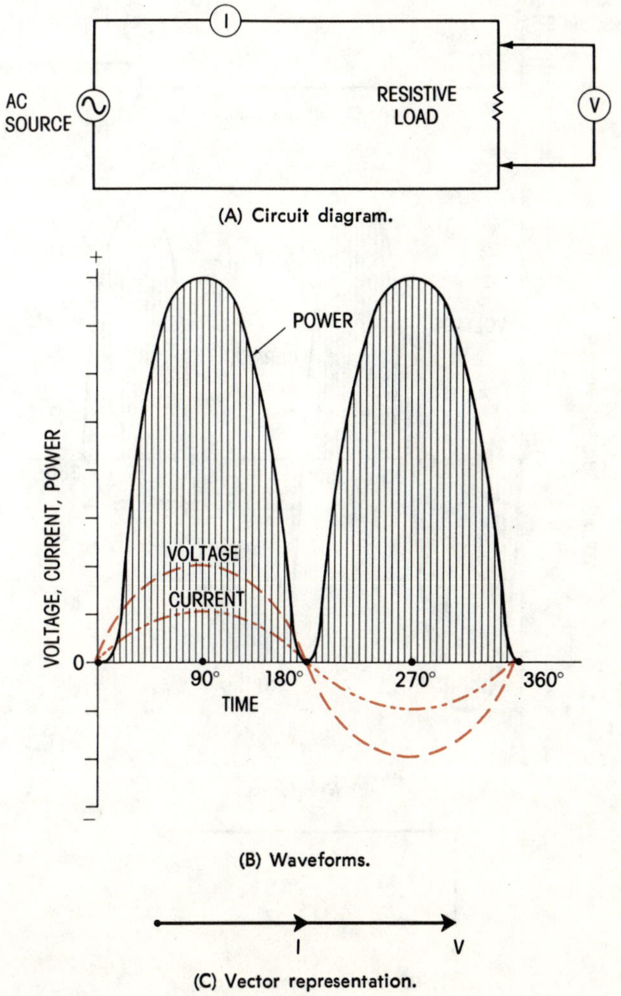

Fig. 5-1. Resistive load connected to an ac source.

purely resistive circuit are in phase. An in-phase relationship exists when the minimum and maximum values of both voltage and current occur at the same time. Also, the power converted by the load is a product of voltage times current ($P = V \times I$). The power curve is also shown in Fig. 5-1B. Thus, when an ac circuit contains only resistance, its behavior is very similar to a dc circuit. Purely resistive loads would seldom be encountered in industrial plants, although some loads are primarily resistive in nature.

Inductive Loads

The property of inductance (L), which is very common in industrial loads, adds more complexity to the relationship between voltage and current in an ac circuit. All motors, generators, and transformers exhibit the property of inductance. This property is evident due to the counterelectromotive force (cemf) produced when a magnetic field is developed around a coil of wire. The magnetic flux produced around coils affects circuit action. Thus, the inductive property (cemf) produced by a magnetic field offers an opposition to change in the current in a circuit.

The opposition to change of current is evident in the diagram of Fig. 5-2B. In an inductive circuit, we can say that the voltage leads the current or that the current lags the voltage. If the circuit were purely inductive (containing no resistance), the voltage would lead the current by 90° (Fig. 5-2B), and no power would be converted in the circuit. However, since all actual circuits have resistance, the inductive characteristic of a load might typically cause the condition shown in Fig. 5-3 to exist. Here the voltage is leading the current by 30°. The angular separation between voltage and current is called the *phase angle*. The phase angle increases as the inductance of the circuit increases. This type of circuit is called a *resistive-inductive* (RL) circuit.

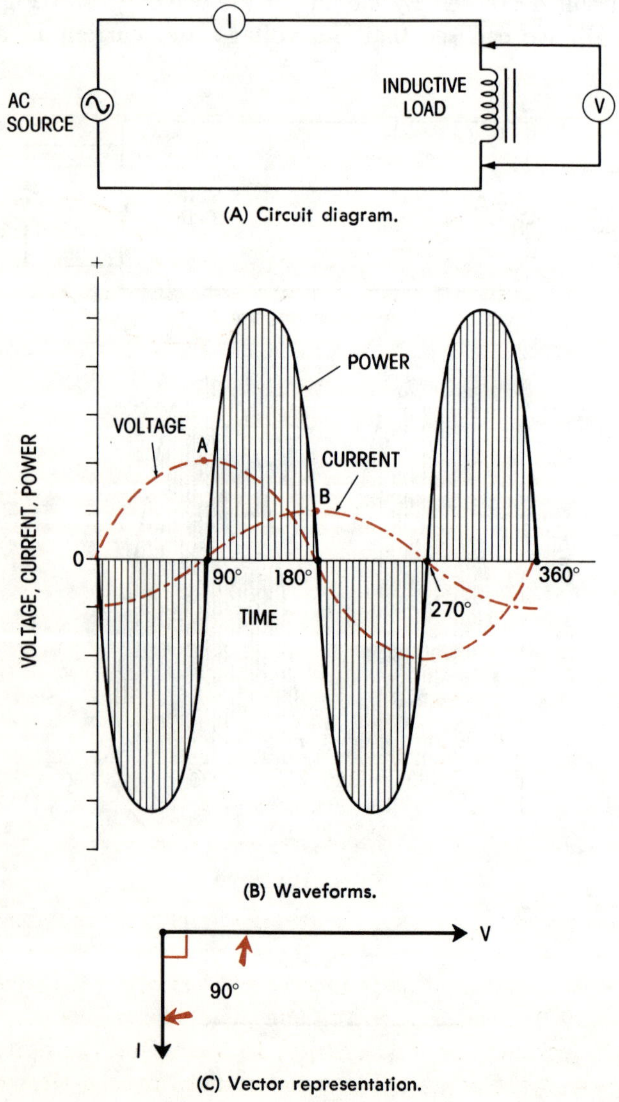

Fig. 5-2. Inductive load connected to an ac source.

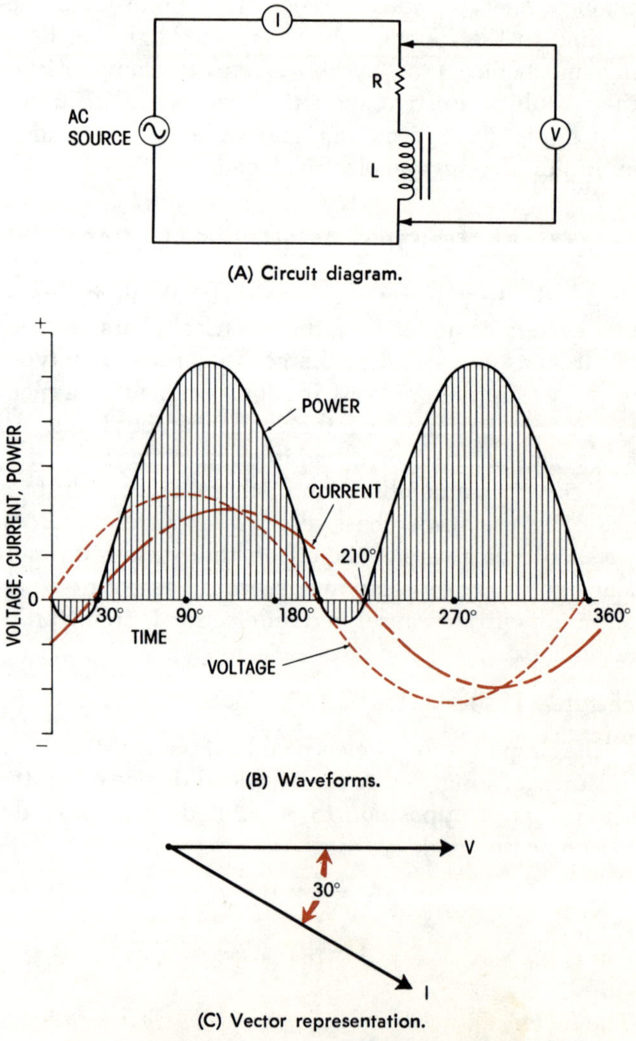

Fig. 5-3. Resistive-inductive (RL) load connected to an ac source.

In terms of power converted, a purely inductive circuit would not convert any power in the load. All ac power would be delivered back to the source. Note points A and B on the waveforms of Fig. 5-2B. These points show that at the peak of each waveform, the corresponding value of the other waveform is zero. The power curves shown are equal and opposite and will cancel each other out. Where both voltage and current are positive, the power is also positive since the product of two positive values is positive. When one value is positive and the other is negative, the product of the two is negative; therefore, the power converted is negative. Negative power means that electrical energy is returned from the load to the source without being converted to another form. Therefore, the power converted in a purely inductive load (90° phase angle) would be equal to zero.

Compare the purely inductive waveforms to those of Fig. 5-3B. In the practical resistive-inductive (RL) circuit, part of the power supplied from the source is converted in the circuit. Only during the intervals from 0° to 30° and from 180° to 210° does negative power result. The remainder of the cycle produces positive power; therefore, most of the electrical energy supplied by the source is converted to another form of energy in the load.

Any inductive circuit exhibits the property of inductance (L), which is an opposition to change in current flowing in a circuit. This property is found in coils of wire, which are called *inductors,* and in rotating machinery and transformer windings. Inductance is also present in electrical power transmission and distribution lines to some extent. The unit of measurement of inductance is the henry (H). A circuit has an inductance of 1 henry if a current changing at a rate of 1 ampere per second produces an induced counterelectromotive force (cemf) of 1 volt.

In an inductive circuit with ac applied, an opposition to current flow is created by the inductance. This type of opposition is known as *inductive reactance* (X_L). The inductive reactance of an ac circuit depends on the inductance (L) of the circuit and the rate of change of current. The frequency of the applied ac establishes the rate of change of current. Inductive reactance is expressed as:

$$X_L = 2\pi fL$$

where,

X_L is the inductive reactance in ohms,
f is the frequency of the ac source in hertz,
L is the inductance of the circuit in henrys.

The factor of 2π (6.28) is the mathematical expression for one sine wave (0° to 360°).

Capacitive Loads

Fig. 5-4 shows a capacitive load connected to an ac source. We know that whenever two conductive materials (plates) are separated by an insulating material (dielectric), the property of capacitance is exhibited. Capacitors have the capability of storing an electrical charge. They have many applications in industrial circuits. They are particularly useful in timing systems and in power distribution systems.

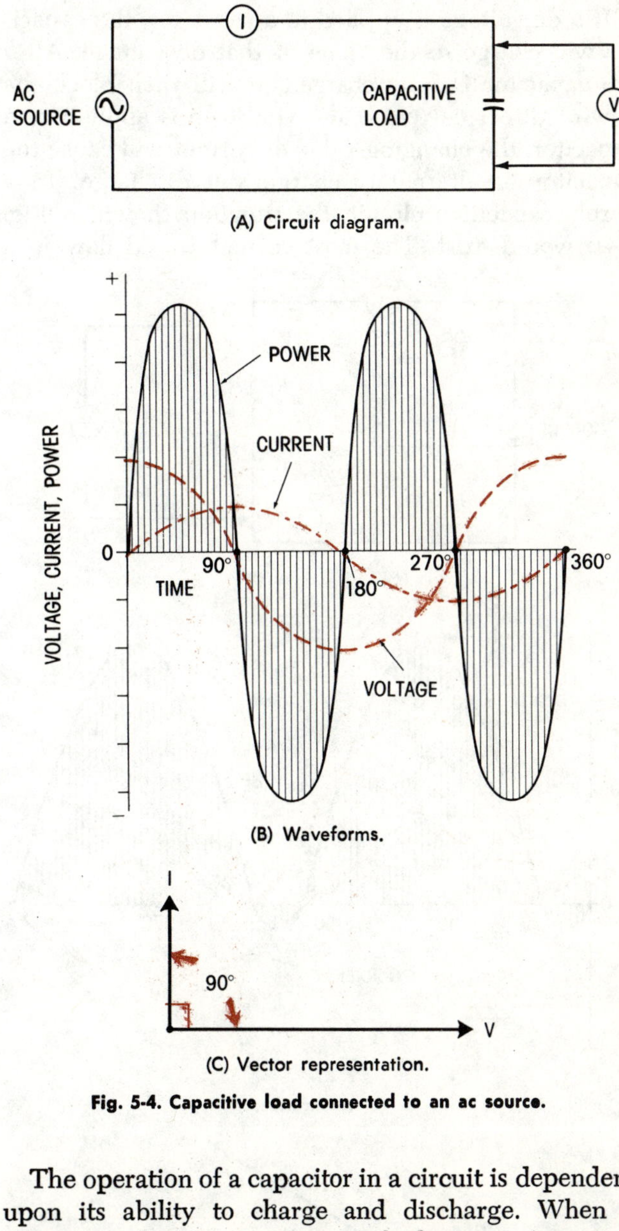

Fig. 5-4. Capacitive load connected to an ac source.

The operation of a capacitor in a circuit is dependent upon its ability to charge and discharge. When a capacitor charges, an excess of electrons (negative charge) is accumulated on one plate, and a deficiency of electrons (positive charge) is created on the other plate. Capacitance (C) is determined by the size of the conductive material (plates) and their separation (determined by the thickness of the dielectric or insulating material). The type of insulating material is also a factor in determining capacitance. Capacitance is directly proportional to the plate size and inversely proportional to the distance between plates. The unit of capacitance is the farad (F). A capacitance of 1 farad results when 1 volt causes a charge of 1 coulomb (a specific mass of electrons) to accumulate on a capacitor. Since the farad (F) is a very large unit, microfarad (μF) values are ordinarily assigned to capacitors.

If a dc voltage is applied to a capacitor, the capacitor will charge to the value of that dc voltage. After the capacitor is fully charged, it will then block the flow of direct current. However, if ac is applied to a capacitor, the changing value of current will cause the capacitor to alternately charge and discharge. In a purely capacitive circuit, the situation shown in Fig. 5-4B would exist. The most current would flow in a capacitive circuit when the voltage changes most rapidly. The most rapid change in voltage occurs at the 0° and 180° positions where the polarity changes. At these positions, maximum current is developed in the circuit. When the rate of change of the voltage is slow, such as near the 90° and 270° positions, a small amount of current flows. In examining Fig. 5-4B, it can be seen that the current leads the voltage by 90° in a purely capacitive circuit (or, the voltage lags the current by 90°). Since a 90° phase angle exists, no power would be converted in this circuit, just as no power was developed in the purely inductive circuit. As shown in Fig. 5-4B, the positive and negative power waveforms will cancel each other out.

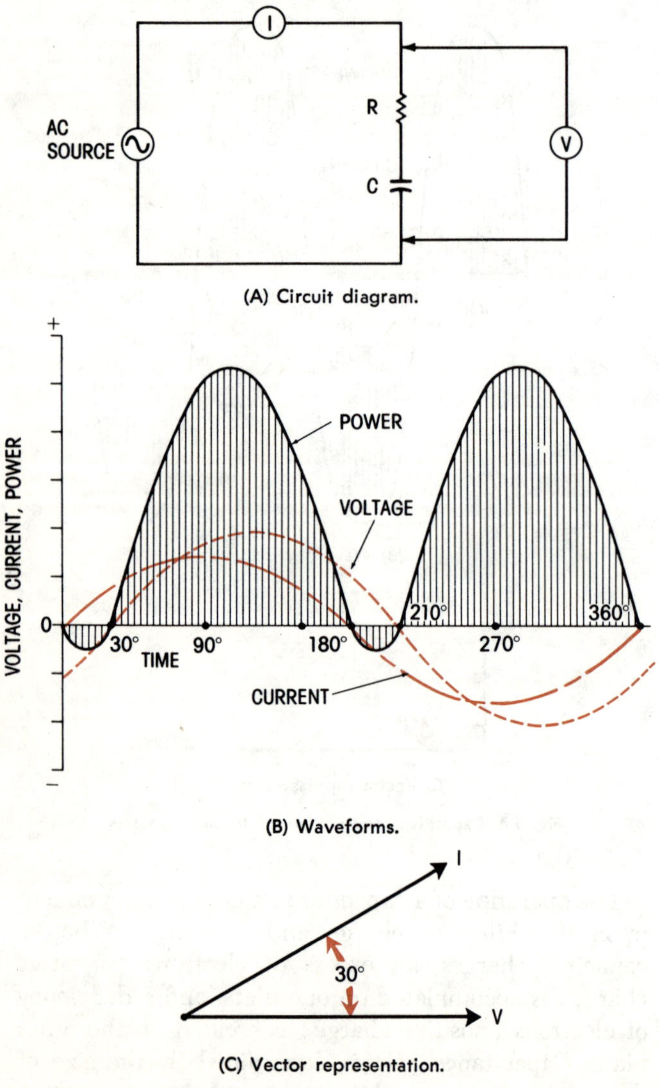

Fig. 5-5. Resistive-capacitive (RC) load connected to an ac source.

Since all circuits contain some resistance, a more practical circuit is the resistive-capacitive (RC) circuit shown in Fig. 5-5. In an RC circuit, the current leads the voltage by some phase angle between 0° and 90°. As capacitance increases, with no corresponding increase in resistance, the phase angle becomes greater. The waveforms of Fig. 5-5B show an RC circuit in which the current leads the voltage by 30°. This circuit is similar to the RL circuit of Fig. 5-3. Power is converted in the circuit except during the 0° to 30° interval, and the 180° to 210° interval. In the RC circuit shown, most of the electrical energy supplied by the source is converted to another form of energy in the load.

Due to the electrostatic field that is developed around a capacitor, an opposition to the flow of alternating current exists. This opposition is known as *capacitive reactance* (X_C). Capacitive reactance is expressed as:

$$X_C = \frac{1}{2\pi fC}$$

where,

X_C is the capacitive reactance in ohms,
f is the frequency of the ac source in hertz,
C is the capacitance in farads.

Again, the factor of 2π (6.28) is the mathematical expression for one sine wave (0° to 360°).

Vector and Phasor Diagrams

In the "C" section of each of the preceding illustrations, a vector diagram was shown for each circuit condition illustrated. The understanding of vector diagrams (sometimes called *phasor* diagrams) is important when dealing with alternating current. Rather than using waveforms to show phase relationships, it is possible to use vector or phasor representation. Vectors are straight lines that have a specific direction and length. They can be used to represent voltage or current values.

A horizontal line is ordinarily drawn when beginning a vector diagram, with its left end as the reference point. A positive direction is considered to be rotation in a counterclockwise direction from the reference point. Note that in the preceding diagrams, the voltage vector was the reference. For the inductive circuits, the current vector was drawn in a clockwise direction from the voltage vector, indicating a lagging condition. A leading condition was shown for the capacitive circuits by a current vector in a counterclockwise direction from the voltage vector.

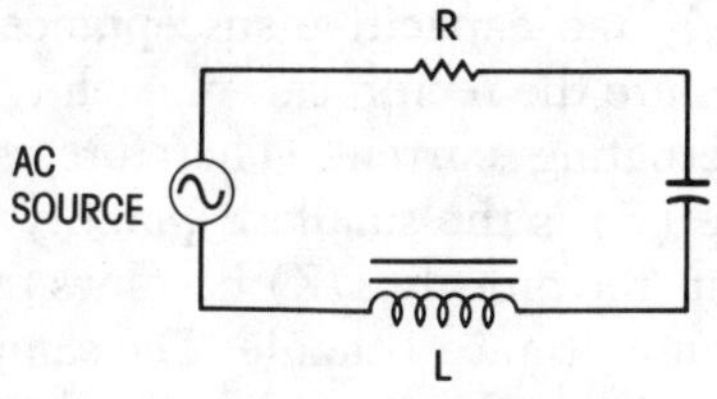

(A) Circuit diagram.

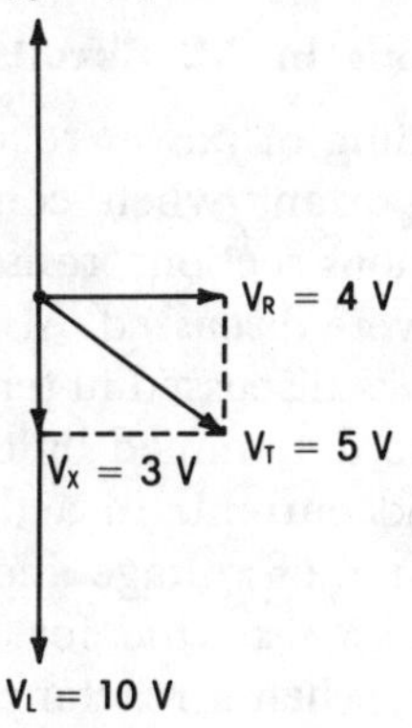

(B) Vector diagram.

$V_X = V_L - V_C$
$= 10\ V - 7\ V$
$= 3\ V$

$V_T = \sqrt{V_R^2 + V_X^2}$
$= \sqrt{4^2 + 3^2}$
$= \sqrt{16 + 9}$
$= \sqrt{25}$
$= 5\ V$

(C) Problem solution.

Fig. 5-6. Voltage vector relationship of a series RLC circuit.

Use of Vectors for Series Circuits

Vectors can be used to compare voltage drops across the components in a series circuit containing resistance, inductance, and capacitance (RLC circuit) as shown in Fig. 5-6. In a series ac circuit, the current is the same in all parts of the circuit, and the voltages must be added by using vectors. In the example, values have been assigned. The voltage across the resistor (V_R) is equal to 4 volts, the voltage across the capacitor (V_C) equals 7 volts, and the voltage across the inductor (V_L) equals 10 volts. We diagram capacitive voltage as leading resistive voltage by 90°, and inductive voltage as lagging resistive voltage by 90°. Since these two values are in direct opposition to one another, they may be subtracted to find the resultant reactive voltage (V_X). By drawing lines parallel to V_R and V_X, the resultant voltage applied to the circuit can be found. Since these vectors form a right triangle, the value of V_T can be expressed as:

$$V_T = \sqrt{V_R^2 + V_X^2}$$

where,

V_T is the total voltage applied to the circuit,
V_R is the voltage across the resistance,
V_X is the total reactive voltage ($V_L - V_C$ or $V_C - V_L$, depending on which is larger).

The solution to this simple problem is shown in Fig. 5-6C.

Use of Vectors for Parallel Circuits

Vector representation is also useful for parallel ac circuit analysis. The voltage in a parallel ac circuit remains the same. The currents through the components in a parallel ac circuit are shown by using vectors. A parallel RLC circuit is shown in Fig. 5-7. The current through the capacitor (I_C) is shown leading the current through the resistor (I_R) by 90°, and the current through the inductor (I_L) is shown lagging I_R by 90°. Since I_L and I_C are 180° out of phase, they are subtracted to find the total reactive current (I_X).

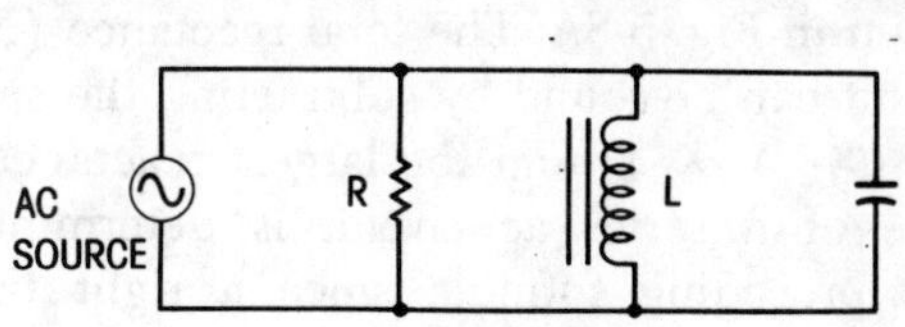

(A) Circuit diagram.

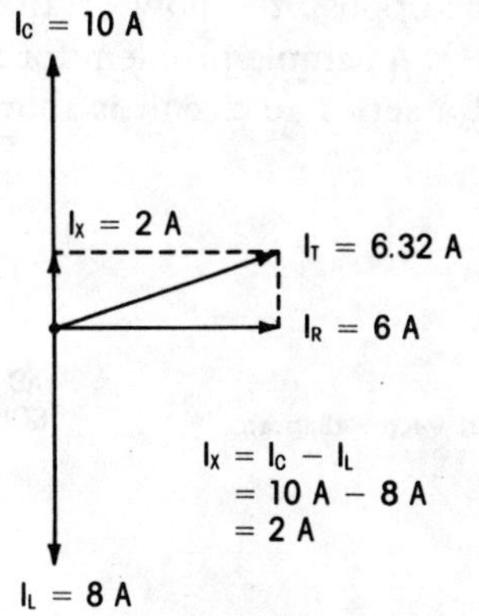

$I_X = I_C - I_L$
$= 10\ A - 8\ A$
$= 2\ A$

(B) Vector diagram.

$I_T = \sqrt{I_R^2 + I_X^2}$
$= \sqrt{6^2 + 2^2}$
$= \sqrt{36 + 4}$
$= \sqrt{40}$
$= 6.32\ A$

(C) Problem solution.

Fig. 5-7. Current vector relationship of a parallel RLC circuit.

By drawing lines parallel to I_R and I_X, the total current of the circuit (I_T) can be found. These vectors form a right triangle; therefore, the total current can be expressed as:

$$I_T = \sqrt{I_R^2 + I_X^2}.$$

A similar method of vector diagramming can be used for voltages in series RL and RC circuits. The method can also be used for currents in parallel RL and RC circuits. RLC circuits were used in the examples to illustrate the method used to find the resultant reactive voltage or current in a circuit.

Impedance in Series AC Circuits

Another application of vectors is for determining the total opposition of an ac circuit to the flow of current. This total opposition is called *impedance* (Z). Resistances and reactances in ac circuits both affect the opposition to current flow. The impedance of an ac circuit is expressed as:

$$Z = \frac{V}{I}$$

or,

$$Z = \sqrt{R^2 + (X_L - X_C)^2}.$$

This formula can be clarified by using the vector diagram shown in Fig. 5-8A. The total reactance (X_T) of an ac circuit can be found by subtracting the smallest reactance (X_L or X_C) from the largest reactance. The impedance of a series ac circuit is determined by using the preceding formula since a right triangle (called an impedance triangle) is formed by the three quantities that oppose the flow of alternating current (see Fig. 5-8B). A sample problem for finding the total impedance of a series ac circuit is shown in Fig. 5-8C.

Impedance in Parallel AC Circuits

When loads are connected in parallel, the calculation of impedance becomes more complex. Fig. 5-9A shows a simple RLC parallel circuit. Since the total impedance in the circuit is smaller than the resistance or reactance, an impedance triangle such as the one shown in Fig. 5-8B cannot be formed. A simple method that can be used to find impedance in parallel ac circuits is the *admittance* triangle shown in Fig. 5-9B. The following quantities are plotted on the triangle: admittance (1/Z), conductance (1/R), inductive susceptance ($1/X_L$), and capacitive susceptance ($1/X_C$). These quantities are the reciprocals of each type of opposition to alternating current. Therefore, since the total impedance (Z) is the smallest quantity in a parallel ac circuit, its reciprocal (1/Z) becomes the largest quantity on the admittance triangle. The sample problem of Fig. 5-9C shows the procedure used to find the total impedance of a parallel RC circuit.

Power Relationships in AC Circuits

The understanding of power relationships in ac circuits is very important when considering industrial loads. In the previous sections, resistive, inductive, and capacitive loads were discussed. Also, power converted in these circuits was discussed in terms of power waveforms, which were determined by the phase angle between voltage and current. In a dc circuit, power is equal to the product of voltage and current ($P = V \times I$). This formula is also true for purely resistive ac circuits. However, when a reactance (either inductive or capacitive) is present in an ac circuit, power is no longer simply a product of voltage and current.

Since reactive loads cause changes in the method used to compute power, the techniques described in

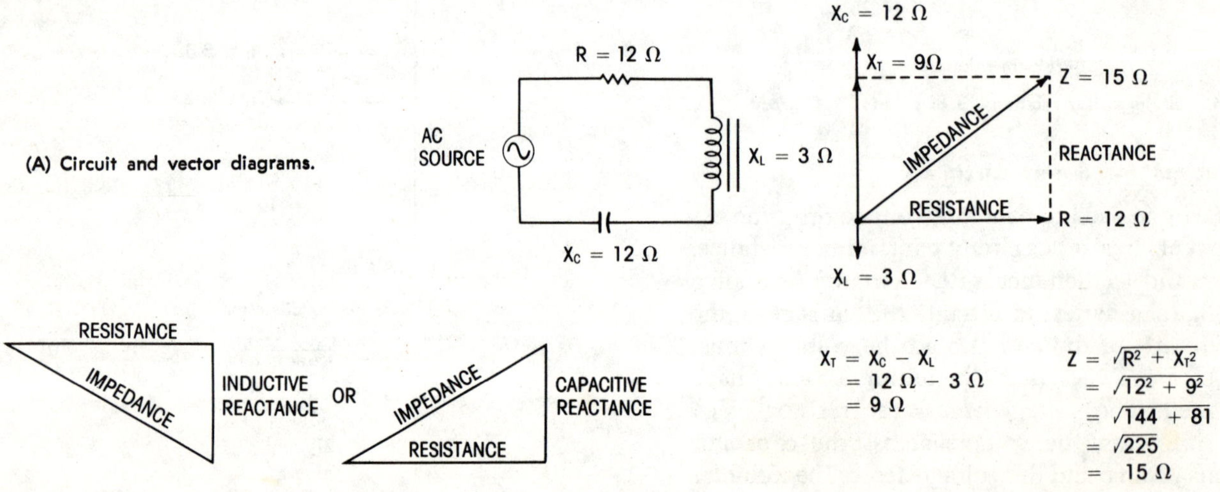

(A) Circuit and vector diagrams.

(B) Impedance triangles.

(C) Problem solution.

Fig. 5-8. Impedance in series ac circuits.

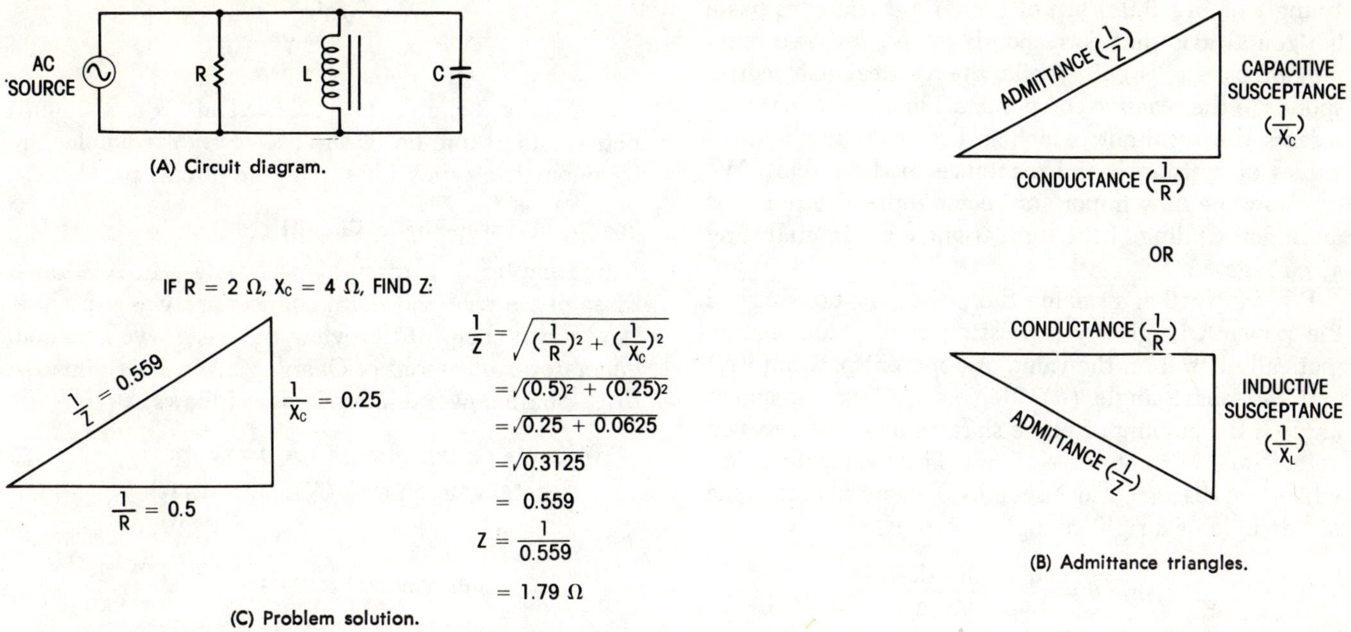

Fig. 5-9. Impedance in parallel ac circuits.

this section express the power relationships in ac circuits. The product of voltage and current is expressed in voltamperes (VA) or kilovoltamperes (kVA), known as *apparent power.* When using meters to measure power in an ac circuit, apparent power is the voltage reading multiplied by the current reading. The actual power that is converted to another form of energy by the load is measured by a wattmeter. This actual power is referred to as *true power.* Ordinarily, it is desirable to know the ratio of true power converted in a circuit to apparent power. This ratio is called the *power factor* and is expressed as:

$$pf = \frac{P}{VA}$$

or,

$$\%\,pf = \frac{P}{VA} \times 100$$

where,

pf is the power factor of the circuit,
P is the true power in watts,
VA is the apparent power in voltamperes.

The maximum pf value is 1.0 or 100% which would be obtained in a purely resistive circuit. This is referred to as *unity power factor.*

The phase angle between voltage and current in an ac circuit determines the power factor. If a purely inductive or capacitive circuit existed, the 90° phase angle would cause a power factor of zero to result. In practical circuits, the power factor varies according to the relative values of resistance and reactance.

The power relationships we have discussed can be simplified by referring to the power triangle shown in Fig. 5-10. There are two components that affect the power relationship in an ac circuit. The in-phase (resistive) component, which results in power conversion in the load, is called *active power.* Active power is the true power of the circuit and is measured in watts. The second component results from an inductive or capacitive reactance and is 90° out of phase with the active power. This component, called *reactive power,* does not produce an energy conversion in the load. Reactive power is measured in voltamperes reactive (var).

The power triangle of Fig. 5-10 shows true power (watts) on the horizontal axis, reactive power (var) at a 90° angle from the true power, and voltamperes (VA) as the longest side (or hypotenuse) of the right triangle. Note the similarity between this right triangle, the voltage triangle for series ac circuits of Fig. 5-6, the current triangle for parallel ac circuits for Fig. 5-7, the impedance triangle of Fig. 5-8, and the admittance

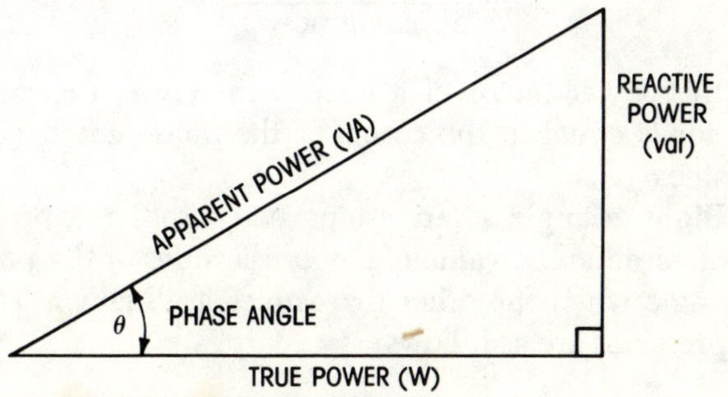

Fig. 5-10. Power triangle.

triangle of Fig. 5-9. Each of these right triangles has a horizontal axis that corresponds to the resistive component of the circuit, while the vertical axis corresponds to the reactive component. The hypotenuse represents the resultant, which is based on the relative values of resistance and reactance in the circuit. We can now see how important vector representation and an understanding of the right triangle are in analyzing ac circuits.

We can further examine the power relationships of the power triangle by expressing each value mathematically based on the value of apparent power (VA) and the phase angle (θ). Remember that the phase angle is the amount of phase shift, in degrees, between voltage and current in the circuit. Trigonometric ratios, which are discussed in Appendix 3, show that the sine of an angle of a right triangle is expressed as:

$$\text{sine } \theta = \frac{\text{adjacent side}}{\text{hypotenuse}}.$$

Since this is true, the phase angle can be expressed as:

$$\text{sine } \theta = \frac{\text{reactive power (var)}}{\text{apparent power (VA)}}.$$

Therefore,

$$\text{var} = \text{VA} \times \text{sine } \theta.$$

We can determine either the phase angle or the var value by using the trigonometric tables.

We also know that the cosine of an angle of a right triangle is expressed as:

$$\text{cosine } \theta = \frac{\text{adjacent side}}{\text{hypotenuse}}.$$

Thus, in terms of the power triangle:

$$\text{cosine } \theta = \frac{\text{true power (W)}}{\text{apparent power (VA)}}.$$

Therefore, true power can be expressed as:

$$\text{W} = \text{VA} \times \text{cosine } \theta.$$

Note that the expression

$$\frac{\text{true power}}{\text{apparent power}}$$

is the power factor of a circuit; therefore, the power factor is equal to the cosine of the phase angle (pf = cosine θ).

Right triangle relationships can also be expressed to determine the value of any of the sides of the power triangle when the other two values are known. These expressions are as follows:

$$\text{VA} = \sqrt{\text{W}^2 + \text{var}^2}$$

$$\text{W} = \sqrt{\text{VA}^2 - \text{var}^2}$$

$$\text{var} = \sqrt{\text{VA}^2 - \text{W}^2}$$

Appendix 3 should be reviewed in order to gain a better understanding of the use of right triangles and trigonometric ratios for solving ac circuit problems.

Power in Three-Phase Circuits

In Chapter 3, a discussion of power in relation to three-phase wye and delta connections was presented. It would be helpful to review the basic wye and delta generator connections of Chapter 3. We can summarize three-phase power relationships as follows:

$$\text{voltamperes per phase } (\text{VA}_\text{P}) = \text{V}_\text{P}\text{I}_\text{P}$$

$$\text{total voltamperes } (\text{VA}_\text{T}) = 3\text{V}_\text{P}\text{I}_\text{P} = 1.73\text{V}_\text{L}\text{I}_\text{L}$$

$$\text{power factor (pf)} = \frac{\text{true power (W)}}{1.73\text{V}_\text{L}\text{I}_\text{L}} = \frac{\text{W}}{3\text{V}_\text{P}\text{I}_\text{P}}$$

$$\text{power per phase } (\text{P}_\text{P}) = \text{V}_\text{P}\text{I}_\text{P} \times \text{pf}$$

$$\text{total power } (\text{P}_\text{T}) = 3\text{V}_\text{P}\text{I}_\text{P} \times \text{pf} = 1.73\text{V}_\text{L}\text{I}_\text{L} \times \text{pf}$$

where,
V_L is the line voltage in volts,
I_L is the line current in amperes,
V_P is the phase voltage in volts,
I_P is the phase current in amperes.

Calculations involving three-phase power are somewhat more complex than for single-phase power calculations. We must keep in mind the difference between phase values and line values to avoid making mistakes.

Power Factor Correction

Since most industries use a large number of electric motors, the industrial plant represents a highly inductive load. This means that industrial power systems operate at a power factor of less than unity (1.0). It is undesirable for an industry to operate at a low power factor, since the utility company would have to supply more electrical power to the industry than is actually used.

A given value of voltamperes (voltage × current) is supplied to an industry. If the power factor of the industry is low, the current must be higher since the power converted by the total industrial load equals: VA × pf. The value of power factor decreases as the reactive power drawn by the industry increases. This is illustrated in Fig. 5-11. We will assume a constant value of true power to see the effect of increases in reactive power drawn by a load. The smallest reactive power shown (var_1) results in the voltampere value of VA_1.

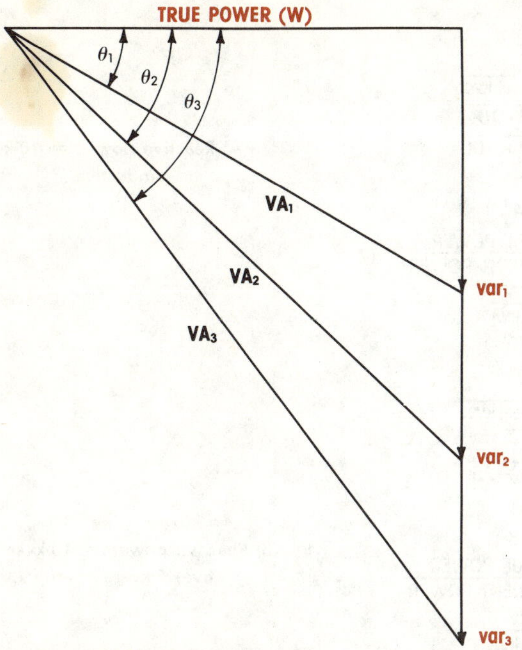

Fig. 5-11. Illustration of the effect of increases in reactive power (var) on apparent power (VA).

As reactive power is increased, as shown by the var_2 and var_3 values, more voltamperes (VA_2 and VA_3) must be drawn from the source. This is true since the voltage component of the supplied voltamperes remains constant. This example represents the same effect as a decrease in power factor, since $pf = W/VA$, and as VA increases, pf will decrease if W remains constant.

Utility companies usually charge industries for operating at power factors below a specified level. It is desirable for industries to "correct" their power factor to avoid such charges and to make more economical use of electrical energy. Two methods can be used to increase the power factor: (1) power-factor-corrective capacitors and (2) three-phase synchronous motors. Since the effect of capacitive reactance is opposite to that of inductive reactance, the reactive effects will counteract one another. Either power-factor-corrective capacitors or three-phase synchronous motors can be used to add the effect of capacitance to an ac power line.

In the example shown in Fig. 5-12, we will assume that both true power and inductive reactive power remain constant at values of 10 kW and 10 kvar. In Fig. 5-12A, the power factor equals 70%. If 5-kvar capacitive reactive power is introduced into the ac power line, the net reactive power becomes 5 kvar (10-kvar inductive minus 5-kvar capacitive), as shown in Fig. 5-12B. With the addition of 5-kvar capacitive, the power factor is increased to 89%. Now, in Fig. 5-12C, if 10-kvar capacitive is added to the ac power line, the total reactive power becomes zero. The true power is now equal to the apparent power; therefore, the power factor is 1.0, or 100%, which is characteristic of a purely resistive circuit. The effect of the increased capacitive reactive power is to increase or "correct" the power factor and thus reduce the current drawn from the ac power lines that supply the industrial loads. It may be beneficial for industries to invest in either power-factor-corrective capacitors or three-phase synchronous motors to correct their power factor.

Utility companies also attempt to correct the power factor of their systems. A certain quantity of inductance is present in most of the power distribution system, including the generator windings, the transformer windings, and the power lines. To counteract the inductive effects, utilities use power-factor-corrective capacitors, such as the pole-mounted units shown in Fig. 5-13. This type of capacitor unit, plus larger banks of capacitors located at substations, can be used to correct the power factor of ac power systems. The use of three-phase synchronous motors to correct the power factor will be discussed later in this chapter.

Load Factor

Another electrical load relationship is becoming increasingly important for industries. This relationship, called *load factor,* expresses the ratio between the average industrial power requirement and the peak power requirement, or:

$$\text{load factor} = \frac{\text{average demand (kW)}}{\text{peak demand (kW)}}.$$

The average demand for an industry is the average electrical power used over a specific time period. The peak demand is the maximum amount of power (kW) used during that time period. The load profile shown in Fig. 5-14 shows a typical industrial demand versus time curve for a working day. Demand peaks that far exceed the average demand cause a decrease in the load factor ratio. Low load factors result in an additional billing charge by the utility company.

Utility companies must design industrial power service to meet their peak demand and ensure that their generating capacity will meet this peak demand. Therefore, it is inefficient electrical design for an industry to operate at a low load factor, since this represents a significant difference between peak demand and average demand. Every industry should attempt to raise its load factor to the maximum level it can. By minimizing the peak demands of industrial plants, power demand control systems and procedures can help increase the efficiency of our nation's electrical power systems.

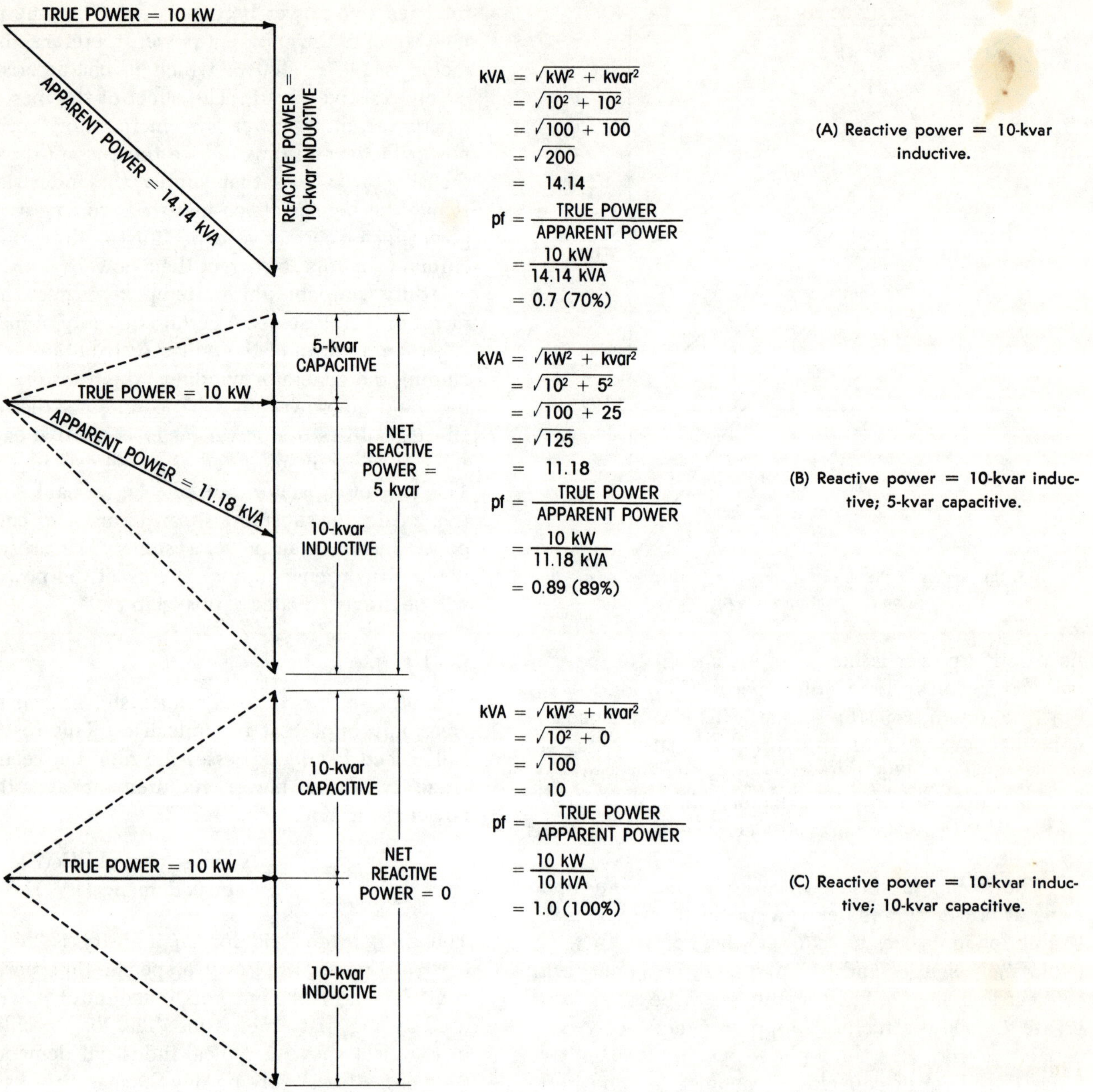

Fig. 5-12. Illustration of the effect of capacitive reactance on an inductive load.

FUNCTIONAL LOADS

Industrial loads are ordinarily referred to by the function they perform. Common types of industrial loads are those that convert electrical energy into light, heat, or mechanical energy. There are various types of lighting, heating, and mechanical loads used in industry. Several of these load devices will be discussed in this section.

Lighting Loads

Industrial lighting systems are designed to create a safe working environment, as well as to produce better accuracy of work done, and more alert employees. There are several types of lighting used in industry. Among these are incandescent, fluorescent, mercury vapor, and several other special-purpose types of lighting.

Incandescent Lighting

For many years, incandescent lighting was the most widely used industrial lighting method. Today, fluorescent lamps are used extensively; however, incandescent lamps are still used for certain applications. The construction of an incandescent lamp is shown in Fig. 5-15. This lamp is simple to install and maintain and costs

Fig. 5-13. Pole-mounted, power-factor-corrective capacitors.

Courtesy McGraw-Edison Co.

less initially; however, it has low efficiency and a short life span.

Incandescent lamps usually have a thin tungsten filament, which is in the shape of a coiled wire. This filament is connected directly to a voltage source (usually 120 volts ac) through the lamp base. When electric current is passed through the filament, the temperature of the filament becomes 3000 to 5000 degrees Fahrenheit. At this temperature range, the tungsten produces a high-intensity white light. When incandescent lamps are manufactured, the air is removed from the glass envelope to prevent the filament from burning, and an inert gas is added.

Fluorescent Lighting

Fluorescent lighting is used extensively in industry today. Fluorescent lamps are tubular bulbs with a filament on each end. There is no electrical connection between the two filaments. The operating principle of the fluorescent lamp is illustrated in Fig. 5-16A. The tube is filled with mercury vapor; thus, when an electric current flows through the two filaments, a continuous arc is formed between them by the mercury vapor. High-velocity electrons passing between the filaments collide with the mercury atoms, producing ultraviolet radiation. The inside of the tube has a phosphor coating which reacts to the ultraviolet radiation to produce visible light. Various types of phosphors are used to produce different colors.

The circuit for a fluorescent lamp is shown in Fig. 5-16B. Note that a thermal starter, which is basically a bimetallic strip and a heater, is connected in series with the filaments. The bimetallic strip remains closed long enough for the filaments to heat and vaporize the

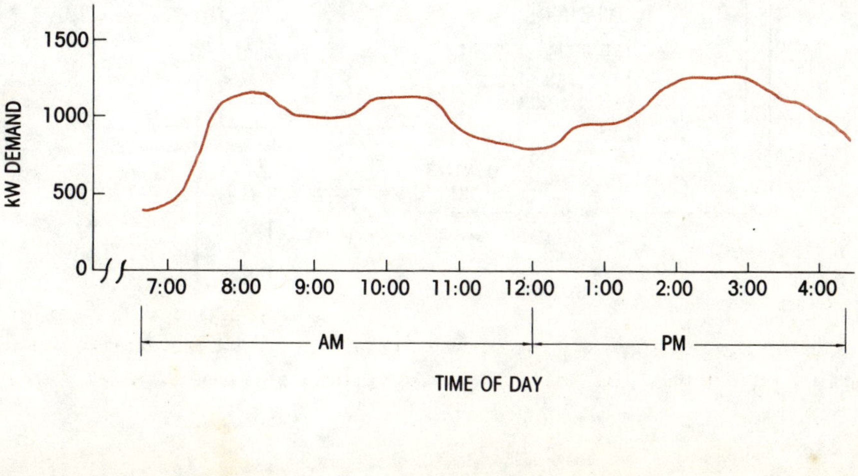

Fig. 5-14. Load profile for an industrial plant.

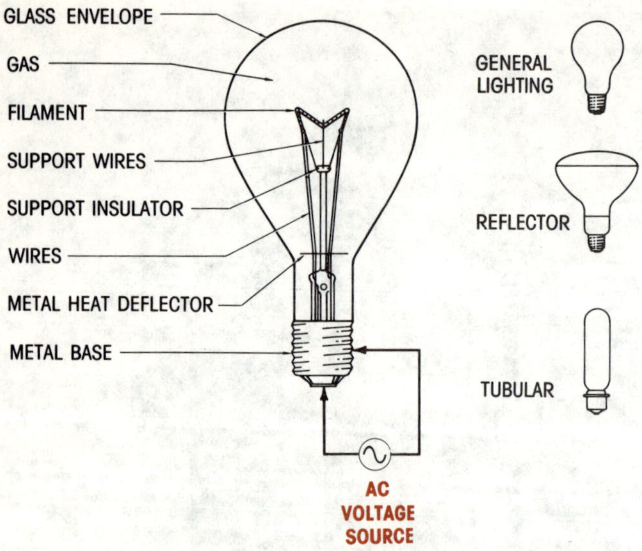

(A) Construction. (B) Types.

Fig. 5-15. The incandescent lamp.

mercury in the tube. The bimetallic strip will then bend due to the heat produced by current flow through the heater. The contact points will then separate and open the filament circuit. A capacitor is connected across the bimetallic switch to reduce contact sparking. Once the contacts of the starter open, a high voltage is momentarily placed between the filaments due to the action of the ballast coil. The ballast coil has many turns of fine wire and thus produces a high counter-electromotive force when the contacts of the starter separate. The high voltage across the filaments causes the mercury to ionize and initiates a current flow across the tube. There are several other methods used to start fluorescent lamps; however, this method illustrates the basic operating principle.

Fluorescent lamps produce more light per watt than incandescent lamps do, so they are cheaper to operate. Since the illumination is produced by a long tube, there is also less glare. The light produced by fluorescent lamps is very similar to natural daylight. The light is whiter and the operating temperature is much less in fluorescent lamps than in incandescent lamps. Various sizes and shapes of fluorescent lamps are available. We may conclude that fluorescent lamps have several advantages for use in industry. They are commonly used on 120- or 277-volt circuits.

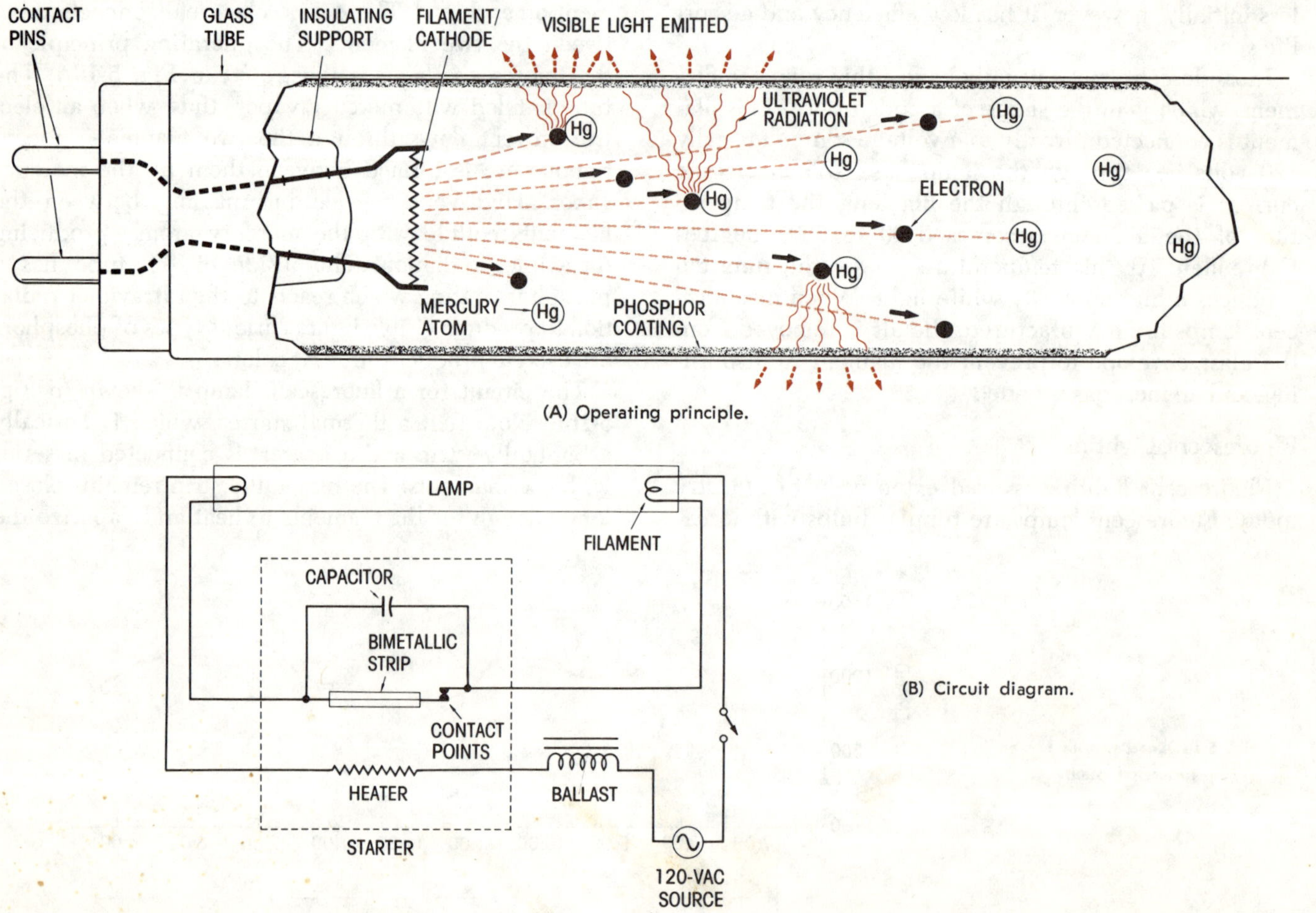

(A) Operating principle.

(B) Circuit diagram.

Fig. 5-16. The fluorescent lamp.

Vapor Lighting

Another popular form of lighting is the vapor type. The mercury vapor lamp is one of the most common vapor-type lamps. Another common type is the sodium vapor lamp. These lamps are filled with a gas that produces a characteristic color. For instance, mercury produces a greenish-blue light and argon produces a bluish-white light. Gases can be mixed to produce various color combinations.

The mercury vapor lamp is shown in Fig. 5-17A. It is made of an inner arc tube placed inside an outer bulb. The inner tube contains mercury. When a voltage is applied between the starting probe and an electrode, an arc is started between them. The arc current is limited by the series resistor; however, it is sufficient to cause the mercury in the inner tube to ionize. Once the mercury has ionized, an intense greenish-blue light is produced.

Mercury vapor lamps are compact, long-lasting, and easy to maintain. They are used in industry to provide a high-intensity light output. At low voltages, mercury is slow to vaporize, so these lamps require a long starting time (from 4 to 8 minutes). Mercury vapor lamps

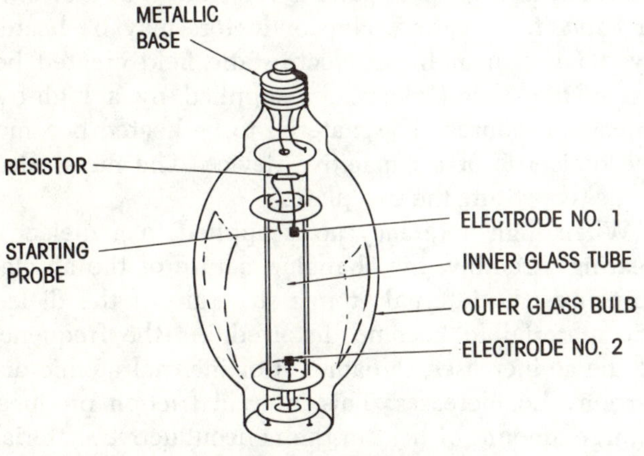

(A) Mercury vapor lamp.

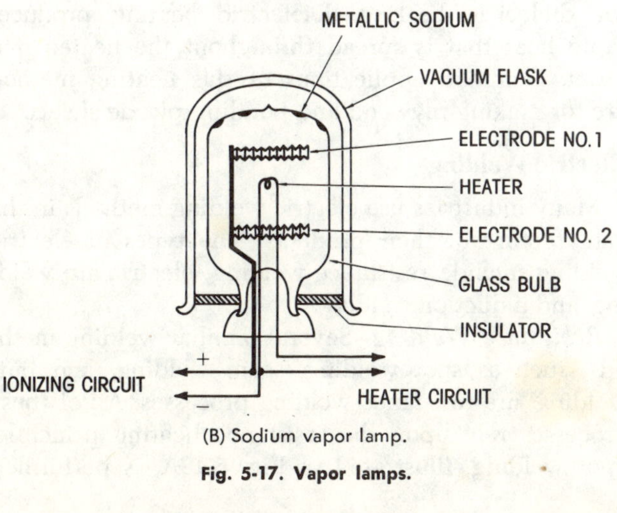

(B) Sodium vapor lamp.

Fig. 5-17. Vapor lamps.

can also be used for outdoor lighting. They are ordinarily used on either 120- or 277-volt circuits.

The sodium vapor lamp, shown in Fig. 5-17B, is popular for outdoor industrial lighting and for highway lighting. This lamp contains some low-pressure neon gas and some sodium. When electric current is passed through the heater, electrons are given off. The ionizing circuit causes a positive charge to be placed on the positive electrodes. As electrons pass from the heater to the positive electrodes, the neon gas is ionized. The ionization of the neon produces enough heat to cause the sodium to ionize. A yellowish light is produced by the sodium vapor. The sodium vapor lamp can produce about three times the candlepower per watt as that of an incandescent lamp.

The three types of lighting loads discussed in this section constitute most of our industrial lighting. Incandescent, fluorescent, and vapor lamps are used for many lighting applications. There are a few other specialized types of lighting that may be used in industry. These include zirconium and xenon arc lights, glow lights, black lights, and infrared lights. The planning involved for proper industrial lighting is quite complex and may involve several types of lamps.

Industrial Heat-Producing Loads

Most loads that are connected to industrial systems produce a certain amount of heat. Heat is produced mainly due to current flow through a resistive device. In many instances, heat represents a power loss (I^2R) in the circuit, since heat energy is not the type of energy that is intended to be produced. Lamps, for instance, produce heat energy as well as light energy. The conversion of electrical energy to heat energy in a light-producing load reduces the efficiency of the load device, since not all of the available energy from the source is converted to light energy. There are, however, several types of loads used in industry that are mainly heat-producing loads. Their primary function is to convert electrical energy into heat energy. These load devices include resistance heating, inductive heating, dielectric (capacitive) heating, and electric welders.

Industrial processes are highly dependent upon heat. Heat is needed to produce chemicals, plastics, and metals, as well as to condition the in-plant environment. Heat is also needed to fasten materials together and to alter their shape. Each of the electrical heating loads mentioned in this section has a specialized industrial application.

Resistance Heating

Heat energy is produced when an electrical current flows through a resistive material. In many instances, the heat energy produced by an electrical current is undesirable; however, certain industrial processes utilize controlled resistance heating. Useful heat can be transferred from a resistive element to a point of utilization by the common methods of heat transfer—convection, radiation, or conduction. A heating element enclosure is needed to control transfer by convection and radiation. For heat transfer by conduction, the heating element is in direct contact with the material to be heated. Actual heat transfer usually involves a combination of these methods.

Fig. 5-18A illustrates the principle of resistance heating. The self-contained heating element uses coiled resistance wire placed inside a heat-conducting material and enclosed in a metal sheath. This principle can be used to heat water, oil, the surrounding atmosphere, or various other media. They may be employed in open air or immersed in the media to be heated. The useful life of resistance elements depends mainly on the operating temperature. As the temperature increases, the heat output also increases. Basically, the heat energy produced depends on the current flow and the resistance of the element (I^2R).

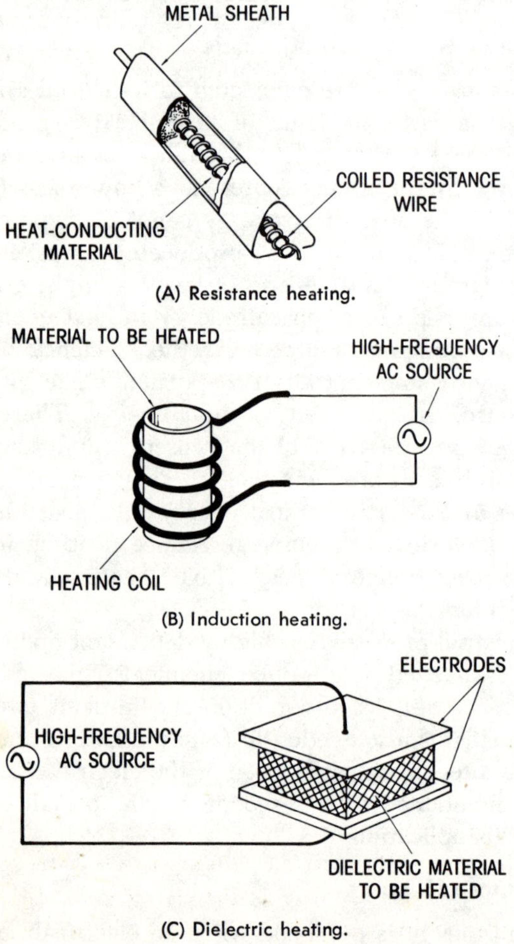

(A) Resistance heating.

(B) Induction heating.

(C) Dielectric heating.

Fig. 5-18. Basic methods of producing heat energy.

Induction Heating

The principle of induction heating is illustrated in Fig. 5-18B. Heat is produced in magnetic materials when they are exposed to an alternating-current field. In the example shown, current is induced into the material to be heated by electromagnetic induction brought about by the application of alternating current to the heating coil. The material to be heated must be a conductor in order for current to be induced. Ordinarily, a high-frequency ac source in the range of 100–500 kHz, is used to produce a higher heat output.

As the magnetic field created by the high-frequency ac source moves across the material to be heated, the induced voltage causes circulating currents (*eddy* currents) to flow in the material. Heat results due to the material's resistance to the flow of eddy currents. The heat produced by this method is rapid and thus advantageous for many industrial processes.

Dielectric (Capacitive) Heating

Induction heating is useful only for conductive materials. Therefore, some other method must be used to heat nonconductive materials. Such a method is illustrated in Fig. 5-18C, and is referred to as *dielectric* or *capacitive* heating. Nonconductors may be heated by placing them in an electrostatic field created between two metal electrodes supplied by a high-frequency ac source. The material to be heated becomes the dielectric of a capacitive device. The metal electrodes constitute the two plates.

When high-frequency ac is applied to a dielectric heating assembly, the changing nature of the applied ac causes the internal atomic structure of the dielectric material to become distorted. As the frequency of the ac increases, the amount of internal atomic distortion also increases. This internal friction produces a large amount of heat in the nonconductive material. Frequencies in the 50-MHz range are ordinarily used for dielectric heating. Dielectric heating produces rapid heat that is spread throughout the heated material. Common applications of this heating method are for making plywood and bonding plastic sheets.

Electric Welding

Many industries use electric welding methods in the manufacture of their products. The types of electric welding include resistance welding, electric-arc welding, and induction welding.

Resistance Welding—Several familiar welding methods, such as spot welding, seam welding, and butt welding, are resistance welding processes. All of these processes rely upon the resistance heating principle. Spot welding, illustrated in Fig. 5-19A, is performed

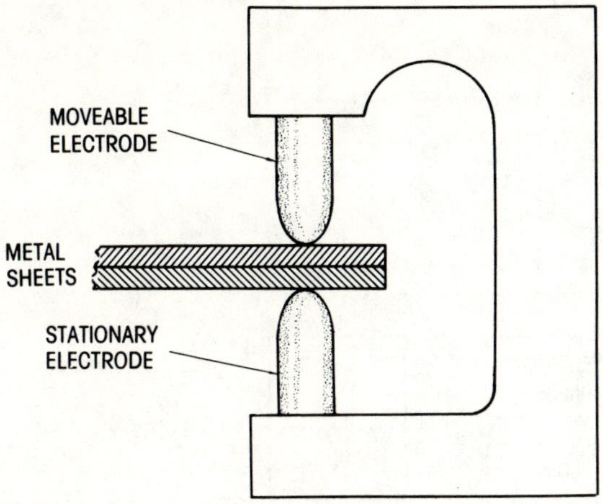

(A) Spot welding.

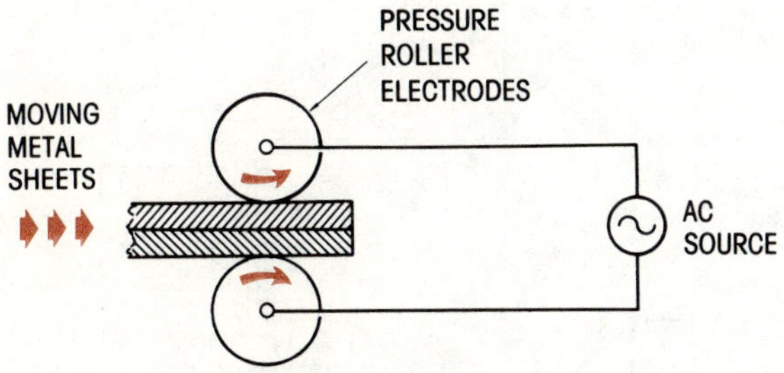

(B) Seam welding.

Fig. 5-19. Resistance welding methods.

on overlapping sheets of metal, usually less than ¼ inch thick. The metal sheets are clamped between two electrodes and an electrical current is passed through the electrodes and the metal sheets. The current is sufficient to cause the metals to fuse together. The instantaneous current through the electrodes is usually in excess of 5000 amperes, while the voltage between the electrodes is less than 2 volts.

Seam welding, illustrated in Fig. 5-19B, is accomplished by passing sheets of metal through two pressure rollers while continuously interrupted current is passed through the electrodes. The operational principle of seam welding is the same as spot welding. Several other similar methods, which are referred to as *butt* welding, *edge* welding, and *projection* welding are also commonly employed in industry.

Electric-Arc Welding—While resistance welding utilizes pressure on the materials to be welded, electric-arc welding produces welded metals by localized heating without pressure, as shown in Fig. 5-20. Carbon electrodes are used for direct-current or alternating-current arc welding of nonferrous metals and alloys. Not all metals can be welded by this process. The welding metal is produced by melting part of the metal to be welded. This metal is then filled into the gap created by the filler material of the electrode. An electric arc is created when the electrode of the welder is brought

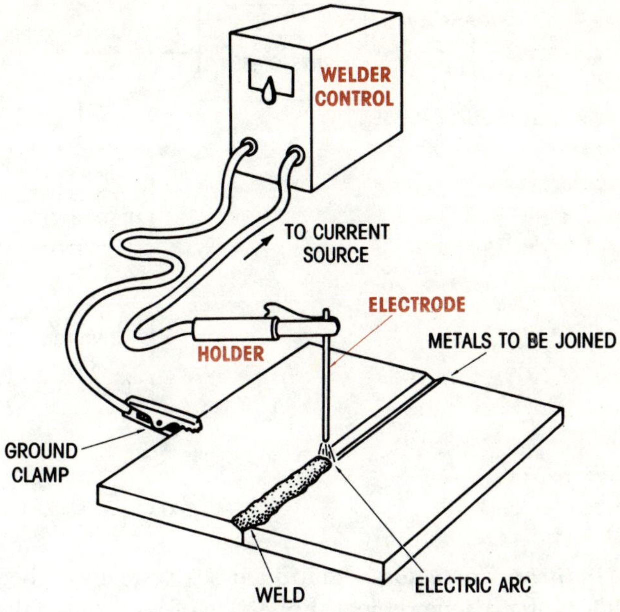

Fig. 5-20. Electric-arc welding.

in contact with the metal to be welded. Various types of electric-arc welders, having various current/voltage ratings, are available.

A smaller amount of current is required for arc welding than for resistance welding. The currents may range from 50 to 200 amperes or higher for some applications. Voltages typically range from 10 to 50 volts. An electric-arc welder may be supplied by an attached generator, a storage battery unit, a step-down transformer, or a rectification unit.

Induction Welding—The induction welding process uses the principle of induction heating to fuse metals together. High-frequency alternating current is applied to a heating coil into which the materials to be welded are placed. Tubular metal is often welded in this way.

Electric Welding Circuits

Electric welders are specialized types of equipment since they use very high amounts of current at low voltage levels. Therefore, their electrical circuits are also specialized. Ignitrons or SCRs are commonly employed to control the starting and stopping of the large currents required by electric welders. The current rating of these devices must be very high, sometimes in the range of 1000 to 100,000 amperes.

Fig. 5-21 illustrates a typical electric-welding system used in industry. The ac power supplied from the branch circuit is either stepped down by a transformer to deliver ac voltage to the welder, or rectified to produce dc voltage for direct-current welders. In either type of machine, an ignitron or SCR contactor is ordinarily used to control the on and off time of the welder.

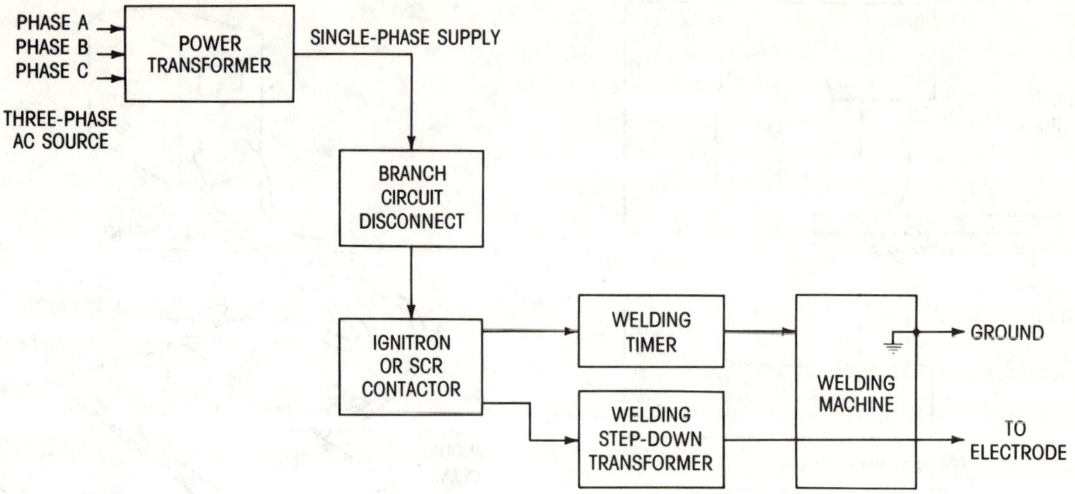

Fig. 5-21. Block diagram of a typical electric welding system.

Ignitron Contactors—Ignitron tubes were described in Chapter 3. One typical application of ignitrons is the control of large currents used in welders. The circuit shown in Fig. 5-22 is an ignitron contactor used to turn a welder on and off. One ignitron will conduct during each alternation of the ac source voltage. Ignitron contactors serve the same purpose in welding circuits that magnetic contactors perform in motor control. They merely serve as switching devices for very large currents. We should keep in mind that ignitrons carry heavy currents only for short periods of time. Contactors are also used with sequence timers to control the welding process.

SCR Contactors—SCR contactors are now replacing

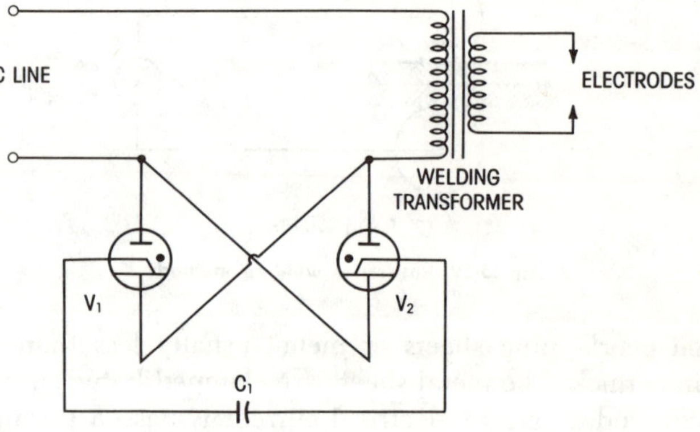

Fig. 5-22. An ignitron contactor circuit.

Fig. 5-23. A solid-state SCR contactor.

Courtesy Honeywell, Inc.

their ignitron counterparts. Since SCRs can now be designed to handle large amounts of current, they are usually more desirable to use than ignitrons. If we replaced the ignitrons in Fig. 5-22 with SCRs, the result would be an SCR contactor circuit. The SCRs are triggered or turned on by pulses supplied by the timing or sequencing circuits of the welder. The SCRs, as well as the ignitrons, are usually cooled by circulating water. An SCR contactor is shown in Fig. 5-23.

Mechanical Loads

Devices that convert electrical energy into mechanical energy represent another broad category of electrical loads. Electric motors fall into this category of load devices. There are many types of motors used in industry today. Most industries depend on motors for their everyday operation. In fact, the motor load is, for many industries, the major power-consuming load. Motors of various sizes are used for processes ranging from precise machinery control to the movement of massive pieces of equipment.

Basic Motor Principles

All motors, regardless of whether they operate from an ac or a dc power source, have several basic characteristics in common. Their basic parts include: (1) a stator, which is the frame and other stationary components; (2) a rotor, which is the rotating shaft and its associated parts; and (3) auxiliary equipment such as a brush/commutator assembly for dc motors, and a starting circuit for single-phase ac motors. The basic parts of a dc motor are shown in Fig. 5-24.

In order to explain the motor principle, we refer to Chapter 3 and the operation of dc generators. The simple dc motor is constructed in the same way as a dc generator. Their basic parts are the same. The function of the motor is to produce mechanical energy in the form of rotary motion. To produce rotary motion, a motor must have an electrical input. Recall from Chapter 3 that generator action is brought about due to a magnetic field, a set of conductors within the field, and relative motion between the two. Motion is similarly produced in a motor due to the interaction of a magnetic field and a set of conductors.

The motor principle is illustrated in Fig. 5-25. In Fig. 5-25A, no current is flowing through the conductors due to the position of the brushes in relation to the commutator. During this condition, no motion is produced in the conductor. When current flows through the conductors, a circular magnetic field is developed around the conductors. The direction of current flow

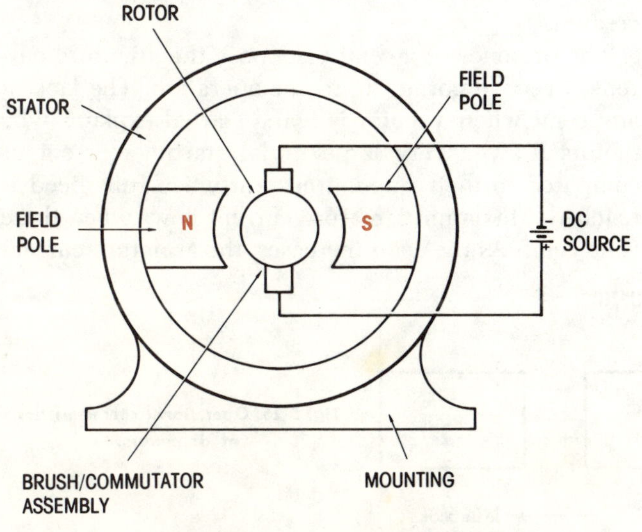

Fig. 5-24. Basic parts of a dc motor.

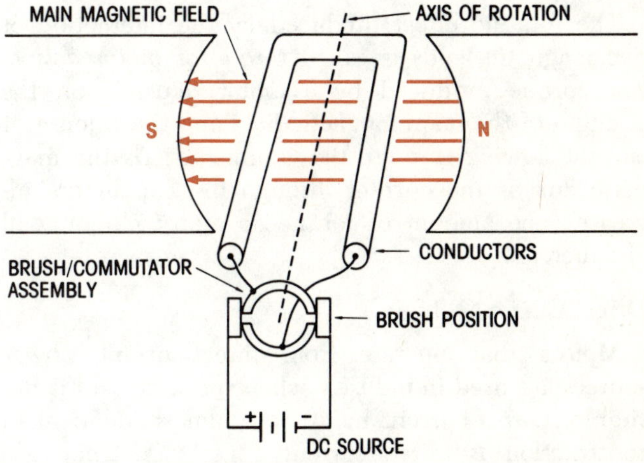

(A) Condition with no current flowing through conductors.

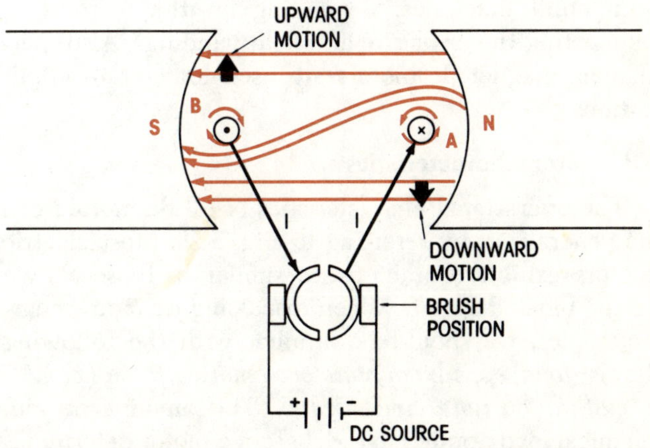

(B) Condition with current flowing through conductors.

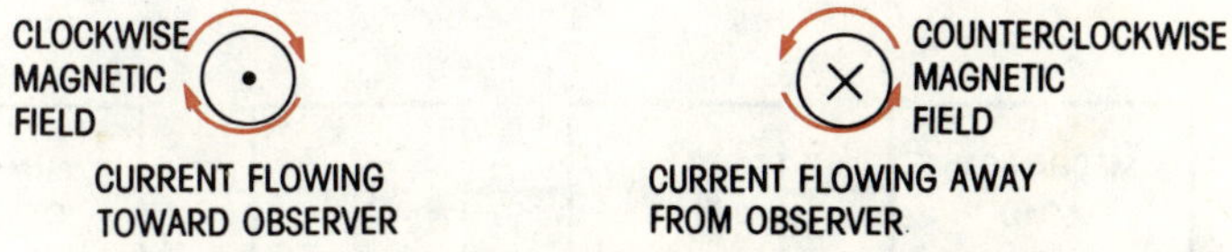

(C) Direction of current flow through conductors determines direction of magnetic field around conductors.

Fig. 5-25. Illustration of the basic motor principle.

determines the direction of the circular magnetic fields, as shown in the cross-sectional diagram of Fig. 5-25C. When current flows through the conductors within the main magnetic field, the field around the conductors interacts with the main field. The interaction of these two magnetic fields results in motion being produced. The circular magnetic field around the conductors causes a compression of the main magnetic flux at points A and B in Fig. 5-25B. This compression causes the magnetic field to produce a reaction in the opposite direction from the compression. Therefore, motion is produced away from points A and B. In actual motor operation, a rotary motion in a clockwise direction would be produced. If we wished to change the direction of motion, we would merely reverse the direction of current flow through the conductors.

The rotating effect produced by the interaction of two magnetic fields is called *torque* or *motor action.* The torque produced by a motor depends on the strength of the main magnetic field and the amount of current flowing through the conductors. As the magnetic flux or the current through the conductors increases, the amount of torque or rotary motion will also increase.

Direct-Current Motors

Motors that operate from direct-current power sources are used in industry when speed control is desirable. Direct-current motors are almost identical in construction to direct-current generators. They are also classified in a similar manner as series, shunt, or compound machines, depending on the method of connecting the armature and field windings. Also, permanent-magnet dc motors are used for certain applications.

DC Motor Characteristics

The operational characteristics of all dc motors can be generalized by referring to Fig. 5-26. Most electric motors exhibit characteristics similar to those shown in the block diagram. When discussing dc motor characteristics, we should be familiar with the following terms: *load, speed, counterelectromotive force (cemf), armature current,* and *torque.* The amount of mechanical load applied to the shaft of a motor determines its operational characteristics. As the mechanical load is increased, the speed of a motor tends to decrease. As the speed decreases, the voltage induced into the conductors of the motor due to generator action (cemf) decreases. The generated voltage, or counterelectromotive force, depends upon the number of rotating conductors and the speed of rotation. Therefore, as the speed of rotation decreases, so does the cemf.

The counterelectromotive force generated by a motor is in opposition to the supply voltage. Therefore, the actual working voltage of a motor can be expressed as:

$$V_T = V_C + I_A R_A$$

where,

V_T is the terminal voltage of the motor in volts,
V_C is the cemf generated by the motor in volts,
$I_A R_A$ is the voltage drop across the armature in volts.

Since the cemf is in opposition to the supply voltage, the actual working voltage of a motor will increase as the cemf decreases. When the working voltage increases, more current will flow through the armature windings. Since the torque of a motor is directly proportional to the armature current, the torque will increase as the armature current increases.

To briefly discuss the opposite situation, if the mechanical load connected to the shaft of a motor decreases, the speed of the motor tends to increase. An increase in speed causes an increase in the cemf. Since the cemf is in opposition to the supply voltage, as the cemf increases, the armature current decreases. A decrease in armature current causes a decrease in torque.

We can see that torque varies with changes in load, but we need to consider each of the steps involved to understand dc motor operation. As the load on a motor is increased, its torque also increases to try to meet the increased load requirement. However, the current drawn by a motor also increases when the load is increased.

The presence of a cemf to oppose the armature current is very important in motor operation. The lack of any cemf when a motor is being started explains why motors draw a very large initial starting current as compared to their running current when full speed is reached. Maximum armature current flows when there is no cemf. As the cemf increases, the armature current

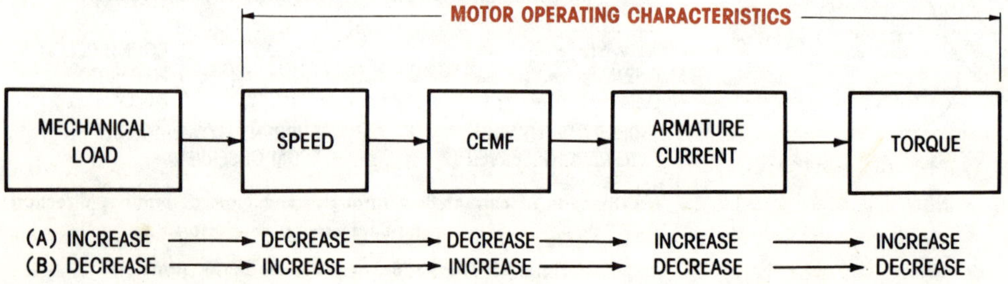

Fig. 5-26. Operational characteristics of dc motors.

decreases. Thus, resistances in series with the armature circuit are often used to compensate for the lack of cemf, and to reduce the starting current of a motor. After the motor has reached full speed, these resistances are by-passed by automatic or manual switching systems in order to allow the motor to produce maximum torque. Keep in mind that the armature current, which directly affects torque, can be expressed as:

$$I_A = \frac{V_T - V_C}{R_A}$$

where,

I_A is the armature current in amperes,
V_T is the terminal voltage of the motor in volts,
V_C is the cemf generated by the motor in volts,
R_A is the armature resistance in ohms.

In determining the functional characteristics of a motor, the torque developed can be expressed as:

$$T = K\Phi I_A$$

where,

T is the torque in foot-pounds,
K is a constant based on physical characteristics (conductor size, frame size, etc.),
Φ is the quantity of magnetic flux between poles,
I_A is the armature current in amperes.

Torque can be measured by several types of motor analysis equipment.

The horsepower rating of a motor is based on the amount of torque produced at the rated full-load values. Horsepower, which is the usual method of rating motors, can be expressed mathematically as:

$$hp = \frac{2\pi ST}{33{,}000} = \frac{ST}{5252}$$

where,

hp is the horsepower rating,
2π is a constant,
S is the speed of the motor in revolutions per minute (r/min),
T is the torque developed by the motor in foot-pounds.

Another dc motor characteristic that we should discuss is armature reaction. Armature reaction was discussed in relation to dc generators in Chapter 3. This effect distorts the main magnetic field in dc motors as well as in dc generators. Similarly, the brushes of a dc motor can be shifted to counteract the effect of armature reaction. However, interpoles are ordinarily used to control armature reaction in dc motors.

The most desirable characteristic of dc motors is their speed-control capability. By varying the applied dc voltage with a rheostat, speed can be varied from zero to the maximum r/min of the motor. Some types of dc motors have more desirable speed characteristics than others. For this reason, we can determine the comparative speed regulation for different types of motors. Speed regulation is expressed as:

$$\%R = \frac{S_{NL} - S_{FL}}{S_{FL}} \times 100$$

where,

%R is the percentage of speed regulation,
S_{NL} is the no-load speed in r/min,
S_{FL} is the rated full-load speed in r/min.

Good speed regulation (low %R) results when a motor has nearly constant speeds under varying load situations.

Types of DC Motors

The types of dc motors commercially available fall basically into four categories: (1) permanent-magnet dc motors, (2) series-wound dc motors, (3) shunt-wound dc motors, and (4) compound-wound dc motors. Each of these motors has different characteristics due to its basic circuit arrangement and physical properties.

Permanent-Magnet DC Motor—The permanent-magnet dc motor, illustrated in Fig. 5-27A, is constructed in the same manner as its dc generator counterpart, which was discussed in Chapter 3. This type of motor is rarely used for industrial applications. The permanent-magnet motor is used only where a low amount of torque is required. When this type of motor is used, the dc power supply is connected directly to the armature conductors through the brush/commutator assembly. The magnetic field is produced by permanent magnets mounted on the stator.

Series-Wound DC Motor—The manner in which the armature and field circuits of a dc motor are connected determines its basic characteristics. Each of the types of dc motors are similar in construction to the type of dc generator that corresponds to it. The only difference, in most cases, is that the generator acts as a voltage source while the motor functions as a mechanical load.

The series-wound motor, illustrated in Fig. 5-27B, has the armature and field circuits connected in a series arrangement. There is only one path for current to flow from the dc voltage source. Therefore, the field is wound of relatively few turns of large diameter wire, giving the field a low resistance. Changes in load applied to the motor shaft cause changes in the current through the field. If the mechanical load increases, the current also increases. The increased current creates a stronger magnetic field. The speed of a series motor varies from very fast at no load, to very slow at heavy loads. Since large currents may flow through the low resistance field, the series motor produces a high torque

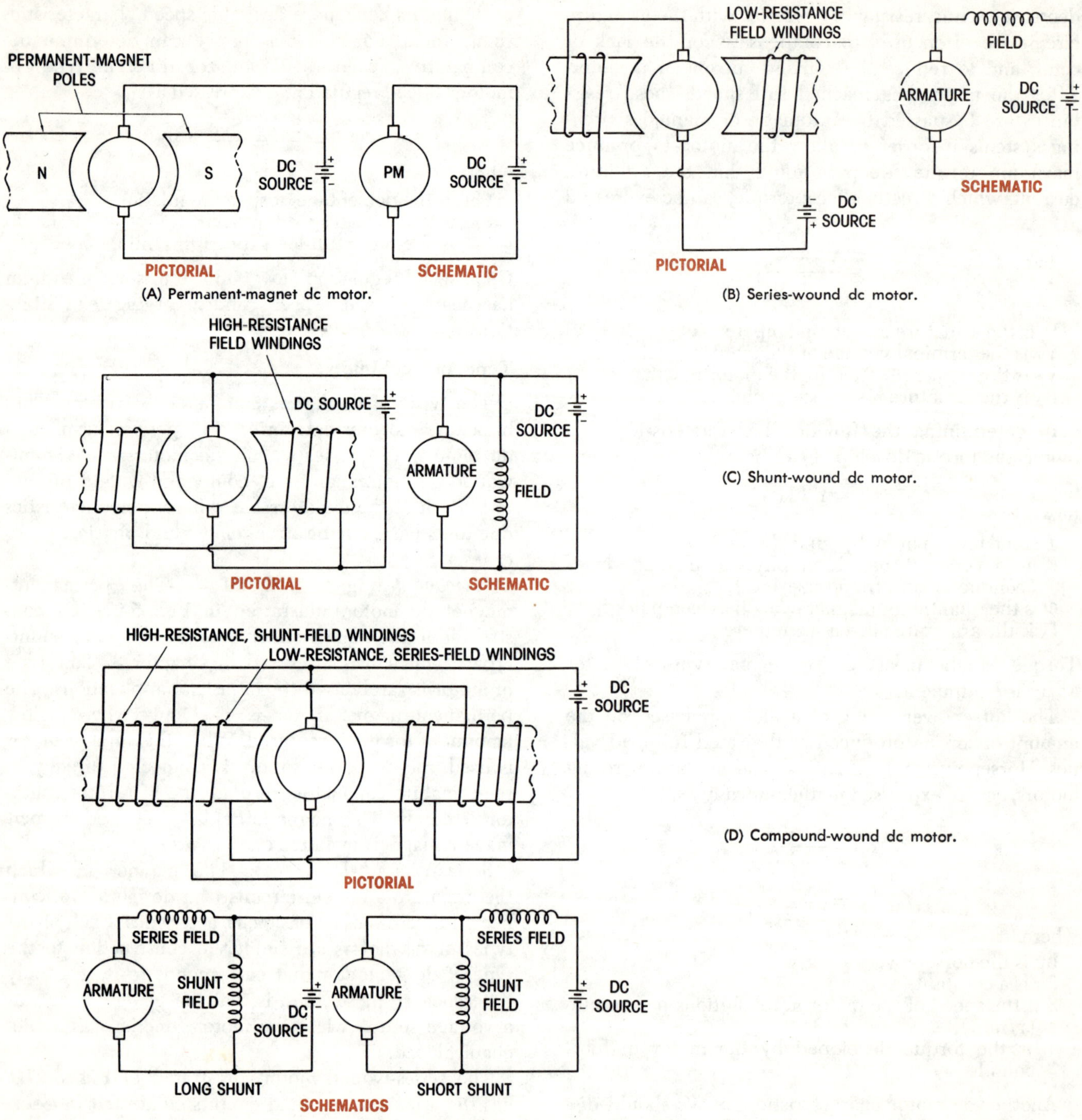

Fig. 5-27. Types of dc motors.

output. Series motors are used where heavy loads must be moved, and speed regulation is not important.

Shunt-Wound DC Motors—Shunt-wound dc motors are more commonly used than any other type of dc motor. As shown in Fig. 5-27C, the shunt-wound dc motor has the field coils connected in parallel with the armature. This type of dc motor has field coils that are wound of many turns of small diameter wire, giving the field a relatively high resistance. Since the field is a high-resistance parallel path, a small amount of current flows through the field. A strong electromagnetic field is produced due to the many turns of wire that form the field windings.

Most of the current drawn by the shunt motor flows in the armature circuit. Since the armature current has little effect on the strength of the field, motor speed is not affected appreciably by variations in load current. The field current, however, can be varied by placing a variable resistance in series with the field windings. Since the current in the field circuit is low, a low-wattage rheostat can be used to vary the speed of the motor. As the field resistance increases, the field current decreases. A decrease in field current reduces the strength of the electromagnetic field. When the field flux is decreased, the armature will rotate faster. Thus the speed of a dc shunt motor can easily be varied by using a field rheostat.

The shunt-wound dc motor has very good speed regulation. The speed does decrease slightly when the load increases, due to the increase of voltage drop across the armature. However, because of its good speed regulation and its ease of speed control, the shunt-wound dc motor is commonly used for industrial applications. Many types of variable-speed machine tools are driven by shunt-wound dc motors.

Compound-Wound DC Motors—The compound-wound dc motor has two sets of field windings, one in series with the armature and one in parallel. This motor combines the desirable characteristics of the series- and shunt-wound motors. It has high torque similar to a series-wound motor, along with good speed regulation similar to a shunt-wound motor. Therefore, when good torque and good speed regulation are needed for an industrial application, the compound-wound dc motor can be used. However, compound-wound motors are expensive.

The characteristics of dc motors should be considered when selecting motors for particular applications. Fig. 5-28 shows comparative graphs that illustrate the relative torque and speed characteristics of dc motors. Fig. 5-29 shows a large industrial dc motor.

A dc motor is designed so that its shaft will rotate in either direction. It is a very simple process to reverse the direction of rotation of any dc motor. By reversing the relationship between the connections of the armature windings and field windings, reversal of rotation is achieved. Usually, this is done by changing the terminal connections where the power source is connected to the motor. Four terminals are ordinarily used for interconnection purposes. They may be labeled A_1 and A_2 for the armature connections, and F_1 and F_2 for the field connections. If either the armature connections or the field connections are reversed, the rotation of the motor will reverse. However, if both the armature and the field connections are reversed, the motor shaft will rotate in its original direction, since the relationship between the armature and the field would still be the same.

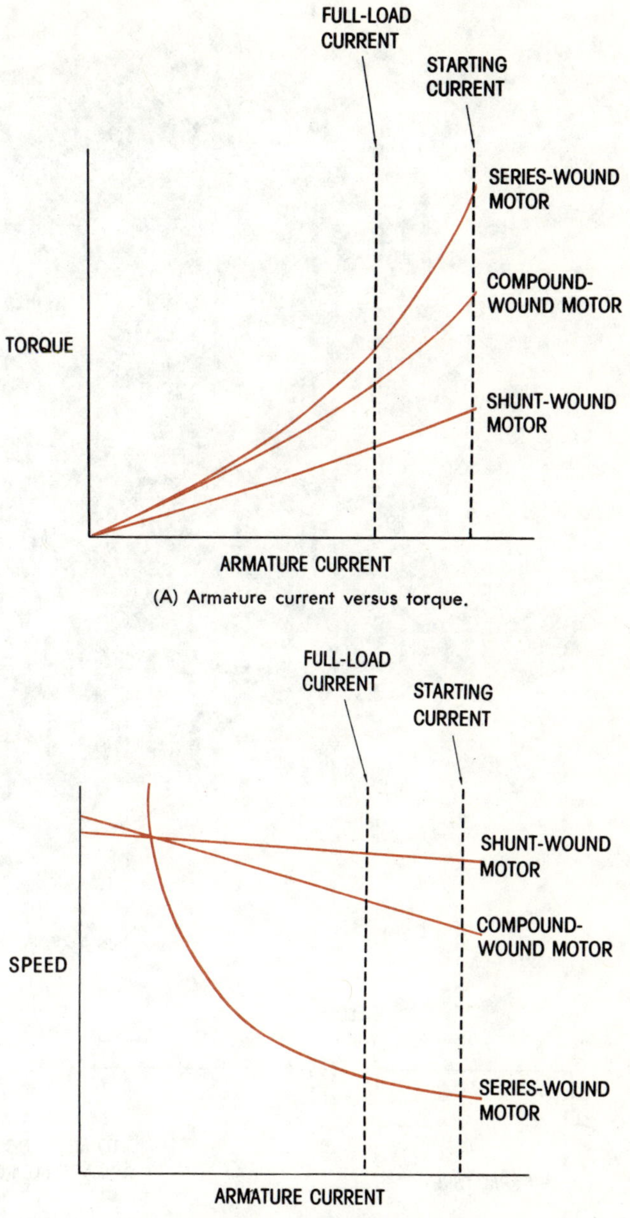

(A) Armature current versus torque.

(B) Armature current versus speed.

Fig. 5-28. Torque and speed characteristics of dc motors.

Single-Phase, Alternating-Current Motors

Another broad classification of mechanical load devices includes single-phase ac motors. These motors, which are common for industrial as well as commercial and residential usage, operate from a single-phase ac power source. There are three basic types of single-phase ac motors. These types are *universal* motors, *induction* motors, and *synchronous* motors.

Universal Motors—Universal motors can be powered by either ac or dc power sources. The universal motor, illustrated in Fig. 5-30, is constructed in the same way as a series-wound dc motor; however, it is designed to operate with either ac or dc applied. The series-wound motor is the only type of dc motor that will operate with ac applied. Shunt-wound motors have windings with inductance values that are too high to function

Courtesy General Electric Co.

Fig. 5-29. A large industrial dc motor.

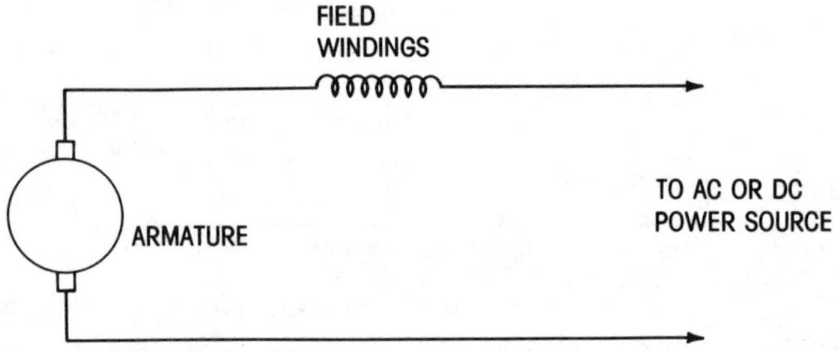

(A) Pictorial diagram.

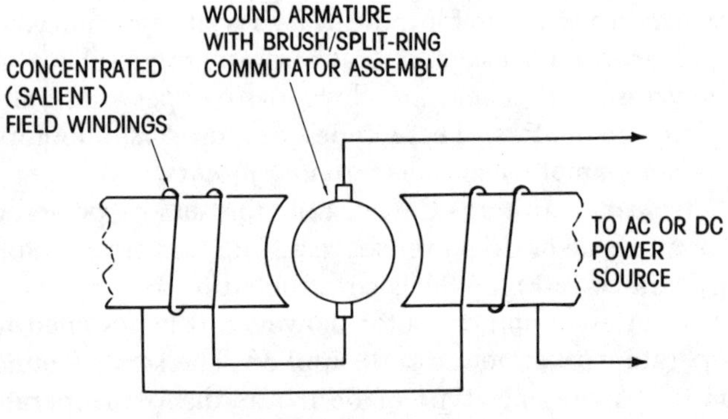

(B) Schematic diagram.

Fig. 5-30. The universal motor.

with ac applied. However, series motors have windings that have low inductance (few turns of large diameter wire), and therefore offer a low impedance to the flow of alternating current. The universal motor is one type of ac motor that has concentrated (or *salient*) field windings. These field windings are similar to those of all dc motors.

The operational principle of the universal motor with ac applied involves the instantaneous change of both field and armature polarities. Since the field windings have low inductance, the reversals of field polarity brought about by the changing nature of alternating current also create reversals of current direction through the armature conductors at the proper time intervals. The universal motor operates in the same manner as a series-wound dc motor except that the field polarity and the direction of armature current change at a rate of 120 times per second when connected to a 60-Hz ac source. The speed and torque characteristics of universal motors are similar to dc series-wound motors. Applications of universal motors in industry are mainly for portable tools and small motor-driven equipment.

Induction Motors—The operational principle of the single-phase ac induction motor is illustrated in Fig. 5-31. The coil symbols along the stator represent the field coils of an induction motor. These coils are energized by an alternating-current source; therefore, their instantaneous polarity changes 120 times per second when 60-Hz ac is applied.

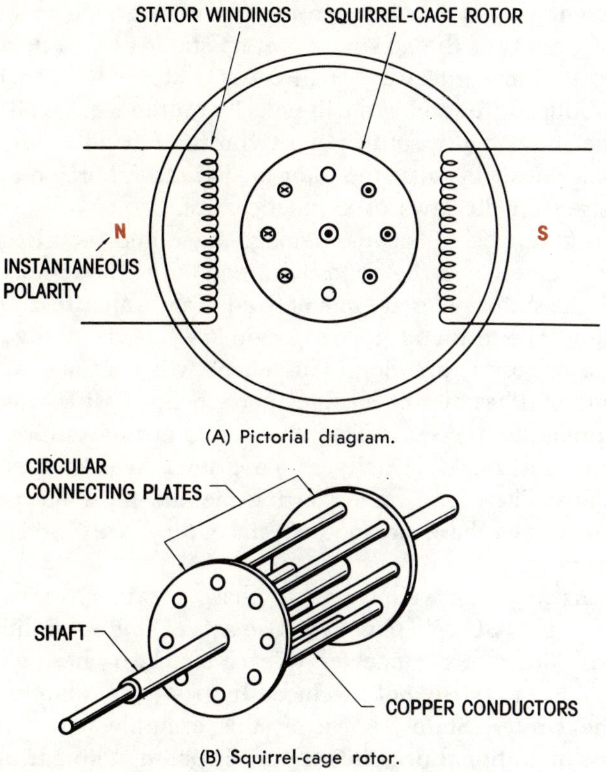

Fig. 5-31. Illustration of the induction principle.

Induction motors have a solid rotor, referred to as a *squirrel-cage* rotor. This type of rotor, which is illustrated in Fig. 5-31B, has large diameter copper conductors soldered at each end to a circular connecting plate. This plate actually short-circuits the individual conductors of the rotor. When current flows in the stator windings, a current is induced in the rotor. This current is developed due to "transformer action" between the stator and rotor. The stator, which has ac voltage applied, acts as a transformer primary. The rotor is similar to a transformer secondary since current is induced into it.

Since the stator polarity changes in step with the applied ac, it develops a rotating magnetic field. The rotor becomes instantaneously polarized due to the induced current flow through the short-circuited copper conductors. The rotor will, therefore, tend to rotate in step with the revolving magnetic field of the stator. If some method of initially starting the rotation is used, the rotor will continue to rotate. However, due to inertia, a rotor must be initially put into motion by some auxiliary method.

It should be pointed out that the speed of an ac induction motor is based on the speed of the rotating magnetic field and the number of stator poles that the motor has. The speed of the rotor will never be as high as the speed of the rotating stator field. If the two speeds were equal, there would be no relative motion between the rotor and stator and, therefore, no induced rotor current and no torque developed. The rotor speed of an induction motor is always somewhat less than the rotating stator field developed by the applied ac voltage.

The speed of the rotating stator field can be expressed as:

$$S = \frac{f \times 120}{n}$$

where,

S is the speed of the rotating stator field in r/min,
f is the frequency of the applied ac voltage in hertz,
n is the number of poles in the stator windings,
120 is a conversion constant.

A two-pole motor operating from a 60-Hz source would have a stator speed of 3600 r/min. The stator speed is also referred to as the *synchronous* speed of a motor.

The difference between the revolving stator speed of an induction motor and the rotor speed is called *slip*. The rotor speed must lag behind the revolving stator speed in order to develop torque. The more the rotor speed lags behind, the more torque is developed. Slip is expressed mathematically as:

$$\%\ \text{slip} = \frac{S_s - S_r}{S_s} \times 100$$

where,

S_s is the synchronous (stator) speed in r/min,
S_r is the rotor speed in r/min.

As the rotor speed becomes closer to the stator speed, the percentage of slip becomes smaller.

Another factor, referred to as *rotor frequency*, affects the operational characteristics of an induction motor under load. As the load on the shaft of the motor increases, the rotor speed tends to decrease. The stator speed, however, is unaffected. When a two-pole induction motor connected to a 60-Hz source operates at 10% slip, the slip would equal 360 r/min (3600 r/min × 10%). Functionally, this means that a revolving stator field sweeps across a rotor conductor 360 times per minute. Current is induced into a rotor conductor each time the stator field revolves past the conductor. As slip is increased, more current is induced into the rotor, causing more torque to be developed. The rotor frequency depends on the amount of slip and can be expressed as:

$$f_r = f_s \times slip$$

where,

f_r is the frequency of the rotor current,
f_s is the frequency of the stator current,
slip is expressed as a decimal.

Rotor frequency affects the operational characteristics of induction motors.

Single-phase ac induction motors are classified according to the method used for starting. Common types of single-phase ac induction motors include: *split-phase* motors, *capacitor* motors, *shaded-pole* motors, and *repulsion* motors.

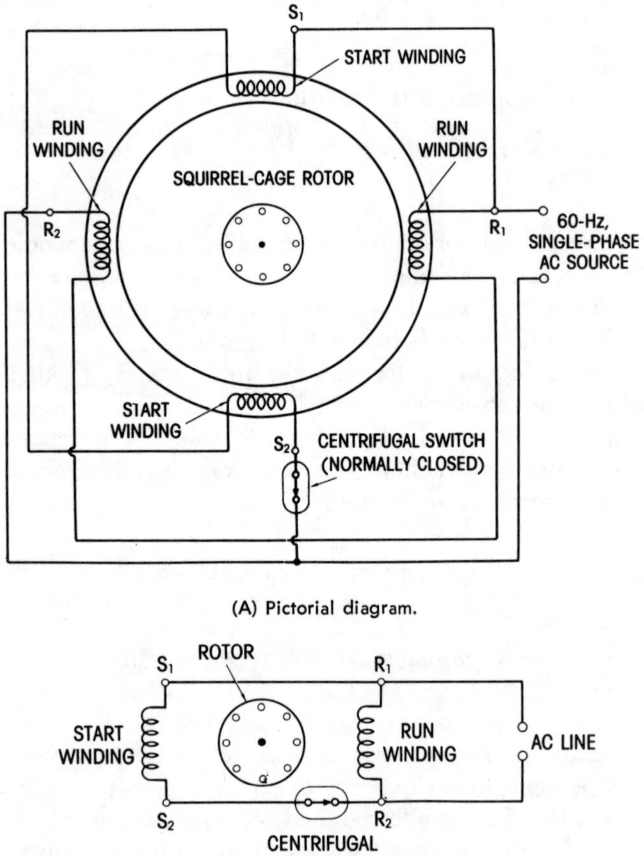

Fig. 5-32. The split-phase induction motor.

Split-Phase Motors—The split-phase ac induction motor, illustrated in Fig. 5-32, has two sets of stator windings. One set, called the *run* winding, is connected directly across the ac line. The other set, called the *start* winding, is also connected across the ac line. However, the start winding is connected in series with a centrifugal switch mounted on the shaft of the motor. The centrifugal switch is in the closed position when the motor is not rotating.

Before discussing the functional principle of the split-phase ac motor, we should understand how rotation is developed by an ac motor. Refer to Fig. 5-33. In Fig. 5-33A, we have a two-pole stator with single-phase ac applied. For purposes of our discussion, a permanent magnet is placed within the stator to represent the squirrel-cage rotor of an induction motor. At time t_0 of the ac sine wave, no stator field is developed. Time interval t_1 will cause a stator field to be produced. Assume a north polarity on the right pole of the stator and a south polarity on the left pole. These polarities will cause the rotor to align itself horizontally based on the laws of magnetic attraction. At time t_2, the stator poles become demagnetized and then begin to magnetize in the opposite direction. At time interval t_3, the stator poles are magnetized in the opposite direction. The rotor will now align itself horizontally in the opposite direction. This effect will continue at a rate of 120 polarity changes per second if 60-Hz ac is applied to the stator. The rotor would not start unless it was positioned initially to be drawn toward a pole piece. Therefore, some starting method must be used for single-phase ac motors, since they are not self-starting.

Assume that we have a two-phase situation as shown in Fig. 5-33B. We now have two sets of stator windings with one phase connected to each set. Two-phase voltage is, of course, not produced by power companies in the United States; however, this example will show the operational principle of a split-phase motor. In the two-phase voltage diagram, when one phase is at minimum value, the other is at maximum.

At time interval t_1, phase 1 is maximum while phase 2 is minimum. Assume that the right stator pole becomes a north polarity and the left stator pole becomes a south polarity. The rotor will align itself horizontally since no polarity is developed in the vertical poles at this time. Now, as we progress to time t_2, phase 1 is minimum and phase 2 is maximum. Assume that the upper stator pole becomes a south polarity and the bottom stator pole becomes a north polarity. The rotor will now align itself vertically, by moving 90°. At time t_3, phase 1 becomes maximum in the opposite direction and phase 2 is minimum. This time interval results in a north pole on the left and a south pole on the right. Thus, the rotor moves 90° farther in a clockwise direction. This effect will continue as two-phase voltage is applied to the stator poles. We can see from the two-phase situation that a direction of rotation is established by the relationship of phase 1 and phase 2, and that an ac motor with two-phase voltage applied would be self-starting. The same is true for a three-

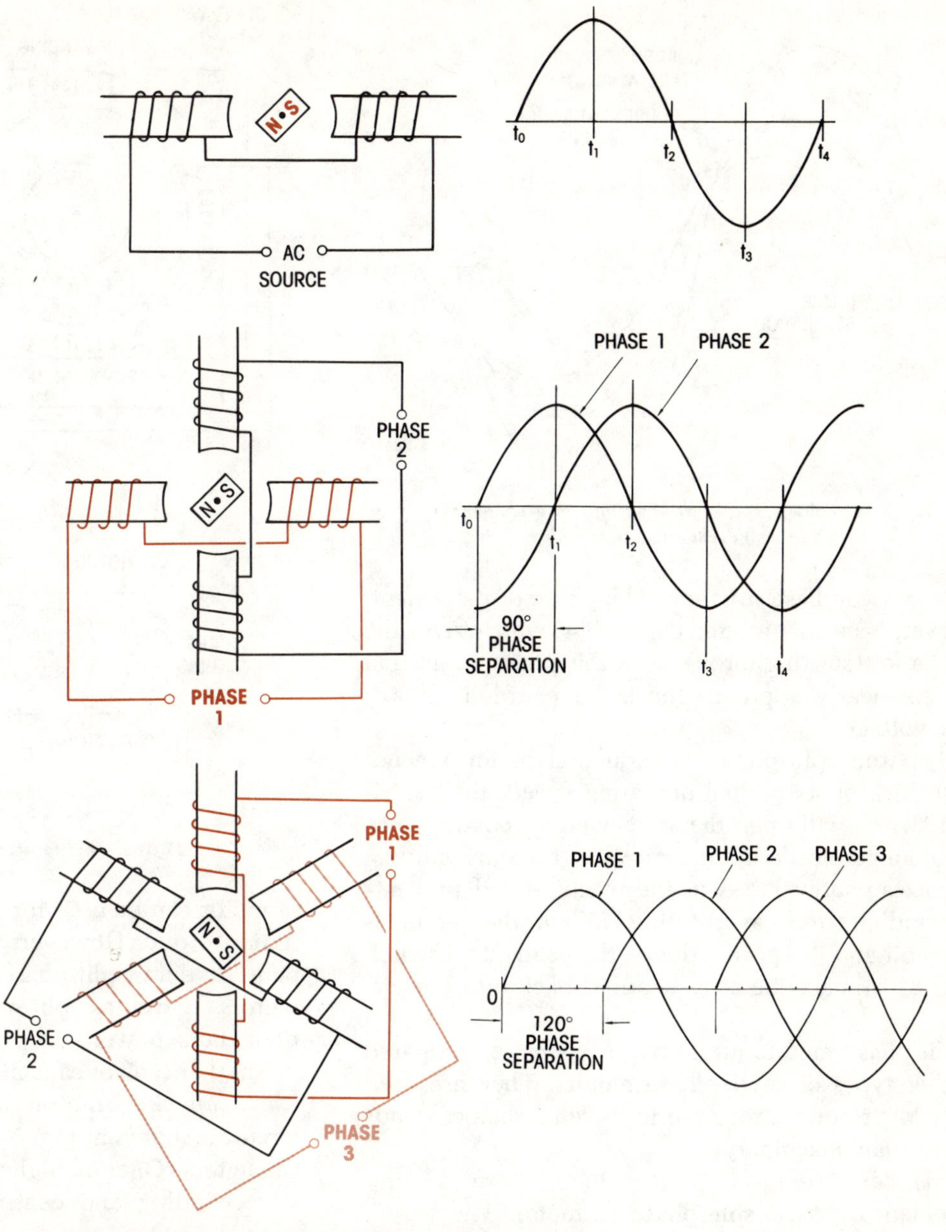

(A) Single-phase situation.

(B) Two-phase situation.

(C) Three-phase situation.

Fig. 5-33. Illustration of how ac motors develop rotation.

phase situation, which is illustrated in Fig. 5-33C. Rotation of the rotor would result due to the 120° phase separation of the three-phase voltage applied to the stator poles. Three-phase induction motors are, therefore, self-starting with no auxiliary starting method required.

Referring back to the split-phase motor of Fig. 5-32, we can see that the purpose of the two sets of windings is to establish a simulated two-phase condition to start the motor. The single-phase voltage applied to this motor is said to be "split" into a two-phase current. A rotating magnetic field is created by phase splitting. The start winding of the split-phase motor is made of relatively few turns of small diameter wire, giving it a high resistance and a low inductance. The run winding is wound with many turns of large diameter wire, causing it to have a lower resistance and a higher inductance. We know that inductance in an ac circuit causes the current to lag the applied ac voltage. The more inductance present, the greater is the lag in current.

When single-phase ac is applied to the stator of a split-phase induction motor, the situation illustrated in Fig. 5-34 will result. Notice that the current in the start winding lags the applied voltage due to its inductance. However, the current in the run winding lags by a greater amount due its higher inductance. The phase separation of the currents in the start and run windings

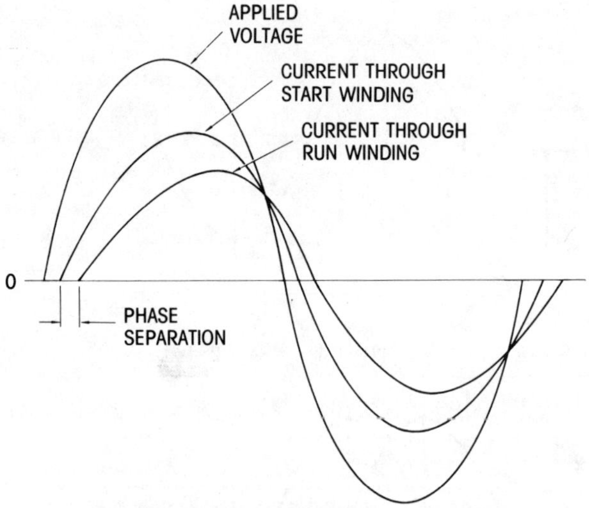

Fig. 5-34. Voltage/current relationships in a split-phase induction motor.

creates a two-phase situation. The phase displacement, however, is usually around 30° or less, which gives the motor a low starting torque since this phase separation does not nearly approach the 90° separation of two-phase voltage.

When the split-phase ac induction motor reaches about 80% of its normal operating speed, the centrifugal switch will open the start winding circuit, since it is no longer needed. The removal of the start winding minimizes energy losses in the machine and prevents the winding from overheating. When the motor is turned off and its speed reduces, the centrifugal switch closes to connect the start winding back into the circuit.

Split-phase motors are fairly inexpensive compared to other types of single-phase motors. They are used where low torque is required to drive mechanical loads such as small machinery.

Capacitor Motors—Capacitor motors are an improvement over the split-phase ac motor. The wiring diagram of a capacitor-start, single-phase induction motor is shown in Fig. 5-35A. Notice that this diagram is the same as the split-phase motor except that a capacitor is placed in series with the start winding. The purpose of the capacitor is to cause the current in the start winding to lead (rather than lag) the applied voltage.

This situation is illustrated in Fig. 5-36. The current in the start winding now leads the applied voltage due to the high value of capacitance in the circuit. Since the run winding is highly inductive, the current through it lags the applied voltage. Notice that the amount of phase separation now approaches 90°, or an actual two-phase situation. The starting torque produced by a capacitor-start induction motor is much greater than for the split-phase motor. Thus, this type

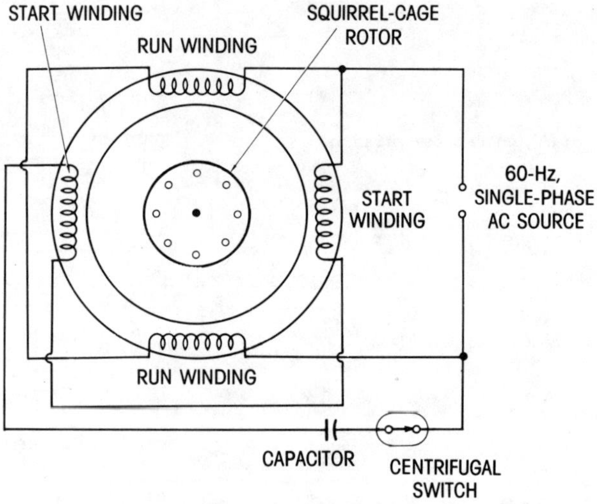

(A) Pictorial diagram.

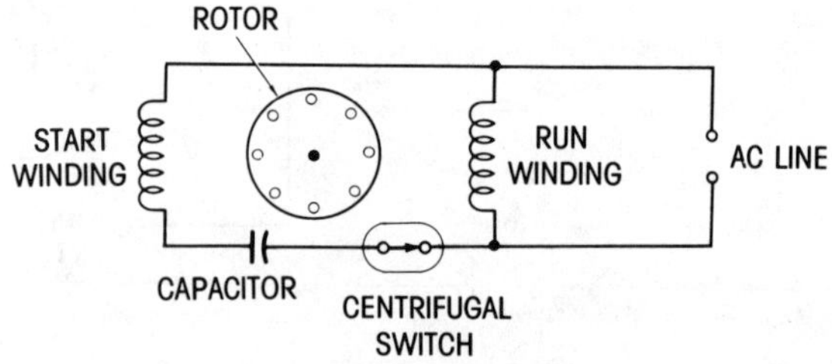

(B) Schematic diagram.

Fig. 5-35. The capacitor-start, single-phase induction motor.

of motor can be used for applications requiring greater initial torque. However, they are somewhat more expensive than split-phase ac motors. Most capacitor motors, as well as split-phase motors, are used in fractional horsepower sizes (less than 1 hp).

Another type of capacitor motor is called a *capacitor-start, capacitor-run,* or *two-value* capacitor motor. Its circuit is shown in Fig. 5-37A. This motor employs two capacitors. One, of higher value, is in series with the start winding and centrifugal switch. The other, of

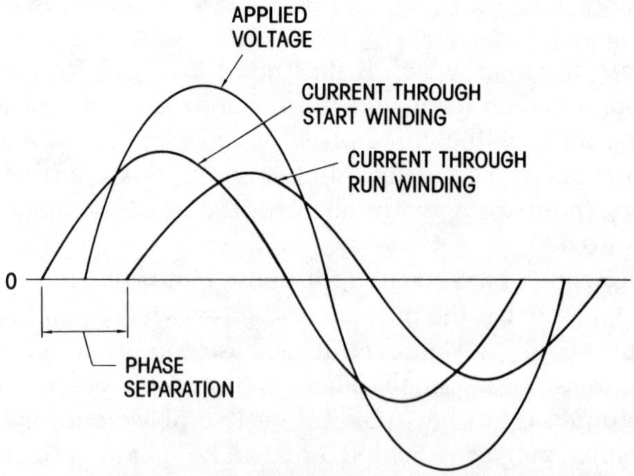

Fig. 5-36. Voltage/current relationships in a capacitor-start, single-phase induction motor.

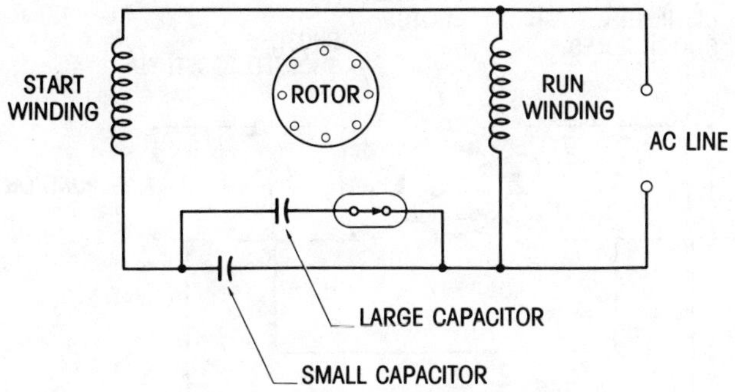

(A) Capacitor-start, capacitor-run motor.

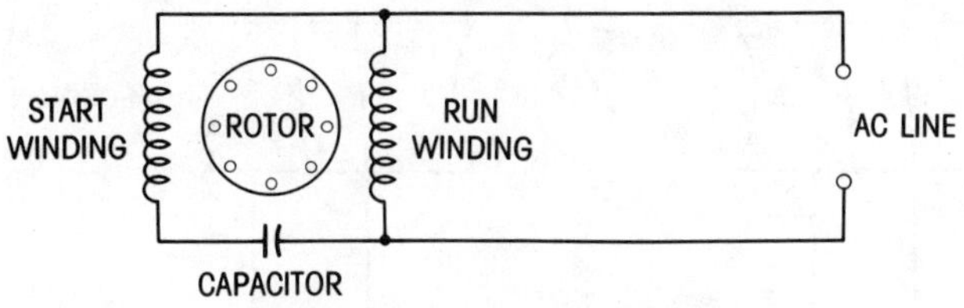

(B) Permanent capacitor motor.

Fig. 5-37. Other types of capacitor motors.

lower value, is in series with the start winding and remains in the circuit during operation. The larger capacitor is used only to increase starting torque, and is removed from the circuit by the centrifugal switch during normal operation. The smaller capacitor and the entire start winding are part of the operational circuit of the motor. The smaller capacitor helps to produce a more constant running torque, as well as quieter operation and improved power factor.

Yet another type of capacitor motor is called a *permanent* capacitor motor. Its circuit is shown in Fig. 5-37B. This motor has no centrifugal switch, so its capacitor is permanently connected into the circuit. These motors are only used for very low torque requirements and are made in small, fractional horsepower sizes.

Both split-phase motors and capacitor motors may have their direction of rotation reversed easily. We simply change the relationship of the start winding and run winding. By reversing either the start winding connections or the run winding connections (but not both), the rotational direction will be reversed.

Shaded-Pole Induction Motors—Another method of producing torque by a simulated two-phase method is called *pole shading*. A shaded-pole ac induction motor is shown in Fig. 5-38. These motors are used for very low torque applications such as fans and blower units. They are low-cost, rugged, and reliable motors that are only made in low horsepower ratings, usually from 1/300 to 1/30 hp.

The operational principle of the shaded-pole motor is shown in Fig. 5-39. The single-phase alternation

Fig. 5-38. A shaded-pole ac induction motor.

shown is for discussion purposes only. The dotted lines represent induced voltage into the shaded section of the field poles. Note the shading coils in the upper right and lower left of the two poles. The shaded pole is encircled by a heavy copper conductor and is actually a part of the main field pole. This closed-loop conductor will have current induced into it when ac is applied to the field.

When ac voltage is applied to the stator windings, the magnetic flux in the main poles induces a voltage into the shaded sections of the poles. Since the shaded section acts like a transformer secondary, its voltage is out of phase with the main field voltage as shown on the waveform diagram. Note the four time intervals shown in sequence in Fig. 5-39. The voltage induced in the shaded pole from the main pole field causes movement of the rotor to continue. Study this illustration to understand the operational principle of the shaded-pole ac induction motor.

The shaded-pole motor is inexpensive since it uses a squirrel-cage rotor and has no auxiliary starting winding or centrifugal mechanism. Industrial application is limited mainly to small fans and blowers.

Repulsion Motors—Another type of ac induction motor is the repulsion motor. This motor was used for many applications, but is now being replaced by other types of single-phase motors. The principle of operation of the repulsion motor is an interesting contrast to other induction motors.

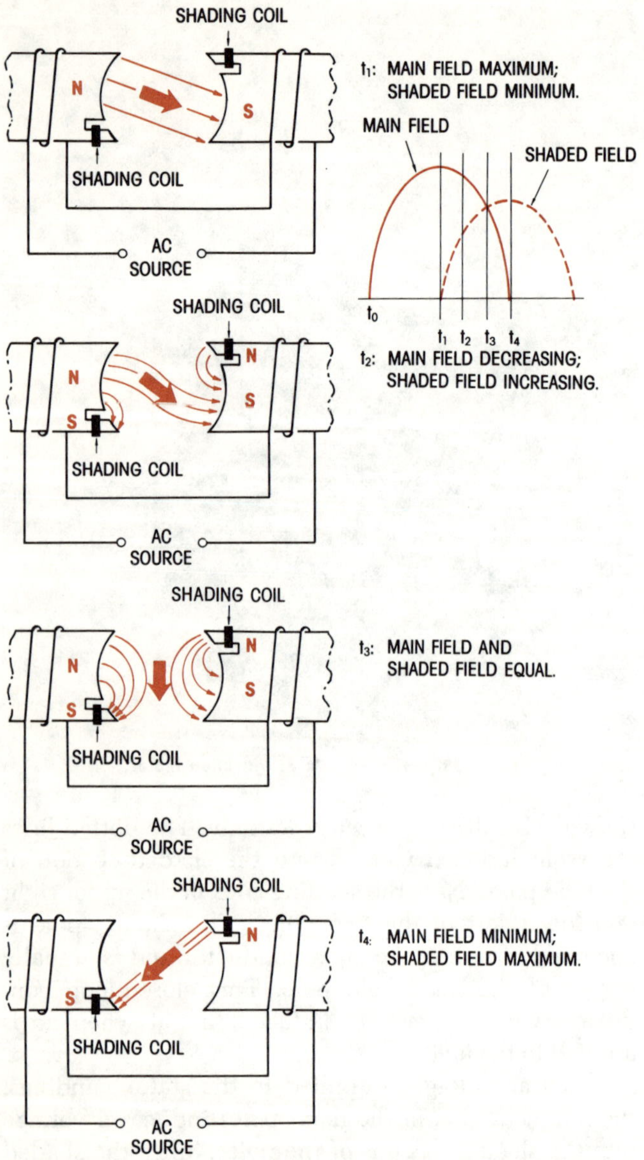

Fig. 5-39. Illustration of the operational principle of the shaded-pole ac induction motor.

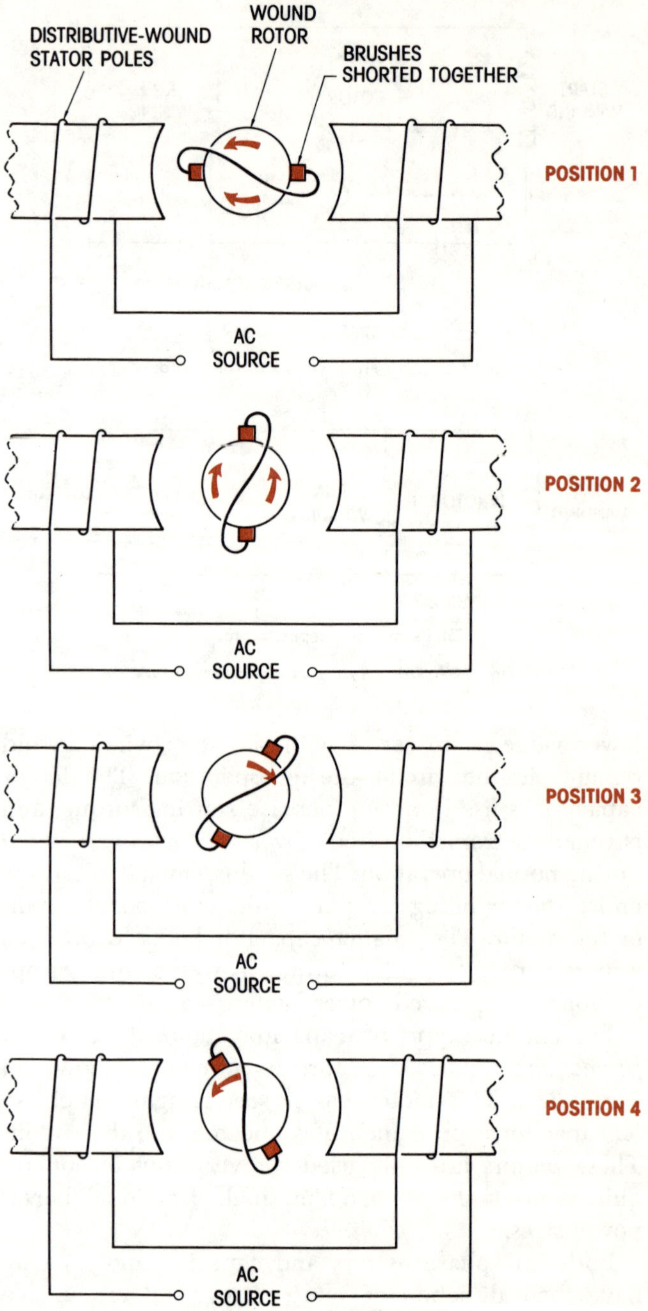

Fig. 5-40. The repulsion motor.

Fig. 5-40 shows the operational principle of the repulsion motor. This motor has a wound rotor that functions similarly to a squirrel-cage rotor. It also has a commutator/brush assembly. The brushes are shorted together to produce an effect similar to the shorted conductors of a squirrel-cage rotor. The position of the brush axis determines the amount of torque developed and the direction of rotation of the repulsion motor.

In position 1 of Fig. 5-40, the brush axis is horizontally aligned with the stator poles. Equal and opposite currents are now induced into both halves of the rotor. Thus, no torque is developed with the brushes in this position. In position 2, the brushes are placed at a 90° angle to the stator field poles. The voltages induced into the rotor again counteract one another and no torque is developed. In position 3, the brush axis is shifted about 60° from the stator poles. The current flow in the armature now causes a magnetic field to exist around the rotor. The rotor field will now follow the revolving stator field in a clockwise direction. As might be expected, when we shift the brush axis in the opposite direction, as shown in position 4, rotation reversal will result. Thus, magnetic repulsion between the stator field and the induced rotor field causes the rotor to turn in the direction of the brush shift.

Repulsion-start induction motors, and some similar types of modified repulsion motors, have very high

(A) Pictorial diagram.

(B) Three-phase waveforms and the resulting stator fields.

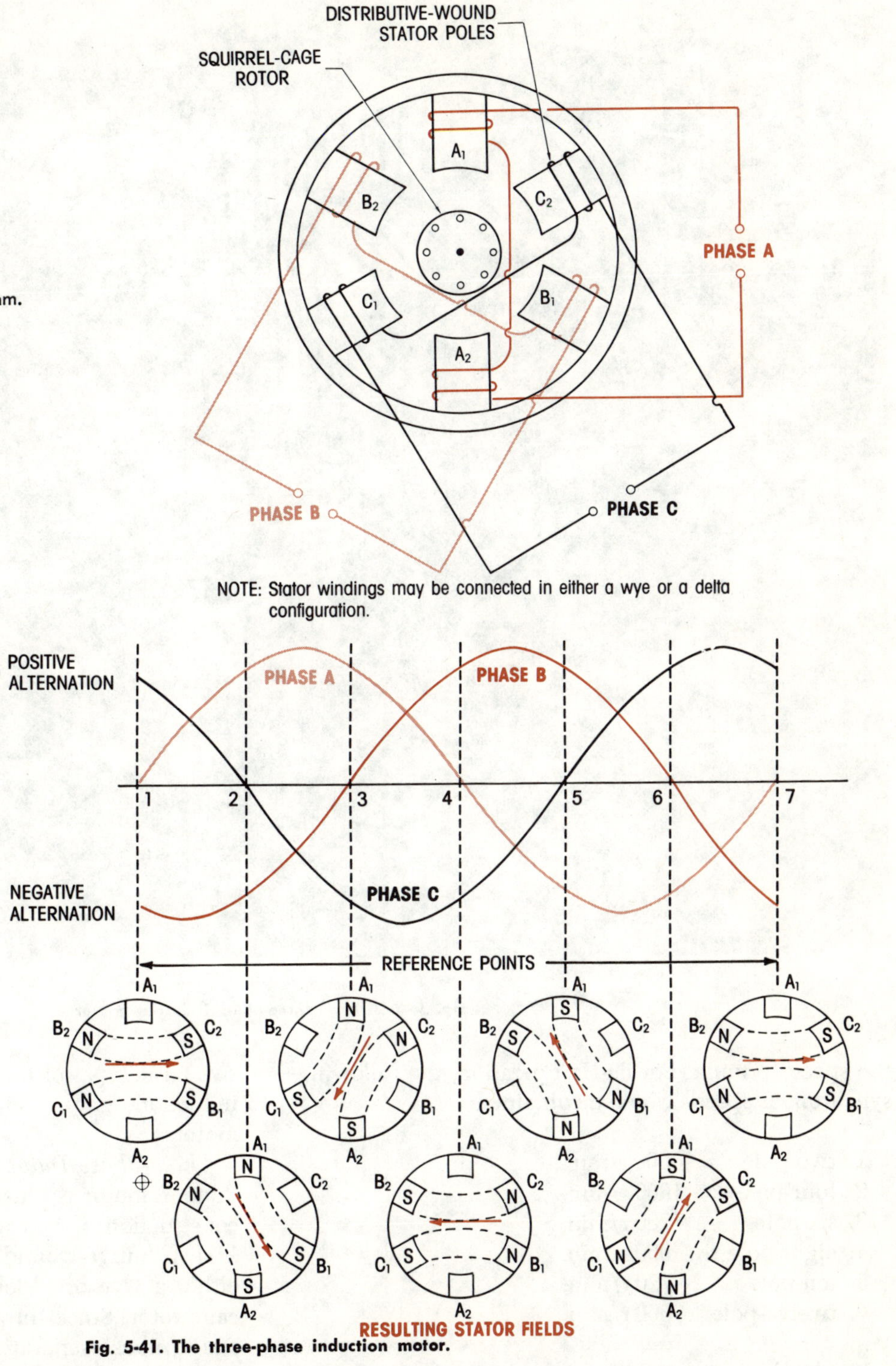

Fig. 5-41. The three-phase induction motor.

starting torque. Their speed can be varied by varying the position of the brush axis. However, the mechanical problems inherent with this type of motor have caused it to become obsolete.

Single-Phase Synchronous Motors—It is often desirable, in timing or clock applications, to use a constant-speed drive motor. Such a motor, which operates from a single-phase ac line, is called a *synchronous* motor. The single-phase synchronous motor has stator windings that are connected across the ac line. The rotor, however, is made of permanent magnetic material. Some external method of starting must be used in order to initiate rotation. Once the rotor is started, it will rotate in synchronism with the revolving stator field, since it does not rely upon the induction principle. The speeds of synchronous motors are based on

Courtesy General Electric Co.

Fig. 5-42. A large three-phase induction motor.

the speed formula. For 60-Hz operation, the following synchronous speeds would be obtained:

1. two-pole 3600 r/min,
2. four-pole 1800 r/min,
3. six-pole 1200 r/min,
4. eight-pole 900 r/min,
5. ten-pole 720 r/min,
6. twelve-pole 600 r/min.

Single-phase synchronous motors are used only for very low torque applications. Such applications include clocks, phonograph drives, and timing devices requiring constant speed.

Three-Phase AC Motors

Three-phase ac motors are often called the "workhorses of industry." Most motors used in industry are operated from three-phase power sources. There are two basic types of three-phase motors: (1) three-phase induction motors, and (2) three-phase synchronous motors.

Three-Phase Induction Motors—The three-phase induction motor is illustrated in Fig. 5-41. Note that the construction of this motor is very simple. It has only a distributive-wound stator, which is connected in either a wye or a delta configuration, and a squirrel-cage rotor. Since three-phase voltage is applied to the stator, phase separation is already established. No external starting mechanisms are needed. Three-phase induction motors are made in a variety of integral horsepower sizes and have good starting and running torque characteristics. A large three-phase induction motor is shown in Fig. 5-42.

The direction of rotation of a three-phase motor of any type can be changed very easily. If any two power lines coming into the stator windings are reversed, the direction of rotation of the shaft will change. Three-

Courtesy Electric Machinery Manufacturing Co.

Fig. 5-43. Large three-phase synchronous motors.

phase induction motors are used for many industrial applications, such as mechanical energy sources for machine tools, pumps, elevators, hoists, conveyors, and many other industrial systems.

Three-Phase Synchronous Motors—Three-phase synchronous motors are unique and very specialized motors. They are considered a constant-speed motor and they can be used to correct the power factor of three-phase systems. Fig. 5-43 shows some large three-phase synchronous motors, and a pictorial diagram of synchronous motor construction is shown in Fig. 5-44.

The three-phase synchronous motor is physically constructed like a three-phase alternator. Direct current is applied to the rotor to produce a rotating electromagnetic field, and the stator windings are connected in either a wye or a delta configuration. The only difference is that three-phase ac power is applied to the synchronous motor, while three-phase power is extracted from the alternator. Thus, the motor acts as an electrical load, while the alternator functions as a source of three-phase power. This relationship should be kept in mind during the following discussion.

The three-phase synchronous motor differs from the three-phase induction motor in that the rotor is wound and is connected through a commutator/brush assembly to a dc power source. Three-phase synchronous motors, in their pure form, have no starting torque. Some external means must be used to initially start the motor. Synchronous motors are constructed so that they will rotate at the same speed as the revolving stator field. We can say that at synchronous speed, rotor speed equals stator speed and the motor has zero slip. Thus, we can determine the speed of a synchronous motor by using the following formula:

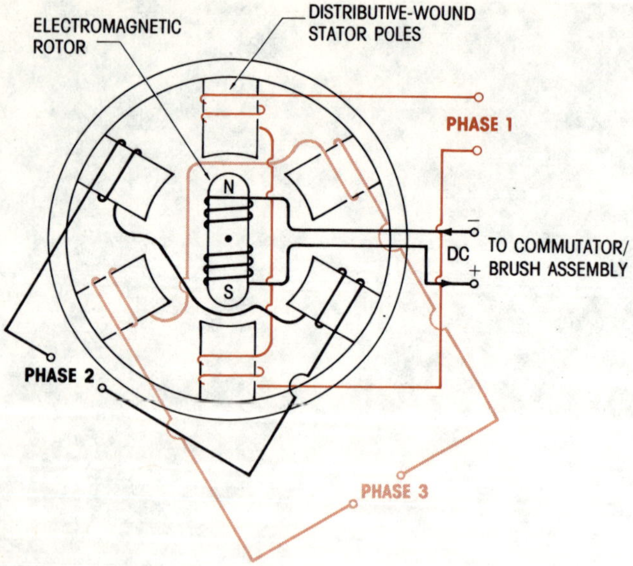

Fig. 5-44. Pictorial diagram of three-phase synchronous motor construction.

$$S = \frac{f \times 120}{n/3}$$

where,

S is the speed of a synchronous motor in r/min,
f is the frequency of the applied ac voltage in hertz,
n/3 is the number of stator poles per phase,
120 is a conversion constant.

Note that this is the same as the formula used to determine the stator speed of a single-phase motor except that the number of poles must be divided by three (the number of phases). A three-phase motor with twelve actual poles would have four poles per phase. Therefore, its stator speed would be 1800 r/min. Synchronous motors have operating speeds that are based on the number of stator poles they have.

Synchronous motors are usually employed in very large horsepower ratings. One method of starting a large synchronous motor is to use a smaller auxiliary dc machine connected to the shaft of the synchronous

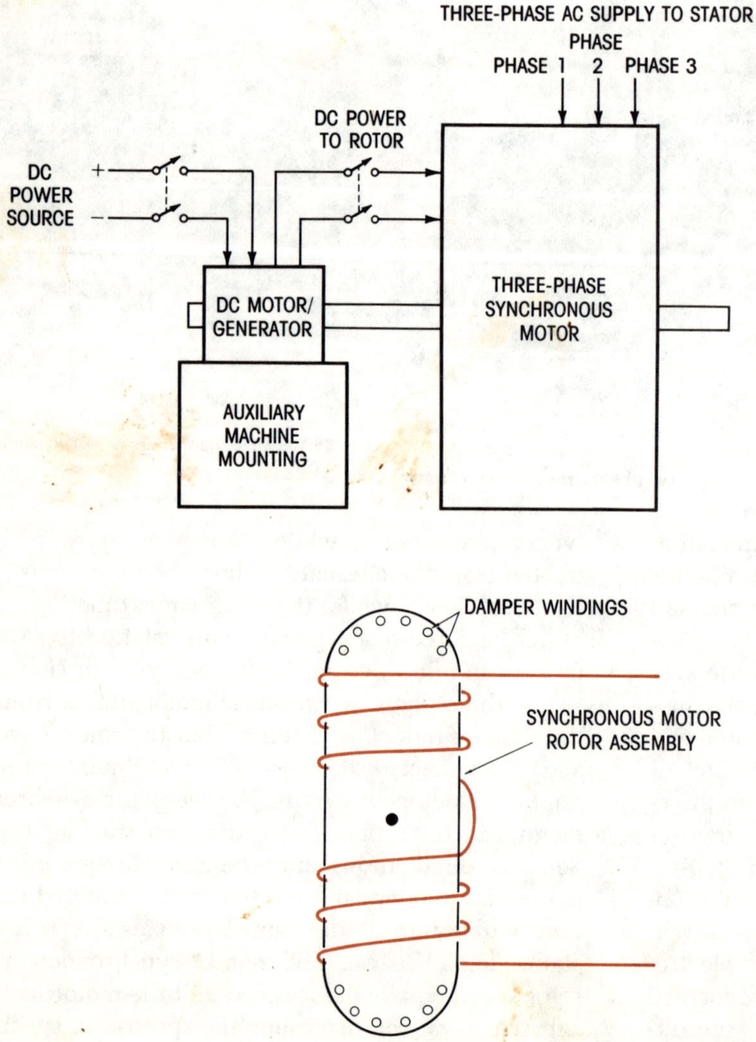

(A) Auxiliary-machine method.

(B) Damper-windings-on-rotor method.

Fig. 5-45. Three-phase synchronous motor starting methods.

motor, as illustrated in Fig. 5-45A. The method of starting would be as follows:

Step 1: Dc power is applied to the auxiliary motor, causing it to increase in speed. Three-phase ac power is applied to the stator.

Step 2: When the speed of rotation reaches a value near the synchronous speed of the motor, the dc power circuit is opened and, at the same time, the terminals of the auxiliary machine are connected across the commutator/brush assembly of the rotor.

Step 3: The auxiliary machine now converts to generator operation and supplies exciter current to the rotor of the synchronous motor, using the motor as its prime mover.

Step 4: Once the rotor is magnetized, it will "lock" in step or synchronize with the revolving stator field.

Step 5: The speed of rotation will remain constant under changes in load condition.

Another starting method is shown in Fig. 5-45B. This method utilizes "damper" windings, which are similar to the conductors of a squirrel-cage rotor. These windings are placed within the laminated iron of the rotor assembly. No auxiliary machine is required when damper windings are used. The starting method used is as follows:

Step 1: Three-phase ac power is applied to the stator windings.

Step 2: The motor will operate as an induction motor due to the "transformer action" of the damper windings.

Step 3: The motor speed will build up so that the rotor speed is somewhat less than the speed of the revolving stator field.

Step 4: Dc power from a rotating dc machine or, more commonly, a rectification system is applied to the commutator/brush assembly of the rotor.

Step 5: The rotor becomes magnetized and builds up speed until rotor speed is equal to stator speed.

Step 6: The speed of rotation remains constant regardless of the load placed on the shaft of the motor.

An outstanding advantage of the three-phase synchronous motor is that it can be connected to a three-phase power system to increase the overall power factor of the system. Power factor correction was discussed previously in this chapter. Three-phase synchronous motors are sometimes used only to correct system power factor. If no load is intended to be connected to the shaft of a three-phase synchronous motor, it is called a *synchronous capacitor*. It is designed to act only as a power factor corrective machine. Of course, it might be beneficial to use this motor as a constant-speed drive connected to a load as well as for power factor correction.

We know from previous discussion that low power factors cannot be tolerated by industries. The expense of installing three-phase synchronous machines could be justified in order to appreciably increase system power factor. To understand how a three-phase synchronous machine operates as a power factor corrective machine, refer to the curves of Fig. 5-46. We know

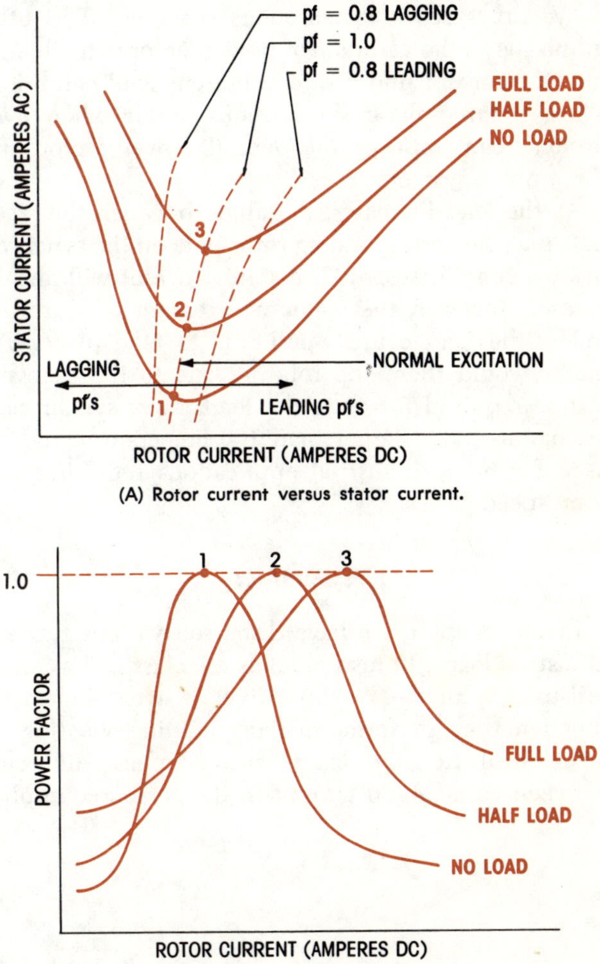

(A) Rotor current versus stator current.

(B) Rotor current versus power factor.

Fig. 5-46. Power relationships of a three-phase synchronous motor.

that the synchronous motor operates at a constant speed. Variation in rotor dc excitation current has no effect on speed. The excitation level will change the power factor at which the machine operates. Three operational conditions may exist, depending on the amount of dc excitation applied to the rotor. These conditions are:

1. Normal excitation—operates at a power factor of 1.0.
2. Under excitation—operates at a lagging power factor (inductive effect).
3. Over excitation—operates at a leading power factor (capacitive effect).

Note the variation of stator current drawn by the synchronous motor as rotor current varies. You should also see that stator current is minimum when power factor equals 1.0 or 100%. The situations shown on the graph indicate no-load, half-load, and full-load conditions with power factors equal to 1.0, 0.8 leading, and 0.8 lagging. These curves are sometimes referred to as *V-curves* for a synchronous machine. The other graph shows the variation of power factor with changes in rotor current under three different load conditions. Thus, a three-phase synchronous motor when over-excited, can improve the overall power factor of a three-phase system.

As the load increases, the angle between the stator pole and the corresponding rotor pole on the synchronous machine increases. The stator current will also increase. However, the motor will remain synchronized unless the load causes "pull-out" to take place. The motor would then stop rotating due to the excessive torque required to rotate the load. Most synchronous motors are rated larger than 100 horsepower and are used for many industrial applications requiring constant speed drives.

SUMMARY

In this chapter, we have discussed various types of industrial loads. Industrial loads are classified as either resistive, inductive, or capacitive; or according to the function they perform, such as lighting, heating, or mechanical. Resistive loads cause no phase difference between voltage and current in the ac source applied to it. However, inductive and capacitive loads cause voltage and current to be out of phase. This out-of-phase situation causes the power factor of a system to be less than unity (1.0). Therefore, ac systems are somewhat more complicated than dc systems in terms of their power relationships. Vector diagrams must be used when making calculations involving ac systems in either series or parallel configurations.

In ac systems, power factor is the relationship of true power and apparent power. Since industries use many electric motors, their power systems are highly inductive. This causes the system power factor to be less than 1.0. It is desirable to keep power factor as high as possible. Power factor corrective equipment, such as capacitors or three-phase synchronous motors, may be used to increase system power factor. Load factor, or the relationship of average power demand and peak power demand, is another consideration of industrial electrical system operation.

Industries use many types of loads. Lighting loads include incandescent, fluorescent, and vapor lighting systems. Heat-producing loads include resistive, inductive, and capacitive systems, as well as electric welders. Welding systems typically are high-current circuits which use ignitrons or SCRs as control devices.

The major type of mechanical load used in industry is the electric motor. Direct-current, single-phase ac, and three-phase ac motors have applications as industrial loads. Direct-current motors are used where speed control is desired. The kinds of dc motors available are permanent-magnet field, series-wound field, shunt-wound field, and compound-wound field. Single-phase motors include the universal motor, split-phase induction motor, capacitor motors, shaded-pole motors, repulsion motors, and synchronous motors. The most popularly used industrial motors are three-phase induction motors. Three-phase synchronous motors are also used where constant speed under changes in load is desired.

Study Questions

1. *What are some types of industrial loads?*
2. *How do resistive, inductive, and capacitive loads differ?*
3. *What is meant by inductive reactance? Capacitive reactance?*
4. *Draw a vector diagram representing the voltages in a series RLC circuit in which V_R = 10 volts, V_L = 20 volts, and V_C = 5 volts.*
5. *Draw a vector diagram representing the currents in a parallel RLC circuit in which V_R = 10 volts, V_L = 20 volts, and V_C = 5 volts.*
6. *What is meant by the impedance of an ac circuit?*
7. *Draw an impedance triangle.*
8. *What is meant by admittance, conductance, and susceptance in a parallel ac circuit?*
9. *Discuss true power, apparent power, reactive power, and power factor of ac circuits.*
10. *Draw a power triangle.*
11. *How may the power factor of an ac system be increased?*
12. *How do incandescent, fluorescent, and vapor lighting systems differ?*
13. *How do resistive, inductive, and capacitive heating methods differ?*
14. *Discuss resistance welding, electric arc welding, and induction welding.*
15. *How are ignitron or SCR contactors used in welding systems?*
16. *Discuss basic motor action or the production of torque in a rotating machine.*
17. *Define: (A) horsepower, (B) torque, (C) speed regulation.*
18. *How do the following types of dc motors differ: permanent magnet, series, parallel, and compound?*
19. *Discuss the induction principle of single-phase motor operation.*
20. *How is the speed of an ac motor determined?*
21. *What is meant by the slip of an ac induction motor?*
22. *Discuss the operation of the following types of motors:*
    a. *Universal motor.*
    b. *Split-phase motor.*
    c. *Capacitor-start, induction motor.*
    d. *Shaded-pole motor.*
    e. *Repulsion motor.*
    f. *Single-phase synchronous motor.*
    g. *Three-phase induction motor.*
    h. *Three-phase synchronous motor.*
23. *How are three-phase synchronous motors commonly started?*
24. *How are three-phase synchronous motors used as power factor corrective devices?*

Problems

1. If the inductance of a motor's windings is 0.5 henry, what is the inductive reactance when connected to a 60-Hz source.
2. When a 200-μF capacitor is connected to a 60-Hz source, how much capacitive reactance does it offer?
3. The voltage drops across the components of a series RLC circuit are as follows: V_R = 15 volts, V_L = 10 volts, V_C = 20 volts. What is the total voltage acting on the circuit? Draw a vector diagram to illustrate these values.
4. The currents through the three paths of a parallel RLC circuit are as follows: I_R = 50 mA, I_L = 80 mA, I_C = 10 mA. What is the total current of the circuit? Draw a current vector diagram.
5. A resistance of 15 ohms is connected in series with an inductive reactance of 10 ohms, with 120 volts, 60 Hz applied. Calculate the following:
   a. Impedance.
   b. Current.
   c. Power factor (R/Z).
   d. Volt-amperes.
   e. True power.
   f. Reactive power.
6. Draw an impedance triangle for the circuit of Problem 5.
7. Draw a power triangle for the circuit of Problem 5.
8. What is the phase angle of the circuit of Problem 5.
9. If a motor converts 1000 watts of electrical power when drawing 12 amperes of current with 120 volts applied, what is its power factor?
10. If a motor converts 2000 watts of electrical power, and the VA of the motor is 2500, what is its power factor and reactive power?
11. If a three-phase motor draws 10 amperes per phase, the line voltage is 240 volts, and the power factor is 0.75, what is the total three-phase power?
12. If an industry has a peak power demand of 1250 megawatts, and an average demand of 880 megawatts, what is its load factor?
13. A shunt-wound dc motor converts 5000 watts when connected to a 120-volt line. If its field resistance equals 100 ohms, find:
    a. Field current.
    b. Line current.
    c. Armature current.
14. If a shunt-wound dc motor draws 60 amperes from a 240-volt source, its field resistance is 40 ohms, and its armature resistance is 0.5 ohm, find:
    a. Field current.
    b. Armature current.
    c. Cemf.
    d. Total power.
15. What is the horsepower of a motor running at 1750 r/min and having a torque of 25 ft-lb?
16. If the no-load speed of a motor is 3550 r/min, and the full-load speed is 3450 r/min, what is its speed regulation?
17. What is the percentage slip of the motor of Problem 16?
18. What is the rotor frequency of the motor of Problem 16?
19. What is the synchronous speed of a 30-pole, 60-Hz, three-phase ac motor?
20. What is the synchronous speed of a 12-pole, 60-Hz, single-phase ac motor?

6

Indicating Systems in Industry

Indicators of many types are used with industrial systems to monitor various quantities. Industrial processes are dependent upon precision indicating systems for everyday operation. Indicating systems are used in industry to measure basic electrical quantities, power-related quantities, and physical quantities. A typical industrial control center that uses a variety of indicating systems is shown in Fig. 6-1. The measurement of several types of electrical and physical quantities will be discussed in this chapter.

Courtesy Leeds & Northrup Co.

Fig. 6-1. A typical industrial control center employing a variety of indicating devices.

BASICS OF INDICATING SYSTEMS

All indicating systems have certain basic characteristics. Usually, a specific quantity is to be monitored either periodically or continuously. Therefore, some type of visual indication of the quantity being monitored must be presented. Several types of instruments for measuring electrical and physical quantities are available. The basic types of indicating systems may be classified as: (1) hand-deflection indicators, (2) com-

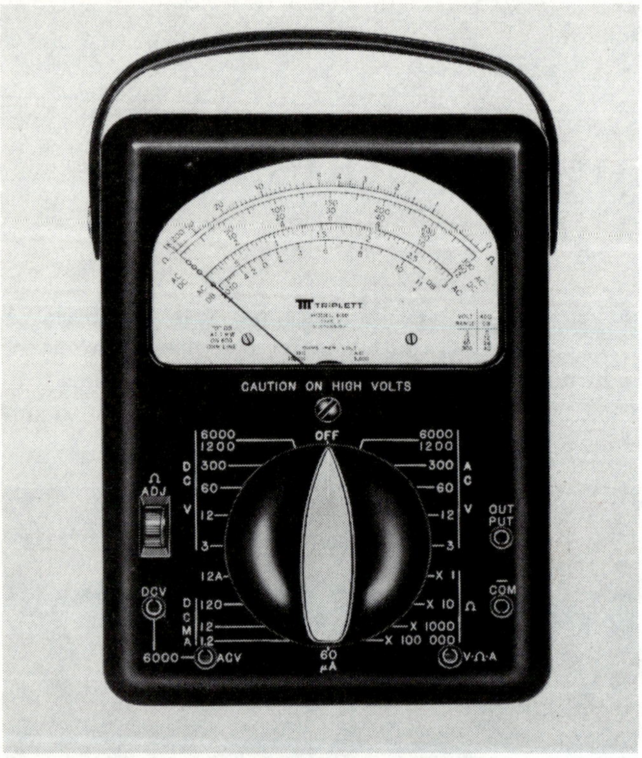

(A) A volt-ohm-milliammeter (vom).

(B) A transistorized volt-ohm-milliammeter (tvom).

(C) A vacuum-tube voltmeter (vtvm).

(D) A transistorized voltmeter (tvm).

Courtesy Triplett Corp.

Fig. 6-2. Typical hand-deflection indicators.

parative indicators, (3) cathode-ray tube indicators, (4) numerical-readout indicators, and (5) chart-recording indicators.

Hand-Deflection Indicators

Indicators that rely on the motion of a hand or pointer assembly are referred to as *hand-deflection* indicators. The volt-ohm-milliameter (vom) is one type of hand-deflection indicator (see Fig. 6-2). The vom is a multifunction instrument that is used for measuring electrical quantities. Single-function hand-deflection indicators are also commonly used to measure electrical or physical quantities. The basic part of an electrical hand-deflection indicator is called a *meter movement.* Physical quantities such as air flow or fluid pressure can also be monitored by hand-deflection meters. The movement of the hand or pointer relative to a calibrated scale indicates the electrical or physical quantity.

Many meters employ the *D'Arsonval* type of meter movement. Construction details and the basic operational principle of this type of movement are shown

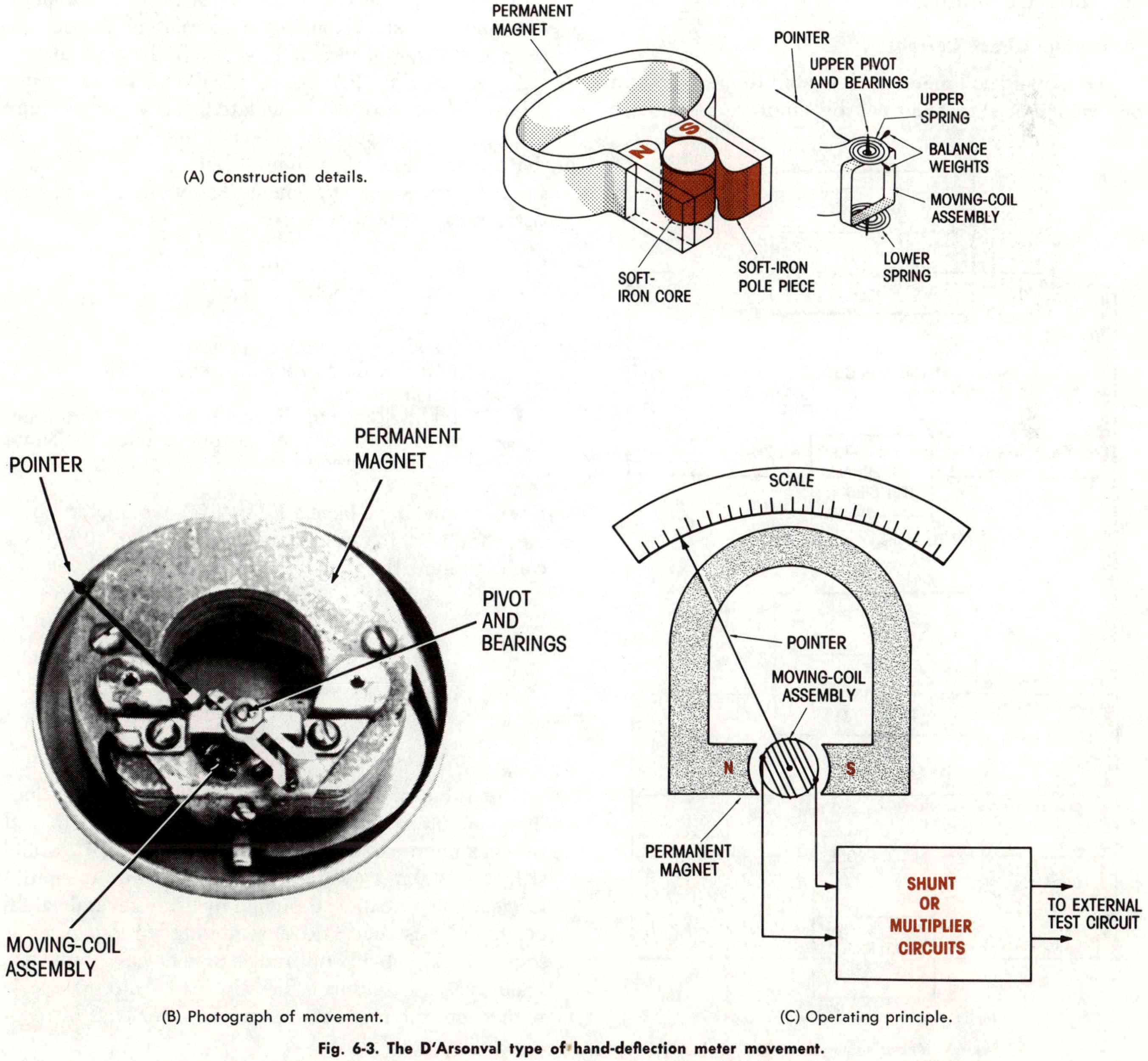

(A) Construction details.

(B) Photograph of movement.

(C) Operating principle.

Fig. 6-3. The D'Arsonval type of hand-deflection meter movement.

in Fig. 6-3. The hand or pointer of the movement remains stationary on the left portion of the calibrated scale until current flows through the electromagnetic coil that is centrally located within a permanent magnetic field. When current flows through the coil, a reaction between the electromagnetic field of the coil and the stationary permanent magnetic field is developed. This reaction causes the hand to deflect toward the right portion of the scale. This basic moving-coil meter movement operates on the same principle as an electric motor. It can be used for either single-function electrical indicators, which measure only one quantity, or for multifunction meters such as a vom or vtvm (vacuum-tube voltmeter). The basic D'Arsonval meter movement can be modified so that it will measure voltage, current, or resistance.

Measuring Direct Current

The moving-coil meter can be used to measure various values of direct current by employing shunt resistors in parallel with the movement. A schematic of the basic movement is illustrated in Fig. 6-4A The coil of the meter movement is not designed to handle high currents. The typical current rating (I_M) of a meter movement ranges from 10 μA to 30 mA. The current rating indicates the amount of current required to cause the pointer to deflect to the full-scale (extreme right) position. When the current value is lowered, more turns of wire must be used for the coil in order to develop sufficient electromagnetic field strength. Therefore, as the current rating (I_M) decreases, the resistance (R_M) of the movement increases. We must know the values of I_M and R_M in order to design a meter circuit that will measure currents higher than the value of I_M.

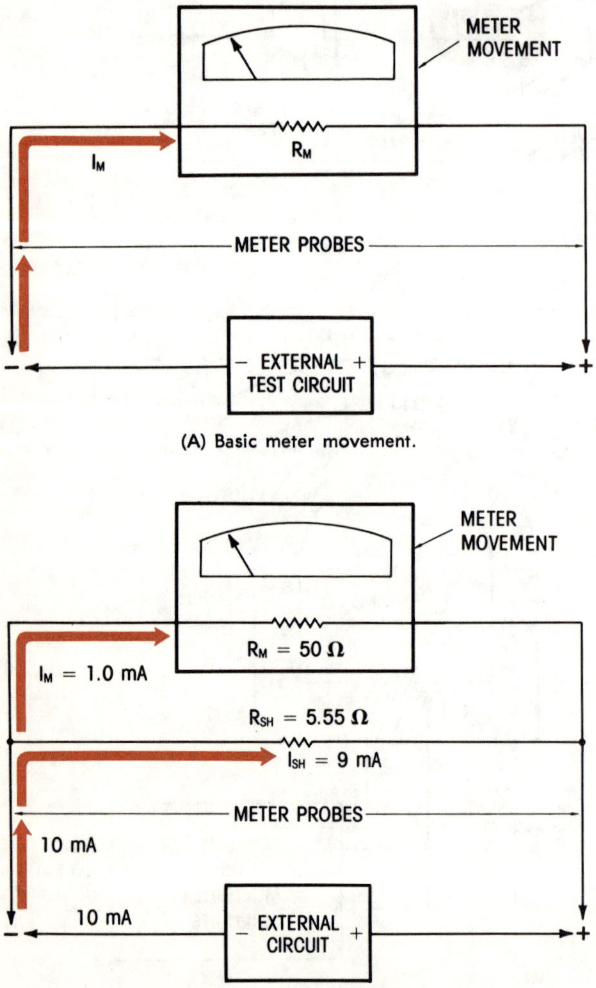

(A) Basic meter movement.

(B) Meter movement with shunt resistor.

Fig. 6-4. Measuring direct current.

If a resistor is placed in parallel with the basic meter movement, as shown in Fig. 6-4B, part of the current from the external circuit is shunted through the resistor. Therefore, this resistor is called a *shunt* resistor (R_{SH}). When properly calculated, the value of shunt resistance will establish meter circuit conditions so that other ranges of current (above I_M) can be measured by the same movement. The value of shunt resistance can be expressed as:

$$R_{SH} = \frac{I_M \times R_M}{I_{SH}}$$

where,

R_{SH} is the shunt resistance in ohms,
I_M is the full-scale current rating of the meter movement in amperes,
R_M is the resistance of the meter movement in ohms,
I_{SH} is the desired value of current through the shunt resistance in amperes.

For example, if we have a 1.0-mA, 50-ohm meter movement and wish to design a meter to measure 10 mA, we can determine the shunt resistance as follows:

$$R_{SH} = \frac{I_M \times R_M}{I_{SH}}$$

$$= \frac{0.001 \times 50}{0.009}$$

$$= 5.55 \text{ ohms.}$$

The value of I_{SH} was set at 9 mA (0.009 A) since that amount of current had to be shunted so that at the maximum current of the extended range, I_M would still equal 1.0 mA (9 mA + 1.0 mA = 10 mA). A multirange ammeter can be designed by using several values of shunt resistance and a switching arrangement as shown in Fig. 6-5. Multirange meters have only one scale; however, either a directly read scale or a scale with a multiplying factor is provided for ease of reading each range.

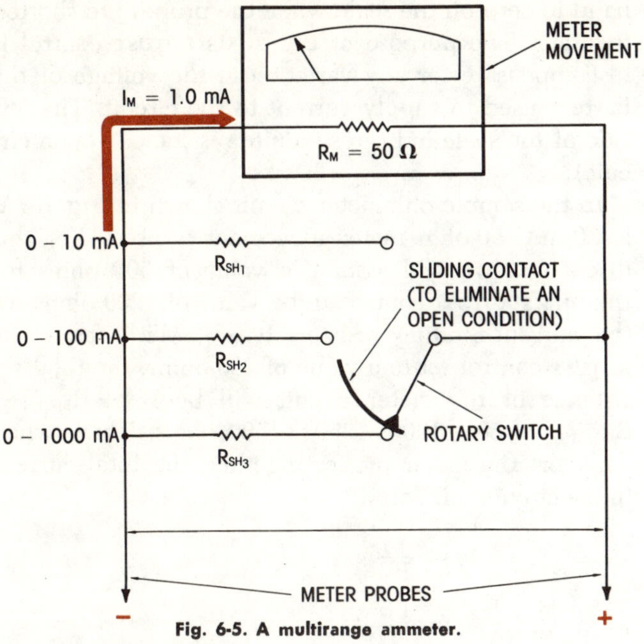

Fig. 6-5. A multirange ammeter.

Measuring DC Voltage

Hand-deflection meter movements can also be used to measure dc voltage. The basic movement actually responds to current through its electromagnetic coil; however, predetermined voltages can also produce accurate readings on a calibrated meter scale. To measure voltage, a resistor is placed in series with the meter movement. This resistor, shown in Fig. 6-6, is called a *multiplier.* The purpose of the multiplier is to adjust the value of current in the meter circuit so that at a predetermined voltage, I_M will cause full-scale deflection on the meter scale. This is true since

$$V_{FS} = I_M \times R_M$$

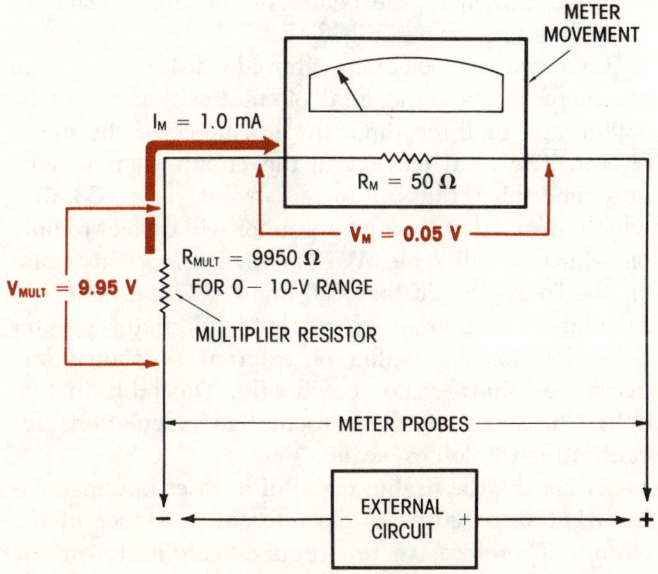

Fig. 6-6. Measuring dc voltage.

where,

V_{FS} is the voltage required for full-scale deflection in volts,
I_M is the current required for full-scale deflection in amperes,
R_M is the resistance of the meter movement in ohms.

Assume that we wish to measure a range of dc voltage from 0 to 10 volts using the 1.0-mA, 50-ohm meter movement of the previous section. With 1.0 mA flowing through the 50-ohm meter movement at full-scale deflection, there must be 0.05 volt applied across the meter terminals, as shown in Fig. 6-6. Thus, in order to measure 10 volts at full-scale deflection, we must add sufficient multiplier resistance in series with the meter to drop 9.95 volts with 1.0 mA current flowing in the circuit, or

$$R_{MULT} = \frac{V_{MULT}}{I_T} = \frac{9.95}{0.001} = 9950 \text{ ohms.}$$

Another way to determine the value of the multiplier resistance is to say that the total resistance in the combined circuit with 10 volts applied is

$$R_T = \frac{V_T}{I_T} = \frac{10}{0.001} = 10{,}000 \text{ ohms.}$$

Since the total resistance of the circuit is also

$$R_T = R_M + R_{MULT},$$

then

$$R_{MULT} = R_T - R_M = 10{,}000 - 50 = 9950 \text{ ohms.}$$

A switching arrangement can be used to provide several voltage ranges using the same meter movement.

An important aspect of meter operation is called *sensitivity*. The sensitivity of a meter increases as the current for full-scale deflection (I_M) decreases. Since a voltmeter is connected in parallel with the circuit under test, part of the circuit current flows through the meter. The current that flows through the meter should be small compared to the circuit current. The more sensitive the meter, the less current it draws from the circuit. Sensitivity is expressed in ohms per volt. Thus, sensitivity is equal to $1/I_M$, where I_M is the current required for full-scale deflection of the meter. A 50-μA movement would have an ohms-per-volt rating of 1/0.000005, or 20,000 ohms/volt. This means that the meter would require 20,000 ohms for a 1-volt range. Sensitivities of meter movements range from as low as 100 ohms/volt to as high as 200,000 ohms/volt. Low-sensitivity meters should not be used for making measurements in circuits that have low currents.

Measuring Resistance

Resistance can also be measured by using the basic D'Arsonval meter movement. One type of ohmmeter circuit is shown in Fig. 6-7A. This circuit is called a *series* ohmmeter since the resistance to be measured is connected in series with the meter circuit. Note that a 1.5-volt cell is used to supply current to the meter circuit. The scale of the meter is calibrated so that the deflection of the pointer will indicate a specific value of resistance. When the meter probes are shorted together (zero external resistance), the meter pointer is adjusted to read full-scale. The full-scale mark on the "ohms" scale, shown in Fig. 6-7B, indicates zero resistance. Therefore, as higher values of resistance are measured between the meter probes, less current will flow through the ohmmeter circuit. The meter pointer will deflect less for higher resistance values. The OHMS ADJUST control is a variable resistance in the ohmmeter circuit that is used to adjust the meter hand to zero on the scale when the probes are shorted together. The purpose of the OHMS ADJUST control is to compensate for any variation in the voltage of the battery used to supply current to the circuit. The left side of the scale indicates infinite resistance (open circuit).

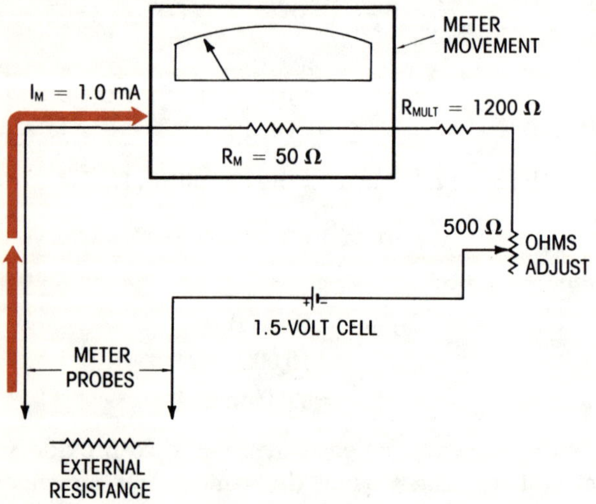

(A) Series ohmmeter circuit.

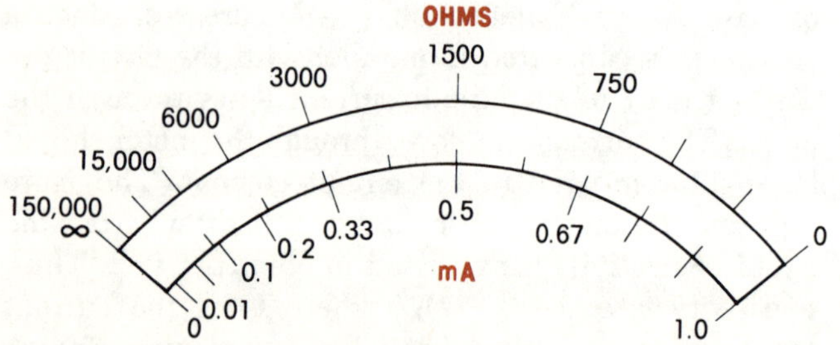

(B) "Ohms" scale calibration.

Fig. 6-7. Measuring resistance.

In the sample ohmmeter circuit shown in Fig. 6-7A, a 1.0-mA, 50-ohm meter movement is used. For this discussion, we will assume a value of 500 ohms for the OHMS ADJUST control and a value of 1200 ohms for the current-limiting resistor, R_{MULT}. With the OHMS ADJUST control set to a value of 250 ohms, the total resistance of the meter circuit will be $R_M + R_{MULT} + R_{OHMS}$, or $50 + 1200 + 250 = 1500$ ohms. Then, when we short the meter probes together, the total current in the circuit will be:

$$I_T = \frac{V_T}{R_T} = \frac{1.5}{1500} = 0.001 \text{ ampere } (1.0 \text{ mA}),$$

which is the current value required for full-scale deflection of the meter. The "ohms" scale is then calibrated to read zero ohms at the point of full-scale deflection.

The external resistance that would cause half-scale deflection of the meter pointer would be equal to the total resistance of the ohmmeter circuit that limits the current to full-scale deflection (1500 ohms). With 1500 ohms external resistance connected between the meter probes, the total circuit resistance will be 3000 ohms. The meter pointer will then deflect to only one-half of full scale, since with twice as much resistance in the circuit, there will be only half as much current. Therefore, we can mark the center of the ohms scale for 1500 ohms as shown in Fig. 6-7B.

If we connect 3000 ohms external resistance between the meter probes, the total circuit resistance will be 4500 ohms, or three times the resistance of the meter circuit. The total current in the circuit then will be only one-third the full-scale current, or 0.033 mA, which means that the meter pointer will deflect to only one-third of full scale. We can then mark this point on the "ohms" scale for 3000 ohms. We can similarly calculate the amount of current and meter pointer deflection for any value of external resistance and mark the "ohms" scale accordingly. The table in Fig. 6-7B illustrates some representative calculations for calibrating the "ohms" scale.

Ordinarily, the usable range of a series ohmmeter is from 1/100 to 100 times the internal resistance of the circuit. Therefore, there are practical limits on the amount of resistance this ohmmeter can measure. To measure lower resistances, the internal resistance of the ohmmeter must be reduced. This can be done by employing a multirange ohmmeter that uses a small battery (usually 1.5 volts) for the ×1 and ×100 ranges, and a larger battery (usually 30 volts) for the ranges of ×1000 and above. To measure very low values of resistance, a *shunt* ohmmeter can be used. The ohmmeter shunt is designed similarly to an ammeter shunt, except that a voltage source and an OHMS ADJUST control are placed in series with the shunt resistance.

We should keep in mind that multifunction meters are usually designed to measure voltage, current, and resistance. Most vom's, tvm's, or vtvm's used today employ one meter movement that is modified by internal circuitry to make many types of measurements. A rotary switch is ordinarily used with multifunction meters to select the proper function and range.

Measuring AC Voltage and Current

It is also possible to modify the basic meter movement so that it will indicate values of ac voltage or current. Either half-wave or full-wave rectifier circuits can be employed to convert alternating current from an external circuit to a direct current that will produce an electromagnetic field in the coil of the meter movement. Half-wave and full-wave rectifier meter circuits are shown in Fig. 6-8. Either of these types of rectifiers can be used to convert ac to dc to cause the pointer of the movement to deflect.

The amount of deflection of the pointer must be considered for proper scale calibration. For half-wave rectification, meter current can be expressed as:

$$I_{M(dc)} = \frac{0.45 \times V_{rms}}{R_T}$$

where,

$I_{M(dc)}$ is the dc meter current in amperes,
0.45 is a conversion factor for converting an ac rms value to an average dc value,
V_{rms} is the maximum ac rms input voltage for the meter range in volts,
R_T is the sum of the meter resistance and the required multiplier resistance in ohms.

The ac sensitivity of a half-wave ac meter is thus only 45% of the dc sensitivity unless some circuit modification is made.

Full-wave rectifiers, such as the bridge circuit of Fig. 6-8B, allow us to have an ac voltage range that is 90% of its dc equivalent. The average dc current for a full-wave rectifier can be expressed as:

$$I_{M(dc)} = \frac{0.9 \times V_{rms}}{R_T}.$$

Rectifier-type ac meters also employ blocking capacitors placed in series with the meter probes to isolate any dc component, which may be contained in the circuit being measured, from the meter circuitry. Rectifier-type meters are designed primarily for ac frequencies in the power range (60 Hz), and for audio frequencies. Other types of meters are used to measure higher frequencies.

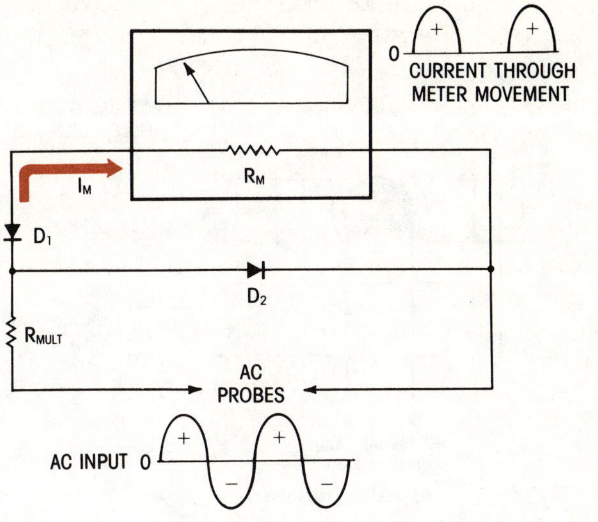

(A) Half-wave ac meter.

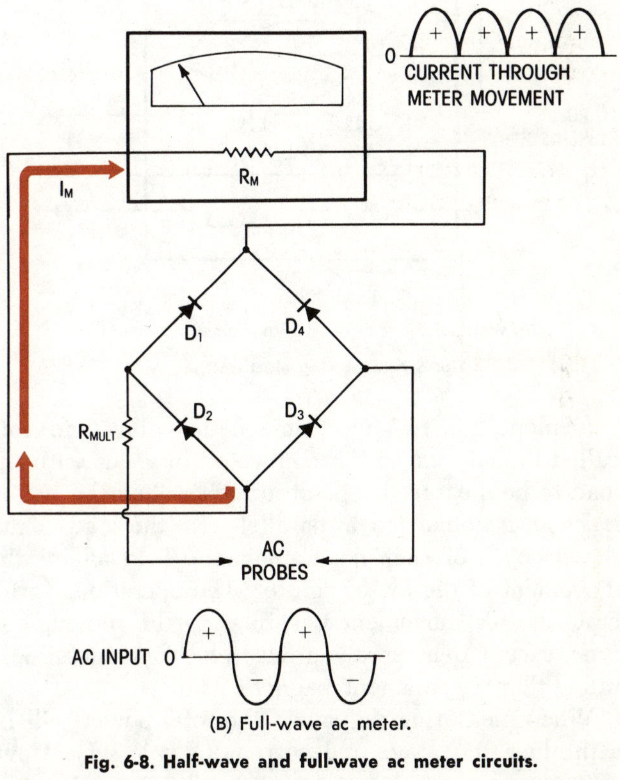

(B) Full-wave ac meter.

Fig. 6-8. Half-wave and full-wave ac meter circuits.

Measuring Electrical Power

Electrical power is measured with a wattmeter. A meter movement referred to as a *dynamometer*, shown in Fig. 6-9, is used for most wattmeters. Note that this

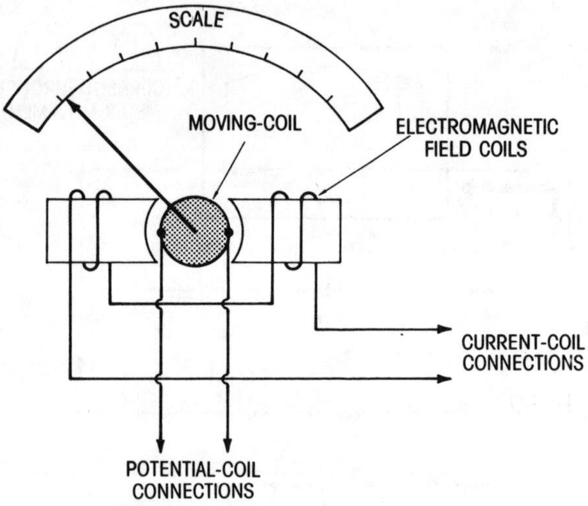

(A) Basic dynamometer movement.

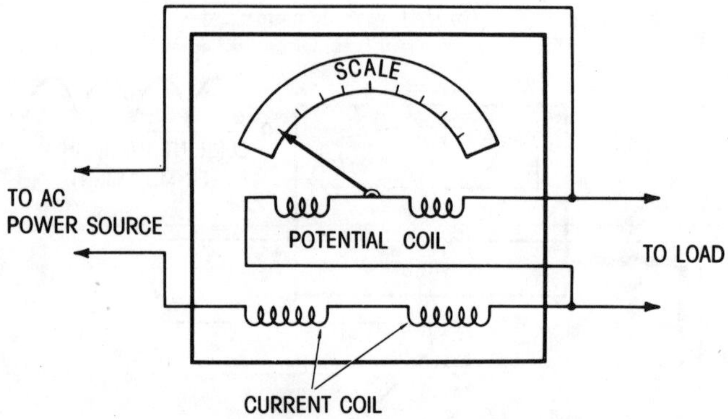

(B) Wattmeter circuit using dynamometer movement.

Fig. 6-9. Measuring electrical power.

movement has two electromagnetic coils. One coil, called the *current* coil, is connected in series with the load to be measured. The other coil, called the *potential* coil, is connected in parallel with the load. Thus, the strength of each electromagnetic field affects the movement of the meter pointer. The operational principle of this movement is similar to the moving-coil type, except that there is a fixed electromagnetic field rather than a permanent magnetic field.

When measuring dc power, the total power will be a product of voltage and current ($P = V \times I$). However, when measuring ac power, we must consider the power factor of the load, since $P = V \times I \times pf$. The true power of an ac circuit can be read directly with a wattmeter. When a load is either inductive or capacitive, recall from Chapter 3 that true power will be less than apparent power ($V \times I$).

Comparative Indicators

Another group of indicators can be classified as *comparative* indicators. Usually, a comparative indicator is designed to compare a known value to some unknown value. The accuracy of a comparative indicator is much better than that of a hand-deflection indicator. Meter-scale calibration is not so important for comparative indicators.

The Wheatstone Bridge

The Wheatstone bridge, shown in Fig. 6-10, is a typical type of comparative indicator. The technique used for Wheatstone-bridge measurement is illustrated in Fig. 6-11. A voltage source is used in conjunction with a sensitive zero-centered meter movement and a resistive bridge circuit. The bridge circuit is completed by adding an unknown external resistance (R_X) to be measured. When resistor R_S is adjusted so that the resistive path formed by R_X and R_S is equal to the path formed by R_1 and R_2, no current will flow through the meter. In this condition, the meter will indicate zero (referred to as a *null*) and the bridge is said to be balanced. The value of R_S is marked on the indicator so that the value of R_X can be determined. Resistors R_1 and R_2 form the "ratio arm" of the indicator. The value of the unknown resistance (R_X) can be expressed mathematically as:

$$R_X = \frac{R_1}{R_2} \times R_S.$$

The Wheatstone bridge will measure values of resistance from 0.01 ohm to 10 megohms with consider-

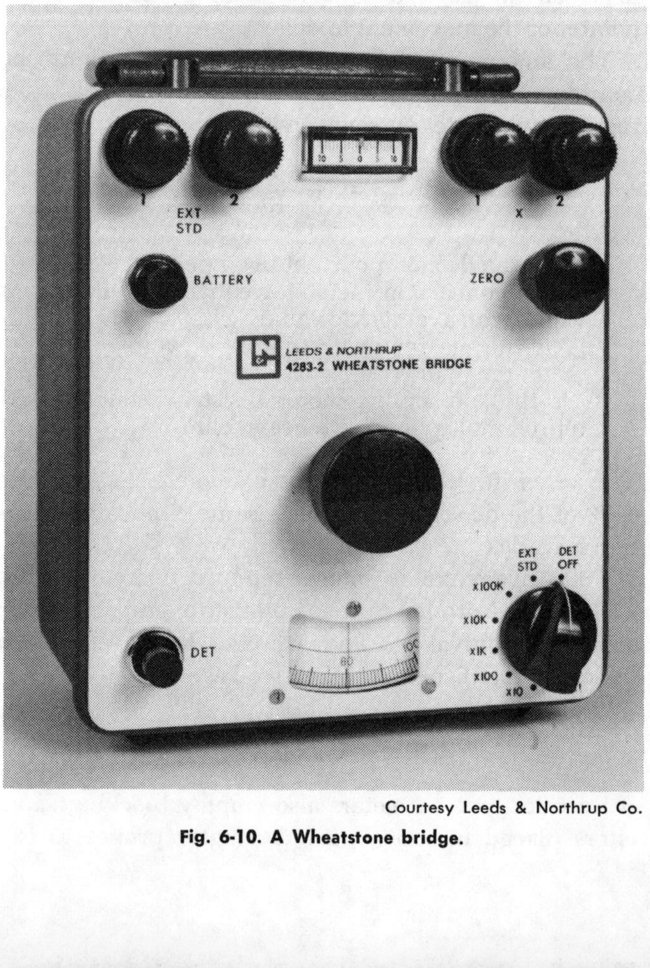

Courtesy Leeds & Northrup Co.

Fig. 6-10. A Wheatstone bridge.

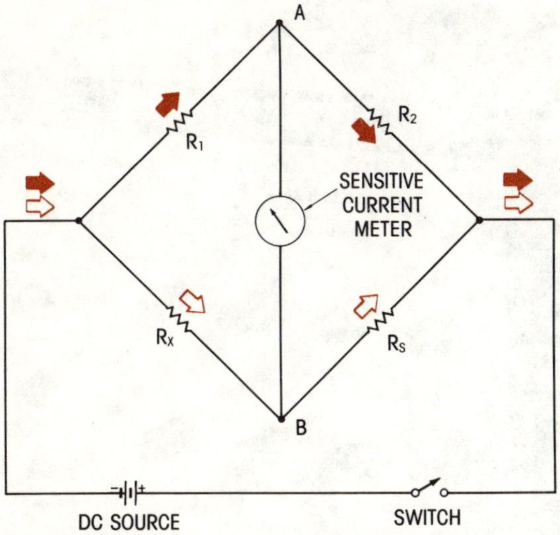

Fig. 6-11. Schematic diagram of a Wheatstone bridge circuit.

able accuracy. Several other types of comparative indicators use the Wheatstone-bridge principle. Comparative indicators compare an unknown quantity with a known quantity contained within the indicator. A standard resistor, shown in Fig. 6-12, can be used to calibrate a Wheatstone bridge.

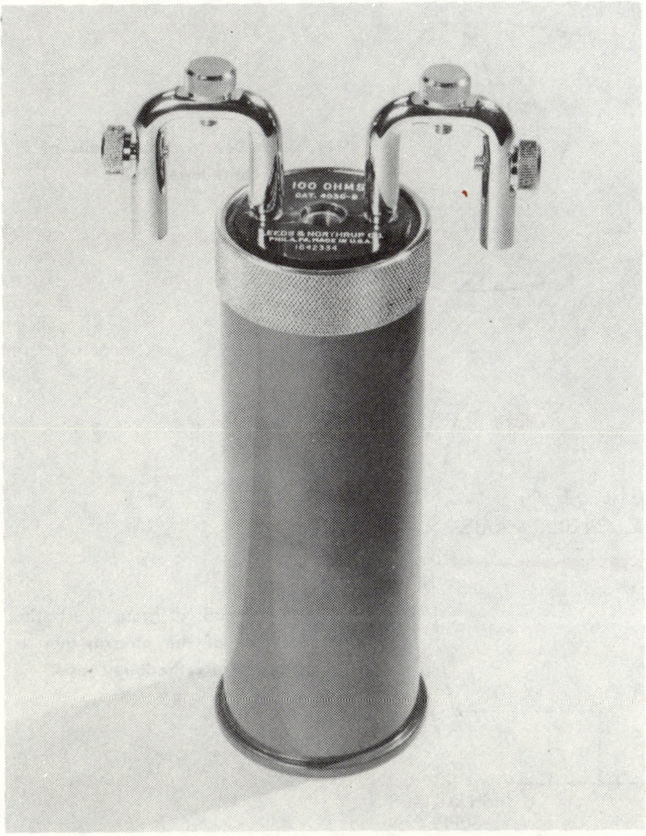

Courtesy Leeds & Northrup Co.

Fig. 6-12. A standard resistor for calibrating Wheatstone bridges.

Potentiometric Comparison

Another technique used with comparative indicators is the *potentiometric* method. Whereas the bridge method is used to measure impedances, the potentiometric method is used to accurately measure voltages. The illustration of Fig. 6-13 shows the simplified po-

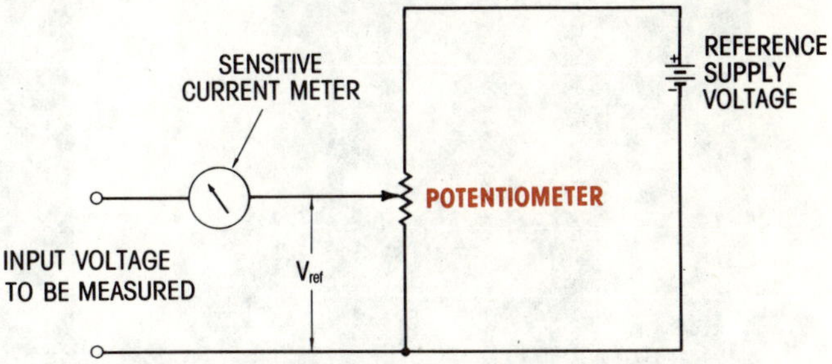

Fig. 6-13. A potentiometric comparison circuit.

tentiometer method. The voltage to be measured is applied to the input terminals of the circuit. The potentiometer, which is connected across the reference supply voltage, forms a voltage divider. When the moveable arm of the potentiometer is adjusted to a voltage (V_{ref}) equal to the voltage being measured, the current through the meter will be zero. When the zero (or *null*) reading on the zero-centered meter is obtained, the value of the unknown voltage can be read from a calibrated scale on the potentiometer.

Since no current is drawn from the voltage being measured during the null condition, the meter can be considered an infinite impedance. Thus, the potentiometric technique can be used for precise calibration of other indicators. Potentiometric indicators can also be used as laboratory voltage standards. Industrial potentiometers either use a standard cell (see Fig. 6-14), which has a precise voltage, or a zener-diode standard reference voltage supply (see Fig. 6-15).

Cathode-Ray Tube (CRT) Display Indicators

The cathode-ray tube (crt) display indicator, usually called an *oscilloscope,* is an important type of indicator. It is possible to monitor the voltages of a system visually by using an oscilloscope. The basic operational part of the oscilloscope is the cathode-ray tube. We should become familiar with the operation of the cathode-ray tube in order to better understand the oscilloscope.

Fig. 6-16 shows the construction details of a crt, while Fig. 6-17 illustrates the operation of the electron-gun assembly in the neck of the crt. A beam of electrons is produced by the cathode of the tube when it is heated by the application of a filament voltage. The electrons are attracted to the positive potential of anode No. 1. The quantity of electrons passing on to

Courtesy Leeds & Northrup Co.

Fig. 6-14. A standard cell.

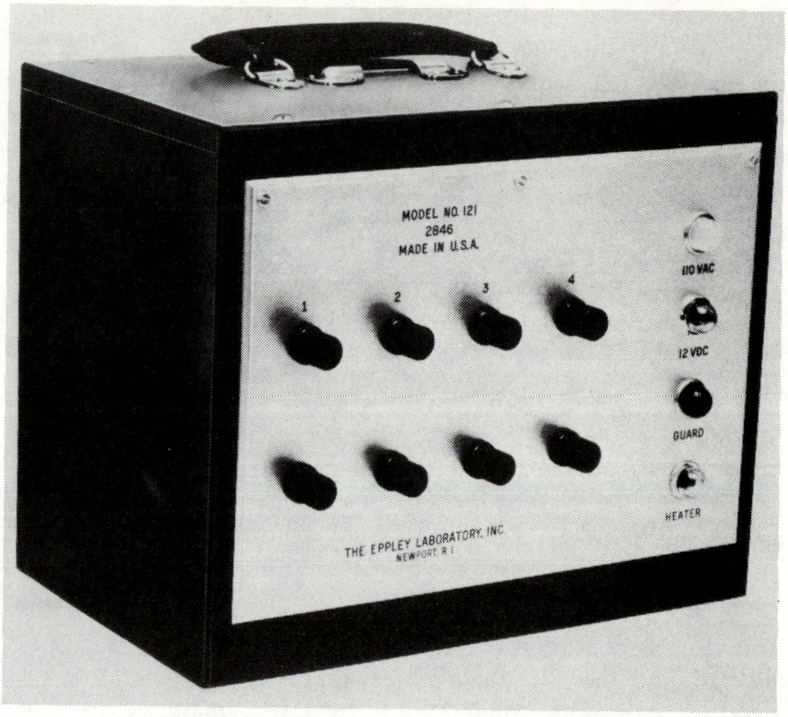

Courtesy Leeds & Northrup Co.

Fig. 6-15. A standard reference voltage supply.

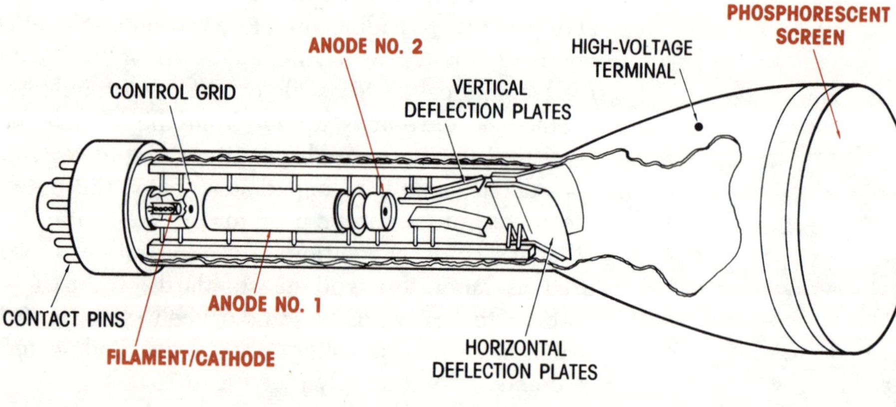

Fig. 6-16. Construction details of a cathode-ray tube.

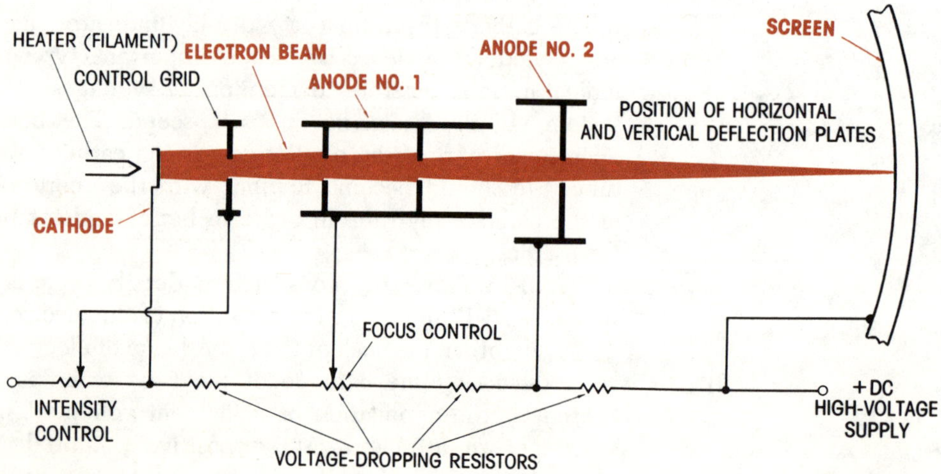

Fig. 6-17. Circuit diagram illustrating the operation of the electron-gun assembly of a cathode-ray tube.

anode No. 1 is determined by the amount of negative bias voltage applied to the control grid. Anode No. 2 is operated at a higher positive potential to further accelerate the electron beam toward the screen of the crt. The difference in potential between anode No. 1 and anode No. 2 determines the point of convergence of the beam on the crt screen. When the electron beam strikes the phosphorescent screen of the crt, light is emitted.

To control the horizontal and vertical movement of the electron beam, deflection plates are used. If no potential is applied to either plate, the electron beam would merely appear as a dot in the center of the crt screen. Electrostatic deflection occurs when a potential is placed on the horizontal and vertical deflection plates. Horizontal deflection usually results from a sawtooth waveform produced by a sweep oscillator, which is part of the oscilloscope circuit. This voltage "sweeps" the electron beam back and forth across the crt screen. The sweep setting of a crt is adjusted to coincide with the frequency range of the voltage to be measured. Vertical deflection takes place in accordance with the magnitude of the applied voltage. The voltage to be measured is applied to the vertical deflection circuit of the oscilloscope (see Fig. 6-18).

Cathode-ray tube display indicators are available in several different types. General-purpose oscilloscopes are used to display simple types of waveforms, and for general electronic servicing. Triggered-sweep oscilloscopes are used where it is desired to apply external voltage to the horizontal circuits to produce horizontal sweep. Other oscilloscopes, classified as laboratory types, have very high sensitivity and good frequency response over a wide range. Crt display indicators can be used to measure ac and dc voltages, frequency and phase relationships, distortion in amplifiers, and various timing and numerical-control applications. Memory and storage-type instruments are also available for more sophisticated measurements.

Numerical-Readout Indicators

Many indicators now in use employ numerical readouts to simplify the measurement process and make more accurate measurements. Instruments such as digital counters, digital multimeters, and digital voltmeters are commonly employed for industrial measurement application (see Fig. 6-19). Numerical-readout indicators rely upon the operation of gating and flip-flop circuitry in order to produce a digital readout of the measured quantity.

Chart-Recording Indicators

The types of indicators discussed previously are used where no permanent record of the measured quantity is needed. Indicators are also employed to provide a permanent record of some quantity that is measured over a specific time period. A typical chart-recording indicator is shown in Fig. 6-20. Types of chart recorders include pen-and-ink recorders, and inkless recorders.

Pen-and-ink recorders use a pen attached to the indicator, which touches a paper chart and leaves a permanent record of the measured quantity on the chart. The charts utilized can either be roll charts that revolve on rollers under the pen mechanism, or circular charts that revolve on an axis under the pen. Chart recorders may use more than one pen to record several different quantities simultaneously.

The pen of a chart recorder is a capillary-tube device that is actually an extension of the basic meter movement. The pen must be connected to a constant source of ink. The pen is moved by the torque exerted

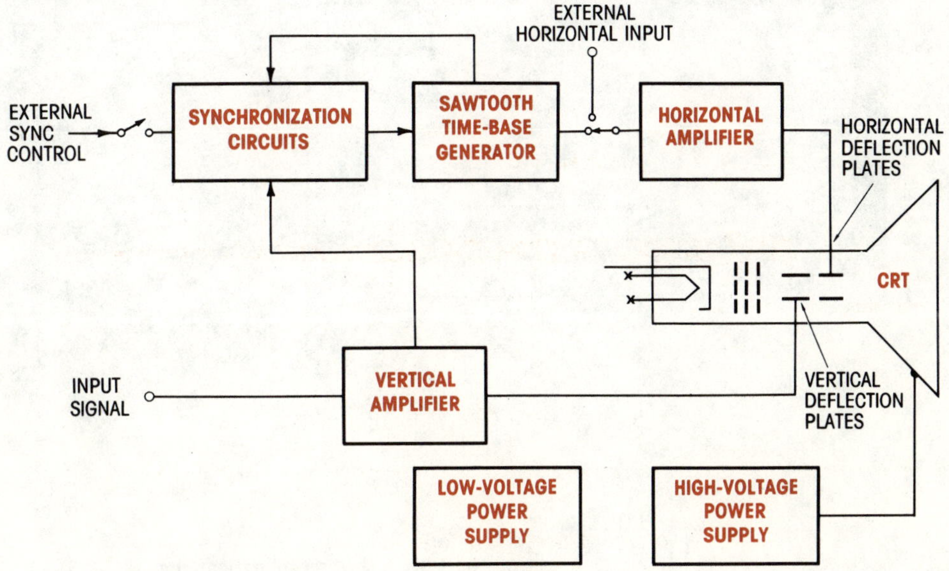

Fig. 6-18. Block diagram of an oscilloscope circuit.

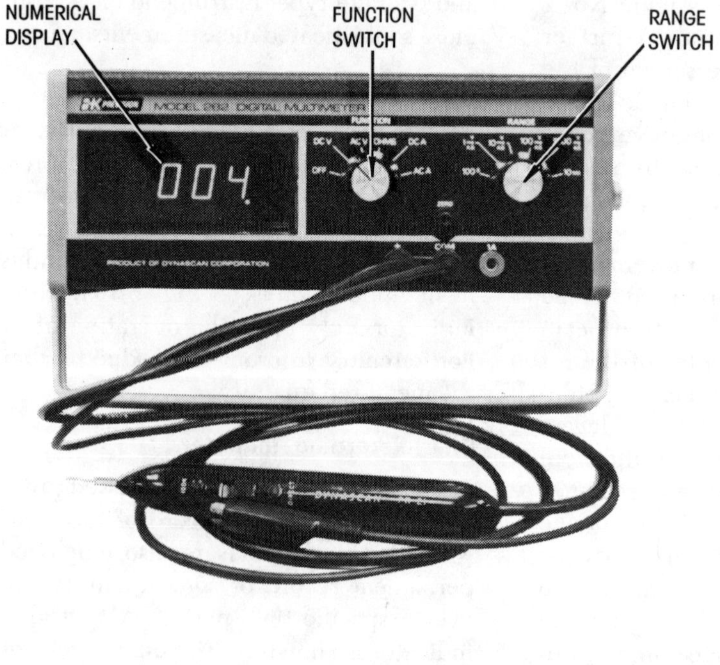

Fig. 6-19. A numerical-readout indicator.

Courtesy Dynascan Corp.

by the meter movement just as the pointer of a hand-deflection meter is moved. The chart used for recording the measured quantity usually contains lines that correspond to the radius of the pen movement. Increments on the chart are marked according to time intervals. The chart must be moved under the pen at a constant speed. Either a spring-drive mechanism, a synchronous ac motor, or a dc servomotor could be used to drive the chart. Recorders are also available that use a single pen to make permanent records of

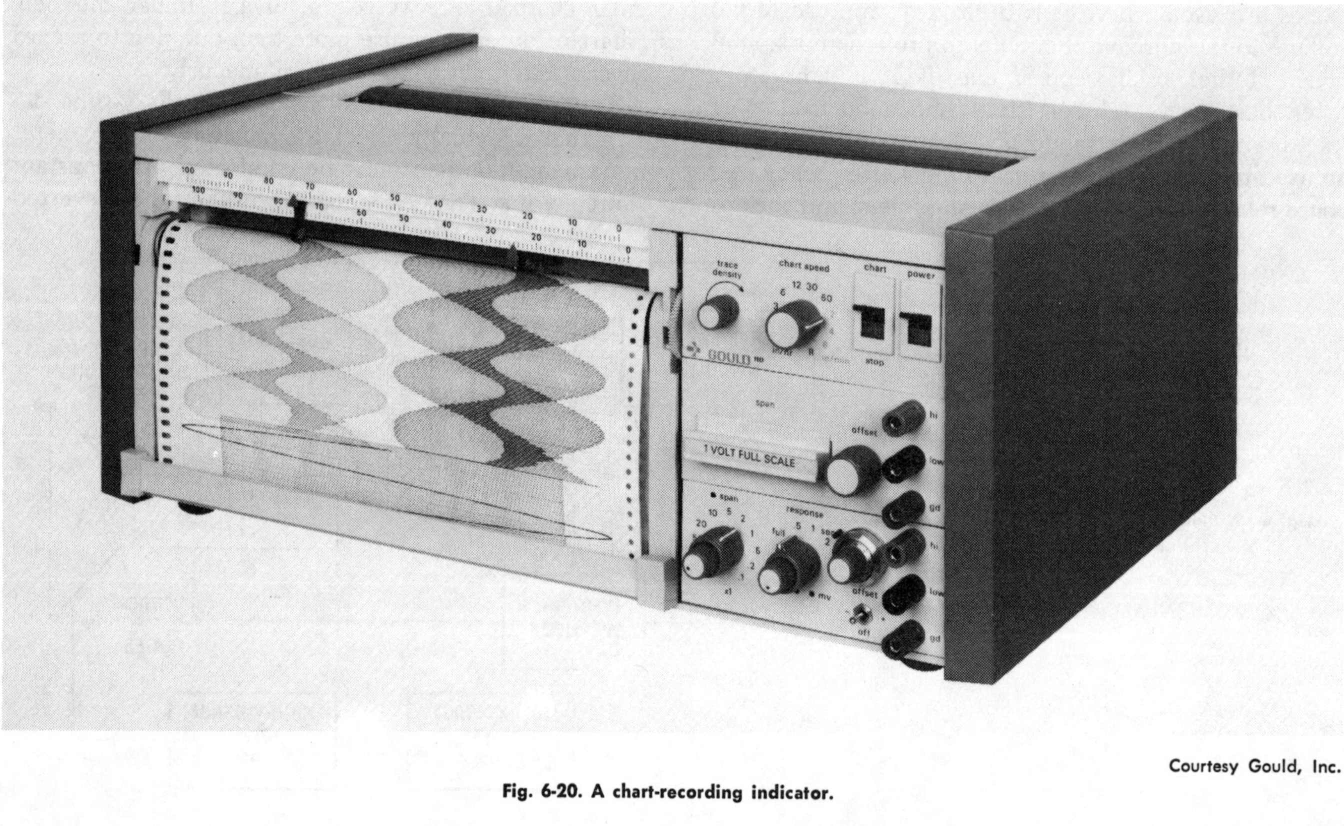

Courtesy Gould, Inc.

Fig. 6-20. A chart-recording indicator.

measured quantities on a single chart. In this case, either coded lines or different colored ink could be used to record the quantities.

Inkless recorders may use a voltage applied to the pen point to produce an impression on a sensitive paper chart. In another process, the pen is heated to cause a trace to be melted along the chart paper. The obvious advantage of inkless recorders is that ink is not required.

Chart-recording indicators for measuring almost any electrical or physical quantity are commercially available. For many industrial applications, the recording system may be located a long distance from the quantity being measured. For accurate process control or system monitoring, a central instrumentation system might be used. Power plants, for instance, ordinarily use chart recorders at a centralized location to monitor the various electrical and physical quantities involved in the power-plant operation.

A specialized example of a chart-recording indicator is the X-Y recorder such as the one shown in Fig. 6-21. An X-Y recorder is used to plot the relationship that two quantities have to one another. Usually some variable quantity is plotted versus time. In other applications, one variable quantity may be plotted versus some other variable quantity. One input is applied to the recorder to cause vertical (X-axis) movement of the indicating pen. The other input is applied to the recorder to cause horizontal (Y-axis) movement of the pen. Thus, the trace produced on the chart shows the relationship of the two quantities connected to the X and Y inputs.

MEASUREMENT OF INDUSTRIAL POWER QUANTITIES

Several types of industrial measurement applications involve indicators used to measure electrical power

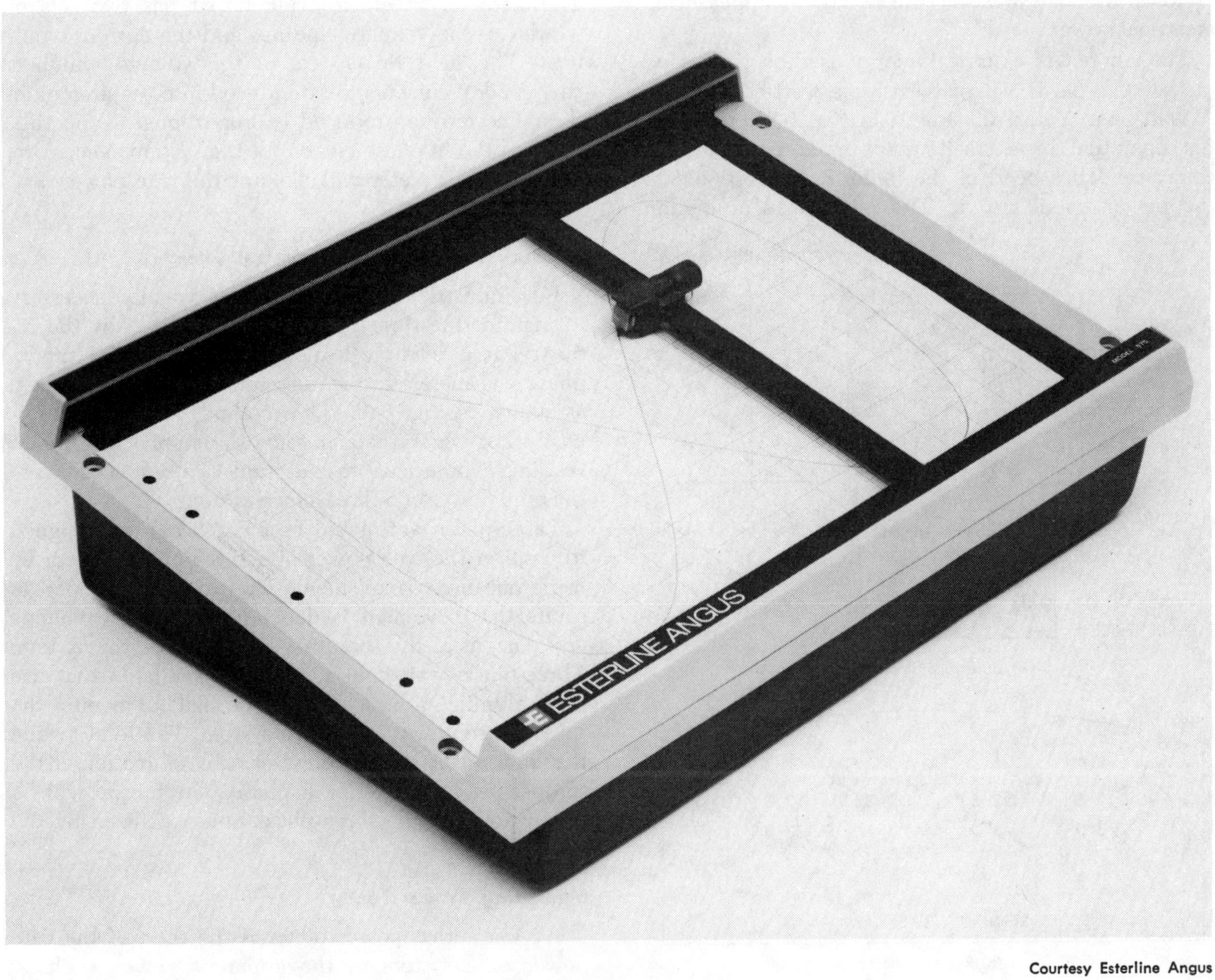

Courtesy Esterline Angus

Fig. 6-21. An X-Y recorder.

and related quantities. The day-to-day operation of industry relies on the use of electrical power. Therefore, in order to use electrical power more effectively, industries must monitor their power system. We have discussed previously the methods used to measure basic electrical quantities. In the following sections, we will discuss the measurement of more specialized quantities.

Measuring Electrical Energy

In a prior section of this chapter, the basic measurement of electrical power was described. Wattmeters with dynamometer movements, which monitor both voltage and current in a circuit, are ordinarily used to measure the electrical power converted by some type of load. For instance, the amount of electrical energy converted into mechanical energy by a motor could be measured by properly connecting a wattmeter to the motor circuit. Recall that the power factor of the load must be considered when measuring electrical power.

The amount of electrical energy used over a period of time can be measured by using a watt-hour meter. A watt-hour meter, illustrated in Fig. 6-22, relies on the operation of a small motor mounted inside its enclosure. The speed of the motor is proportional to the power applied to it. The rotor is an aluminum disk that is connected to a register, which usually indicates the number of kilowatt-hours of electrical energy used. Fig. 6-22B shows the dial-type of register that is frequently used on watt-hour meters. Other types have a direct numerical readout of kilowatt-hours used.

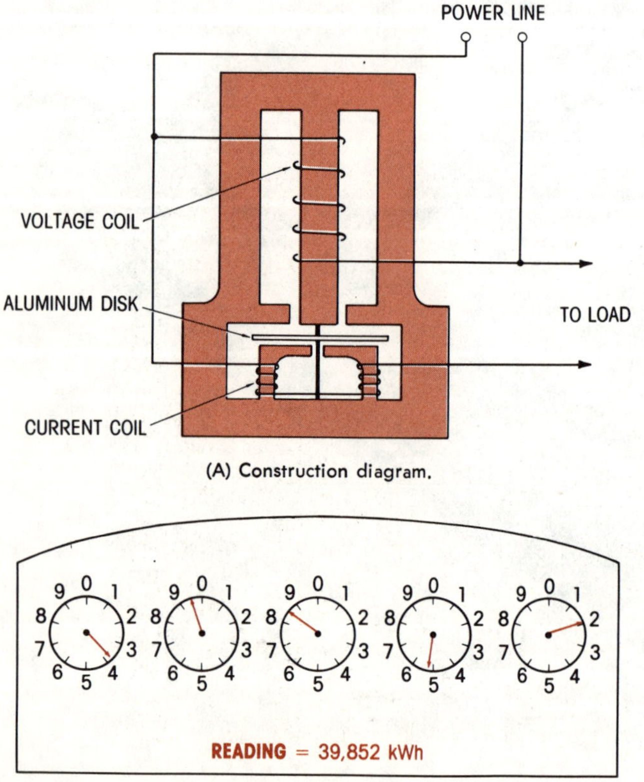

(A) Construction diagram.

(B) Register dials.

Fig. 6-22. The watt-hour meter.

The watt-hour meter is connected between the voltage feeder lines and the branch circuits of an industrial power system. In this way, all electrical energy used by an industry must pass through the kWh meter. The same type of system is used for home service entrances.

The electrical operation of a watt-hour meter is somewhat similar to the conventional wattmeter. A potential coil is connected across the power lines to monitor voltage, while a current coil is placed in series with the line to measure current. The voltage and current of the system affect the movement of the aluminum-disk rotor. The operation of the watt-hour meter may be considered as similar to an ac induction motor. The stator is an electromagnet that has two sets of windings—the voltage windings and the current windings. The field developed in the voltage windings causes eddy currents to be induced into the aluminum disk. The torque produced is proportional to the voltage and the in-phase current of the system. Therefore, the watt-hour meter will monitor the true power converted in a system.

Measuring Three-Phase Electrical Energy

For industrial applications, it is usually necessary to monitor the three-phase energy utilized by the industry. It is possible to use a combination of single-phase wattmeters to measure total three-phase power, as shown in Fig. 6-23. The methods shown are ordinarily not very practical since the sum of the meter readings would have to be found in order to calculate the total power of a three-phase system.

Three-phase wattmeters (see Fig. 6-24) are designed to monitor the true power of a three-phase system by using one meter movement. The torque developed to rotate the movement is dependent upon the voltage and current of the three phases acting on the system. Three-phase watt-hour meters are available that use three aluminum disks on the same shaft to monitor the three-phase power used by a system. The total torque that rotates the shaft is therefore a combination of the torques developed by each phase. The torque will be proportional to the three-phase power utilized by the system.

Measuring Power Factor

We know that power factor is the ratio of the true power of a system to the apparent power (volts × amperes). To determine power factor we could use a

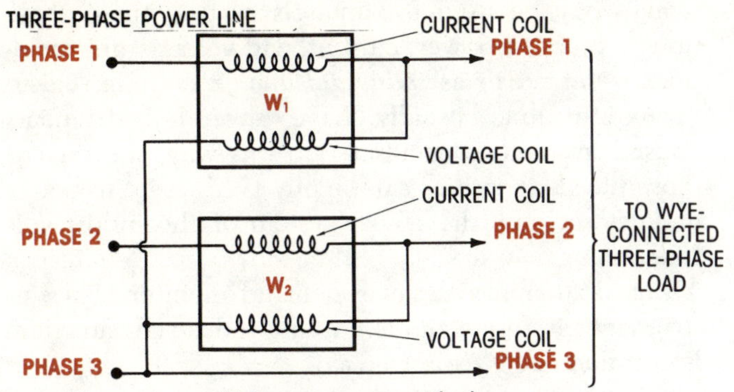

(A) Two-wattmeter method.

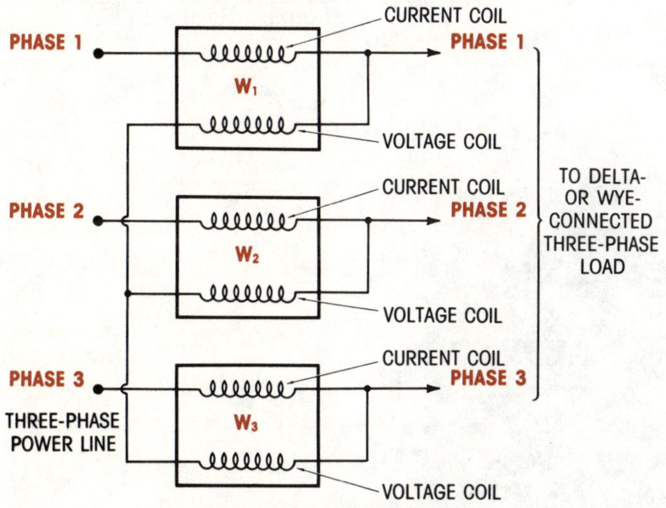

(B) Three-wattmeter method.

Fig. 6-23. Using single-phase wattmeters to measure three-phase power.

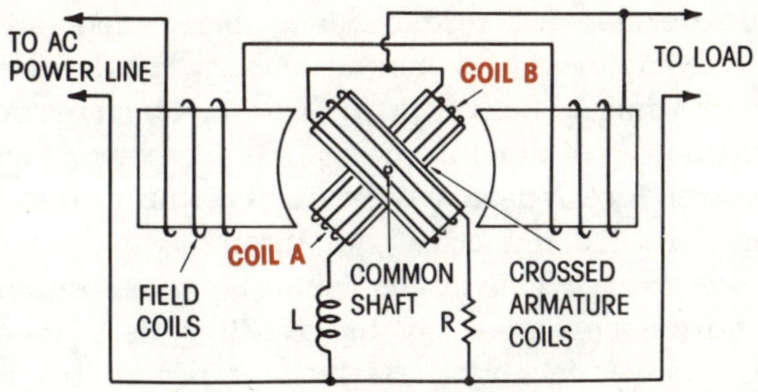

Fig. 6-25. Schematic diagram of a power-factor meter circuit.

wattmeter, a voltmeter, and an ammeter and use the relationship: pf = W/VA. However, it would be more convenient to use a power-factor meter in situations where power factor must be monitored.

The principle of a power-factor meter is illustrated in Fig. 6-25. The power-factor meter is similar to a wattmeter, except that it has two armature coils that will rotate due to their electromagnetic field strengths. The armature coils are mounted on the same shaft so that their alignment is about 90° apart. One coil is connected across the ac line in series with a resistance, while the other coil is connected across the line in series with an inductance. The resistive path through the coil reacts to produce a flux proportional to the in-phase component of the power. The inductive path reacts in proportion to the out-of-phase component of the power.

If a unity (1.0) power-factor load is connected to the meter, the current in the resistive path through coil A

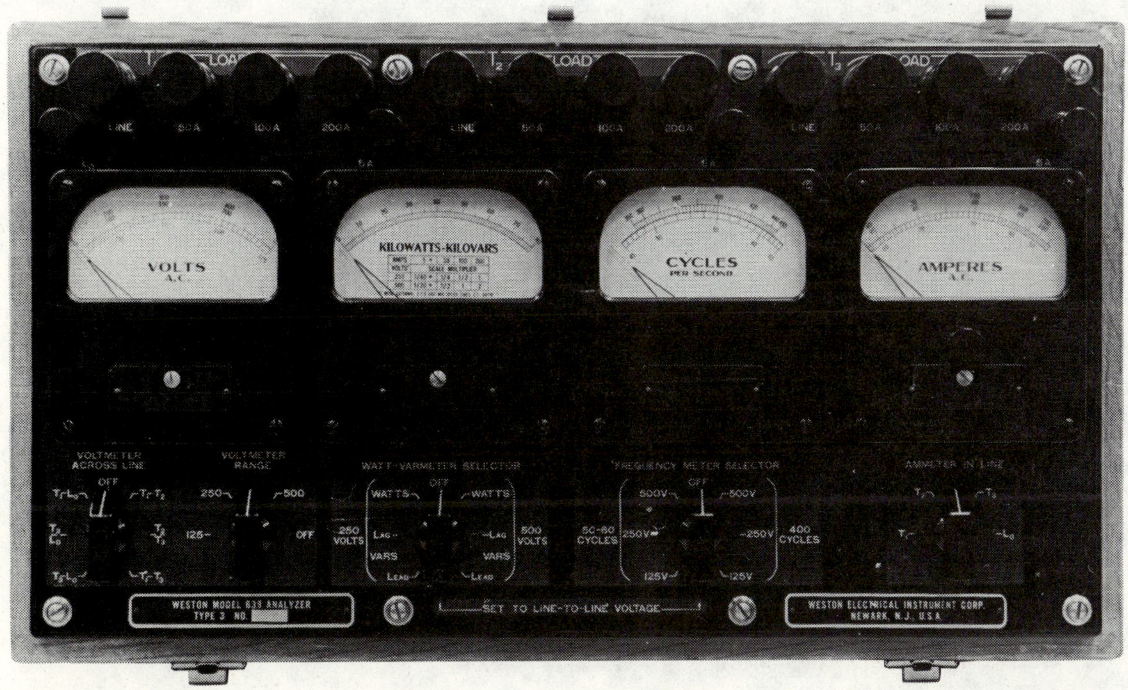

Courtesy Weston Instruments, Inc.

Fig. 6-24. An industrial analyzer that monitors true three-phase power, reactive power, line current, line voltage, and line frequency.

would develop full torque. Since there is no out-of-phase component, no torque would be developed through the inductive path. The meter movement would now indicate full-scale, or unity power factor. As power factor decreases below 1.0, the torque developed by the inductive path through coil B would become greater. This torque would be in opposition to the torque developed by the resistive path. Therefore, a power factor of less than 1.0 would be indicated. The scale must be calibrated to measure power-factor ranges from zero to unity.

Various other types of instruments are commercially available for measuring three-phase power and associated quantities. Power-monitoring instruments, which may be used for analysis of true three-phase power, reactive power, current, and voltage, are sometimes referred to as *industrial analyzers* (Fig. 6-24). Such indicators usually have several instruments housed in a single enclosure. In this way, the related three-phase quantities can be observed easily to better understand how the power system of the industry is functioning.

Fig. 6-26 shows a power-demand monitor. This instrument performs an important industrial function. Power demand is expressed as:

$$\text{power demand} = \frac{\text{peak power used (kW)}}{\text{average power used (kW)}}.$$

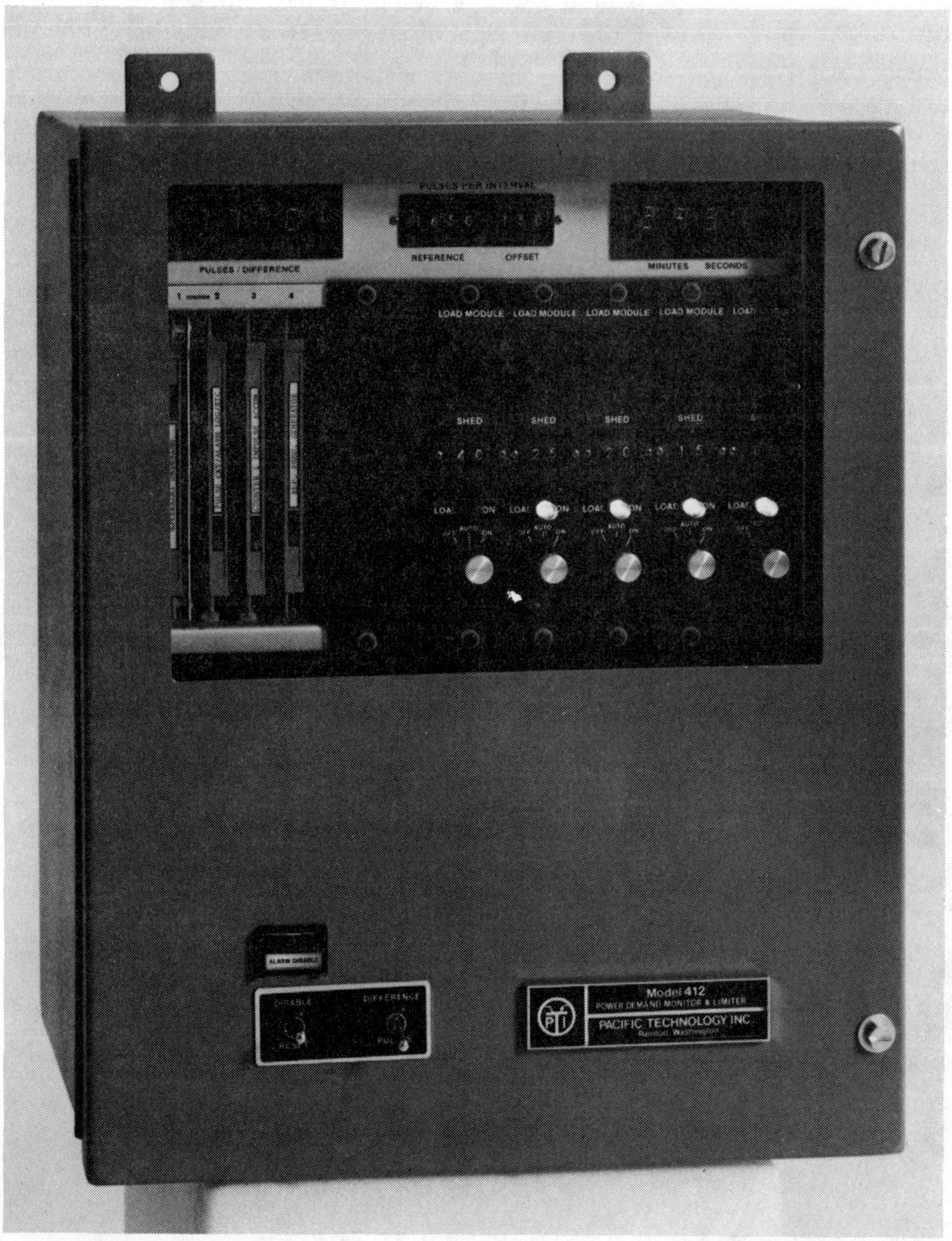

Fig. 6-26. A power-demand monitor.

Courtesy Pacific Technology, Inc.

This ratio is important because it indicates the amount above the average power consumption that a utility company must supply an industry. Power demand is usually calculated over a 15-, 30-, or 60-minute interval and then converted into longer periods of time.

Industries may be penalized by the utility company if their peak power demand far exceeds their average demand. Power-demand monitors help industries to better utilize their electrical power. A high peak demand means that the equipment for the industrial power distribution must be rated higher. The closer that the peak demand approaches the value of the average demand, the more efficient the industrial power system is in terms of power utilization.

Measuring Frequency

Another very important power measurement is frequency. The frequency of the power source must remain stable or the operation of many types of industrial equipment will be affected. Frequency refers to the number of cycles of voltage or current that occur in a given period of time. The international unit of measurement for frequency is the hertz (Hz), which means cycles per second. A table of frequency bands is shown in Fig. 6-27. The standard power frequency in the United States is 60 Hz.

Frequency can be measured with several different types of indicators. An electronic counter is one type of frequency indicator. Vibrating-reed frequency indicators are common for measuring power frequencies (see Fig. 6-28). An oscilloscope can also be used to measure frequency. Graphic recording instruments can be used to provide a visual display of frequency for different time periods. The power industry could use this method to monitor the frequency output of its alternators.

Specialized Indicators for Industrial Power Applications

There are a variety of indicators used specifically for industrial power applications. Some involve the direct measurement of power-related quantities, while others are used for testing or troubleshooting industrial equipment. In the following sections we will discuss the synchroscope, ground-fault indicators, megohmmeters, clamp-type current meters, and telemetering systems as representative types of specialized power indicating systems.

Synchroscope

The major application of the synchroscope is in power plants. Most power plants have more than one alternator. In order to connect two or more alternators onto the same ac line, the following conditions must be met: (1) their voltage outputs must be equal, (2) their frequencies must be equal, (3) their voltages must be in phase, and (4) the phase sequence of the voltages must be the same.

Voltage output levels can be checked easily with a voltmeter of the transformer-type used to monitor high voltages. Frequency is adjusted by varying the speed of the alternator, and is also easy to monitor. The phase sequence is established on each alternator when it is installed and connected to the power system. The additional factor, which must also be monitored before paralleling alternators, is to assure that the voltages of

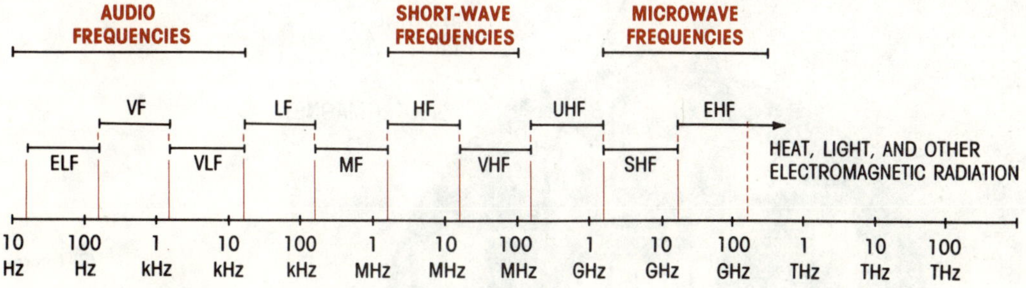

BAND		FREQUENCY RANGE		
EXTREMELY LOW FREQUENCY	(ELF)	30 Hz	TO	300 Hz
VOICE FREQUENCY	(VF)	300 Hz	TO	3 kHz
VERY-LOW FREQUENCY	(VLF)	3 kHz	TO	30 kHz
LOW FREQUENCY	(LF)	30 kHz	TO	300 kHz
MEDIUM FREQUENCY	(MF)	300 kHz	TO	3 MHz
HIGH FREQUENCY	(HF)	3 MHz	TO	30 MHz
VERY-HIGH FREQUENCY	(VHF)	30 MHz	TO	300 MHz
ULTRA-HIGH FREQUENCY	(UHF)	300 MHz	TO	3 GHz
SUPER-HIGH FREQUENCY	(SHF)	3 GHz	TO	30 GHz
EXTREMELY HIGH FREQUENCY	(EHF)	30 GHz	TO	300 GHz

Fig. 6-27. Classification of frequency bands.

(A) External view.

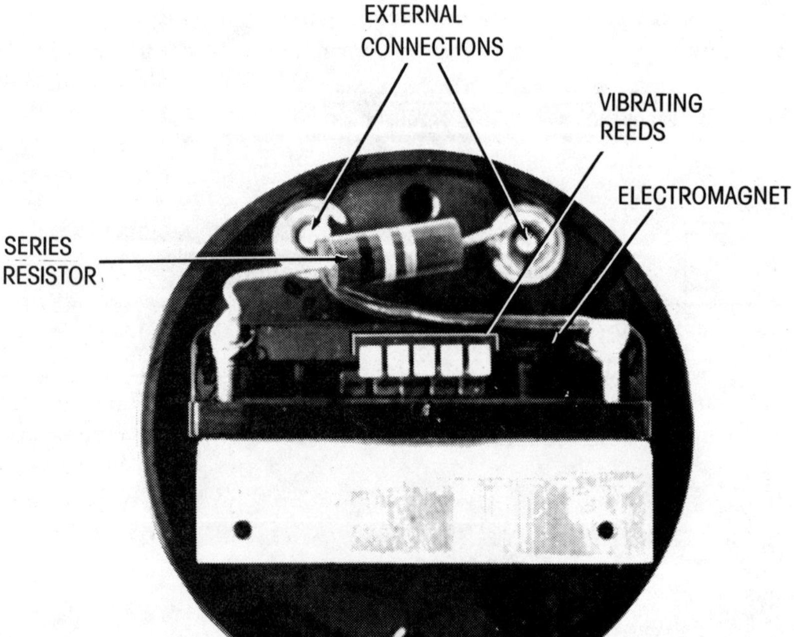

(B) Internal view.

Fig. 6-28. A vibrating-reed frequency indicator.

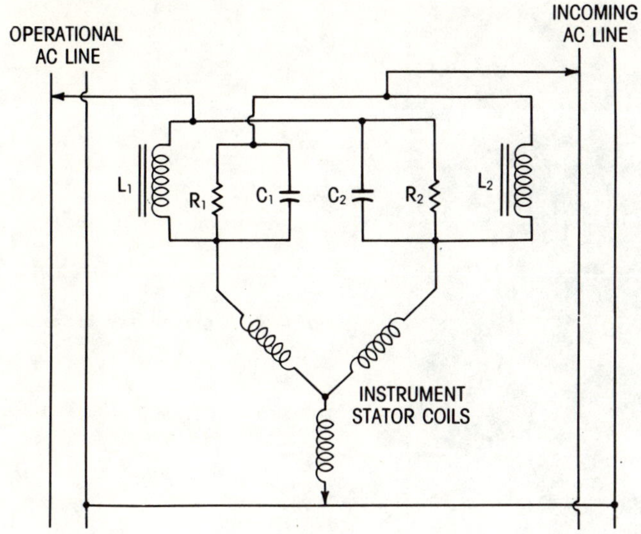

(A) Schematic diagram.

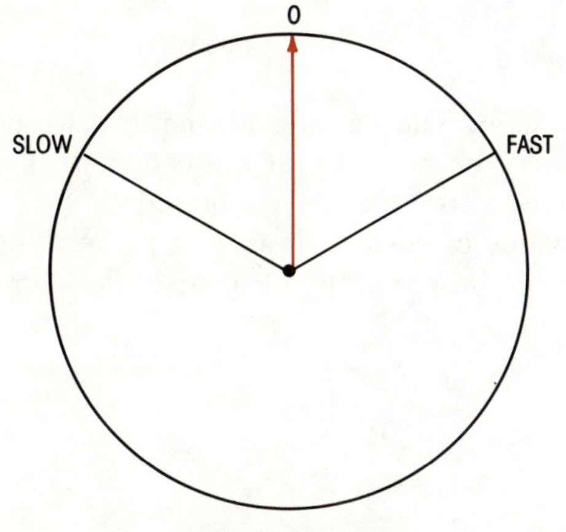

(B) Indicator scale.

Fig. 6-29. The synchroscope.

the alternator to be paralleled with the other alternators is in phase. This is done with a synchroscope (see Fig. 6-29A).

The synchroscope is used to measure the relationship between the phase of the system and the alternator to be put "on line". It also indicates whether the alternator is running faster or slower than the system to which it is being connected. The basic design of the indicator utilizes a phase-comparative network of two RLC circuits connected between the operating system and the alternator to be paralleled. The indicator scale shows whether the incoming alternator is running too slow or too fast (see Fig. 6-29B). When an in-phase relationship exists, along with the other three factors, the alternator can be connected with the system that is already operational. The addition of another alternator will allow a higher power capacity to be produced by the power system.

Ground-Fault Indicators

The ground-fault indicator is used to locate faulty system or equipment grounding conditions. Industrial power systems must have equipment that is properly grounded. Grounding was previously discussed in Chapter 4.

A ground-fault indicator, such as the one shown in Fig. 6-30, can be used to check for faulty grounding at various points in the industrial power system. Several conditions can exist that might be hazardous. These faulty wiring methods include: (1) hot and neutral wires reversed, (2) open equipment ground wire, (3) open neutral wire, (4) open hot wire, (5) hot and equipment ground reversed, or (6) hot wire on the neutral terminal and the neutral not connected. Each of these conditions would present a serious problem for the industrial power system. Although proper wiring eliminates most of these problems, a periodic check with a ground-fault indicator would assure safe and efficient electrical wiring.

Megohmmeters

A megohmmeter, such as the one shown in Fig. 6-31, is used to measure high resistances which extend beyond the range of a typical ohmmeter. These indicators are used primarily for checking the quality of insulation on electrical power equipment in industry. The quality of insulation of equipment varies with age, moisture content, and applied voltage. The megohmmeter is similar to a typical ohmmeter except that it uses a hand-crank, permanent-magnet dc generator as a voltage source rather than a battery. The dc generator is cranked by the operator while making an insulation test. Fig. 6-32 shows a diagram of a megohmmeter circuit. This circuit is essentially the same as any series ohmmeter with the exception of the dc generator used as a voltage source.

Periodic insulation tests should be made on all industrial power equipment. As insulation breaks down with age, the equipment becomes malfunctional. A good method is to develop a periodic schedule for checking and recording insulation resistance so that we can predict when a piece of equipment needs to be replaced or repaired. A resistance-versus-time graph can be made and the trend on the graph noted. A downward trend (decrease in insulation resistance) over a time period indicates that a problem in insulation exists.

Clamp-Type Meters

Clamp-type meters, such as the one shown in Fig. 6-33, are popular for measuring current in power lines. This indicator can be used for periodic checks of

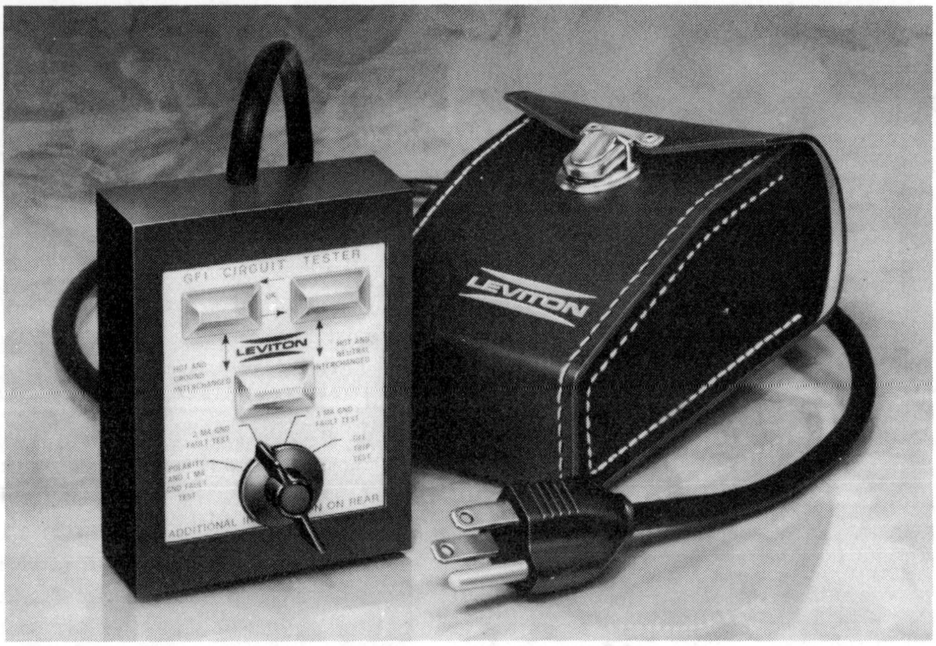

Fig. 6-30. A ground-fault indicator.

Courtesy Leviton Manufacturing Co., Inc.

current by clamping it around a power line. It is an easy-to-use and convenient maintenance and testing instrument. The simplified circuit of a clamp-type current meter is shown in Fig. 6-34. Current flow through a conductor creates a magnetic field around the conductor. The varying magnetic field induces a current into the iron core of the clamp portion of the meter. The meter scale is calibrated so that when a specific value of current flows in a power line, it will be indicated on the scale. Of course, the current flow

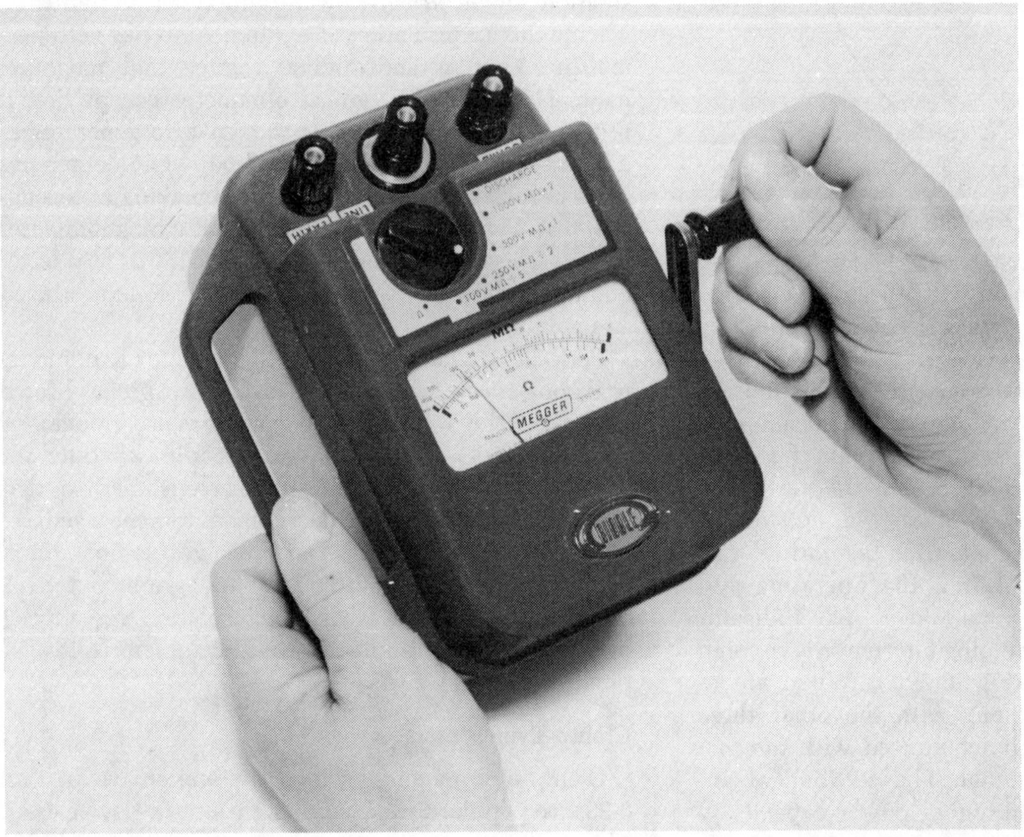

Fig. 6-31. A megohmmeter.

Courtesy James G. Biddle Co.

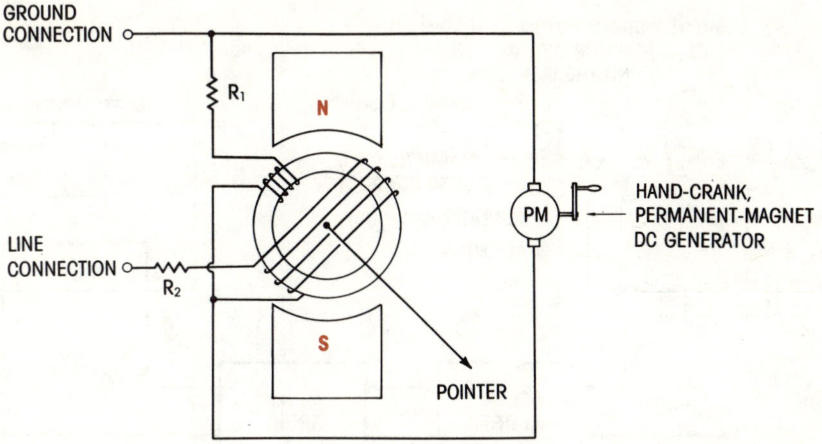

Fig. 6-32. Circuit diagram of a megohmmeter.

in the power line is proportional to the current induced into the iron core of the clamp-type meter. Ordinarily, the clamp-type meter also has a voltage and resistance function which utilizes external test leads.

Telemetering Systems

When a quantity being measured is indicated at a location some distance from its transducer or sensing element, the measurement process is referred to as *telemetering*. Many types of indicating systems fit this definition; however, telemetering systems are usually for long-distance measurement or for centralized measurement. For instance, many industries group their indicating systems together to facilitate process control. Another example of telemetering is the centralized monitoring of electrical power by utility companies on a regional basis. These systems are similar to other measuring systems except that a transmitter/receiver communication process is usually involved.

Many types of electrical and physical quantities can be monitored by using telemetering systems. The most common transmission media for telemetering systems are: (1) wire, such as telephone or telegraph lines; (2) superimposed signals, which are 30- to 200-kHz signals carried on power distribution lines; and (3) radio-frequency signals, mainly from am, fm, and phase-modulation transmitters. A type of telemetering system is shown in Fig. 6-35. In the system shown, a dc voltage from the transducer could be used to modulate an am or fm transmitter. The rf signal is then received and converted back into a dc voltage to activate some end device. The end device, which may be located a considerable distance from the transducer, could be a chart recorder, a hand-deflection indicator, or possibly a process controller. Digital telemetering is also used since binary signals are well suited for data transmission. In this system, the transducer output is converted to binary code for transmission.

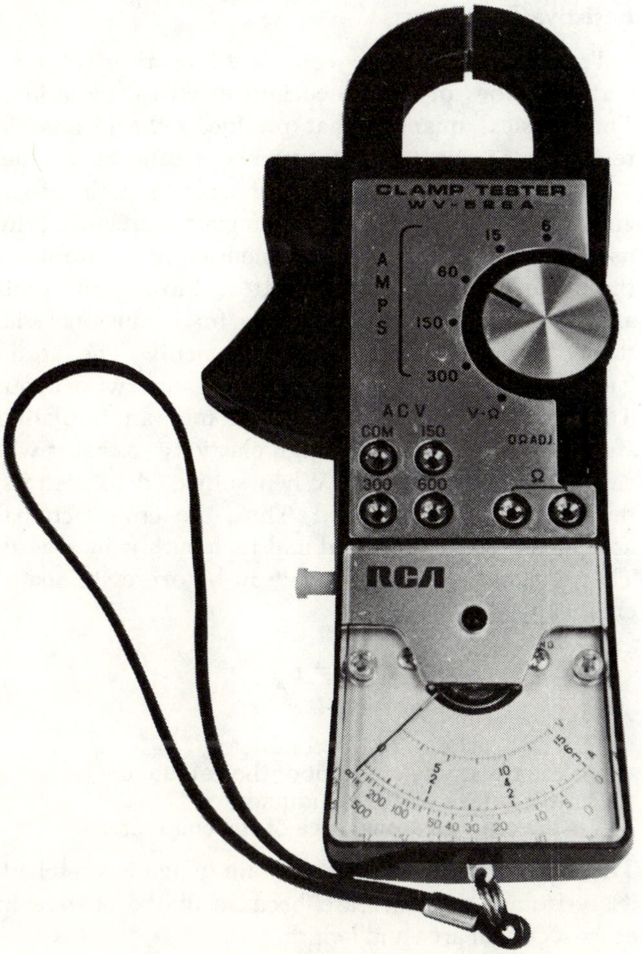

Courtesy RCA Corp.

Fig. 6-33. A clamp-type current meter.

INDICATING SYSTEMS FOR MEASURING PHYSICAL QUANTITIES

Many industrial processes must be monitored and controlled by indicating systems that measure physical quantities. Such physical quantities as temperature, pressure, flow, and motion must be monitored in conjunction with industrial operations.

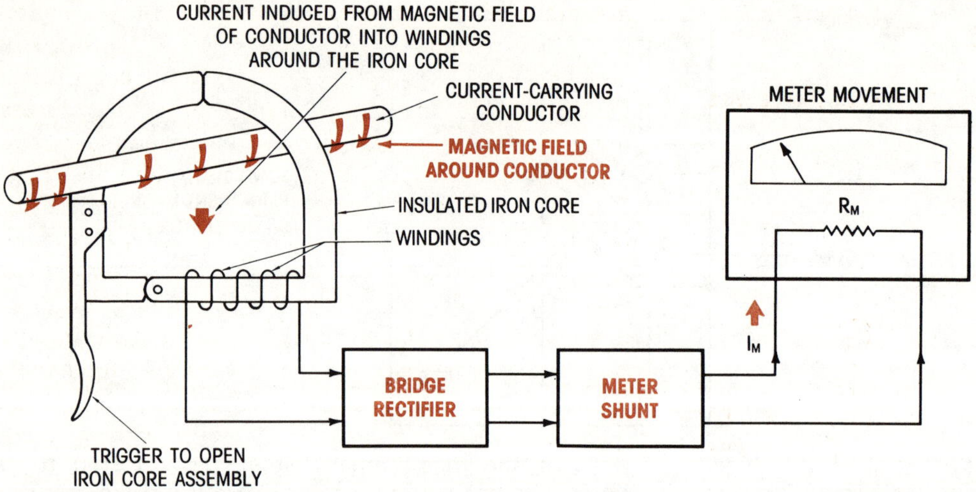

Fig. 6-34. Circuit diagram of a clamp-type current meter.

The previous sections of this chapter have dealt with indicating systems for measuring electrical quantities such as voltage, current, resistance, and power. In the following sections, we will discuss several indicating systems that are used to measure physical quantities. The measurement of physical quantities is becoming more important than ever in industry due to the increased precision of process control and automation.

Transducers

Most physical quantities are measured by first being converted to electrical quantities. The conversion of physical quantities to electrical quantities ordinarily simplifies the measurement process. Devices that convert one form of energy into another form are referred to as *transducers*. Therefore, transducers are used to convert physical quantities to electrical quantities. For example, a thermocouple is a transducer that converts heat energy into electrical energy. Also, a microphone is a transducer that converts sound energy into electrical energy. Numerous other examples of transducers are used in our homes as well as in industry. There are several basic types of transducers that may be used with indicating systems. Common classifications of transducers include resistive, capacitive, inductive, piezoelectric, thermoelectric, optoelectric, and other types. These types, with some modifications for specific indicating systems, may be used for converting most physical quantities into electrical quantities for ease of measurement.

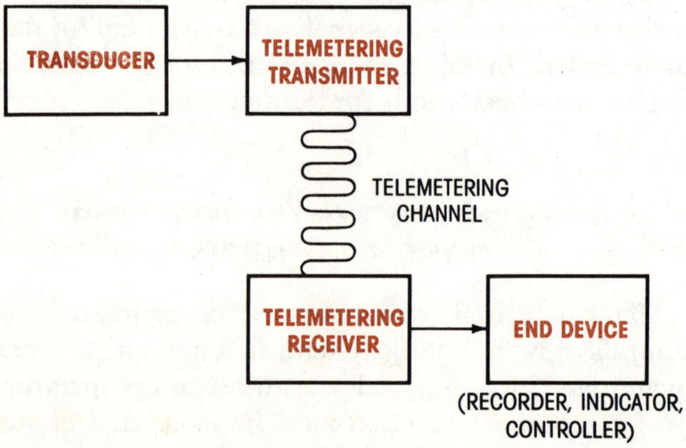

Fig. 6-35. Block diagram of a basic telemetering system.

Resistive Transducers

Transducers that are considered as resistive convert variations of resistance into electrical variations. The physical quantity that produced the change in resistance can then be measured on a calibrated scale.

A common type of resistive transducer is the strain gauge shown in Fig. 6-36A. Strain gauges are ordinarily used to measure a change in dimension of a material as the material is subjected to a stress. Most applications are confined to laboratory usage for testing material strength, or as calibrating instruments. The strain gauge itself is constructed of fine gauge wire about 0.001 inch in diameter mounted into an insulating strip. The wire used has a high elasticity so that it will easily change dimension. When subjected to a stress, the wire will be stretched. Thus, the cross-sectional area of the wire is reduced and its length is increased. The resistance of a conductor can be expressed mathematically as:

$$R = \rho \frac{l}{A}$$

where,

R is the resistance of the conductor,
ρ is the resistivity constant of the conductor,
l is the length of the conductor,
A is the cross-sectional area of the conductor.

Therefore, as the wire of the strain gauge is stretched, its resistance will increase because of the change in cross-sectional area and length.

Another type of resistive transducer utilizes the potentiometer principle, as shown in Figs. 6-36B and C.

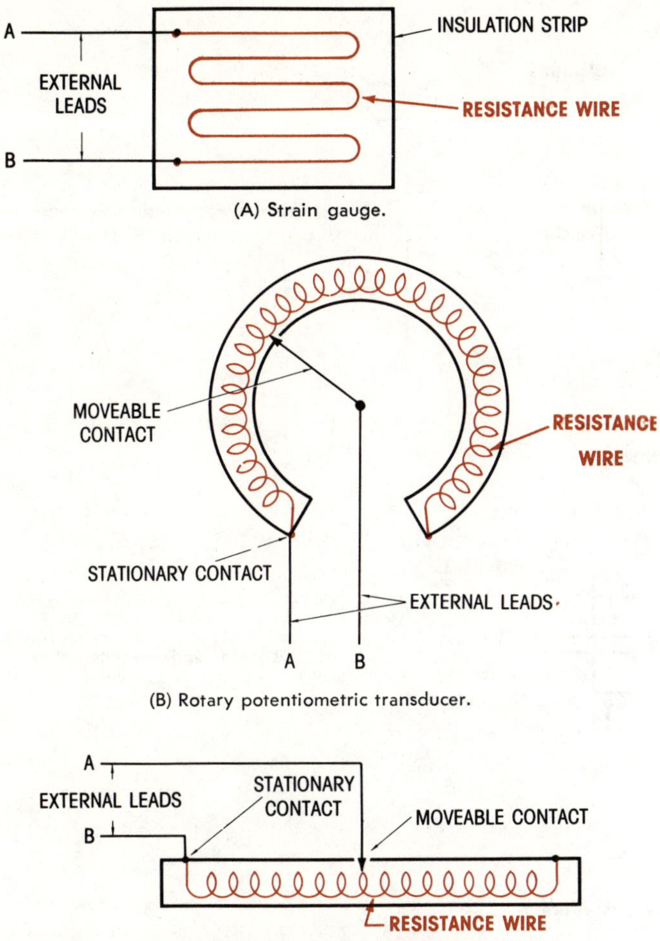

(A) Strain gauge.

(B) Rotary potentiometric transducer.

(C) Flat potentiometric transducer.

Fig. 6-36. Resistive transducers.

This type of transducer changes resistance when the position of its movable contact is changed. By increasing the length of wire between terminals A and B, the resistance between those two points is increased. This principle is often used to measure physical displacement by allowing the displacement to cause movement of the sliding contact.

Another application of resistive transducers is for electrochemical measurements. The resistivity of various chemical solutions changes due to a change of state that may be brought about by the quantity to be measured. For instance, pH (alkalinity or acidity) can be measured by passing an electrical current through a solution containing that which is to be measured. The pH of the solution will cause the resistivity to vary, thus varying the current in the external circuit calibrated to measure pH.

Capacitive Transducers

Capacitive transducers rely on a change in capacitance brought about by the change of some physical quantity. Capacitance exists when two conductive materials (plates) are separated by a dielectric material. We can increase capacitance by increasing the area of the plates or by decreasing the thickness of the dielectric. One application of capacitive transducers, illustrated in Fig. 6-37, is for measuring fluid pressure. This type of transducer is placed into a fluid line. Plate 1 of the capacitor is a conductive diaphragm inserted into the fluid line to sense any variation in fluid pressure. It is electrically connected to the housing. Plate 2 is mounted adjacent to plate 1 and is initially adjusted to calibrate the indicator scale. Plate 2 is held in position, while plate 1 will vary in position due to changes in pressure. When the pressure of the fluid in the line increases, plate 1 will move closer to plate 2. Since the distance between capacitor plates is decreased, the capacitance between terminals A and B will increase. Also, when pressure decreases, capacitance will decrease. Thus, variations in pressure cause changes in capacitance.

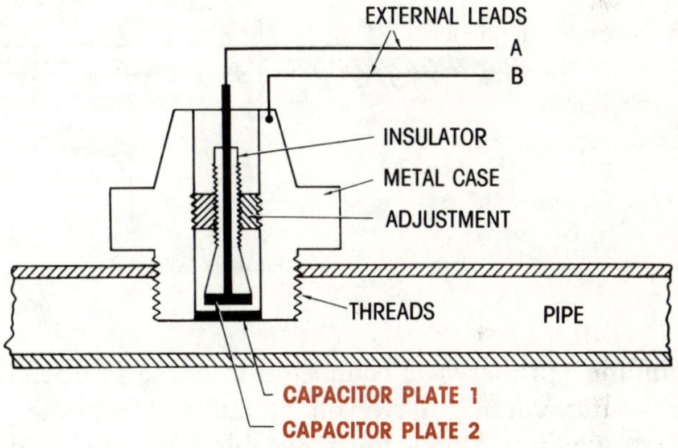

Fig. 6-37. A capacitive transducer.

Inductive Transducers

Inductance can also be varied to cause an electrical change. Usually, inductive transducers such as the one shown in Fig. 6-38A have a stationary coil and a moveable core. The moveable core can be connected to some physical variable whose movement is to be measured. As the core changes position within the stationary coil, the inductance of the coil will vary. The current flow through the coil will vary inversely with the inductive reactance of the coil, since $X_L = 2\pi fL$ and $I = V/X_L$. An ammeter connected in series with the ac line can be calibrated to indicate the position of the core, thus using its relative position to produce an electrical variation.

An often-used type of inductive transducer is the linear variable-differential transformer (LVDT). The principle of operation is illustrated in Fig. 6-38B. A moveable metal core is placed within an enclosure that has three windings wrapped around it. The center

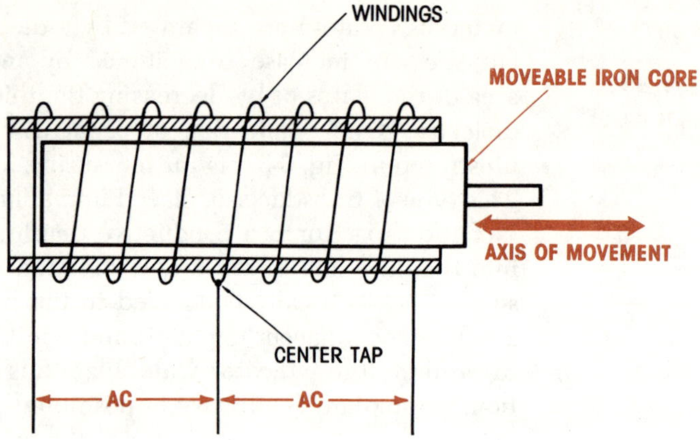

(A) Moveable-core inductive transducer.

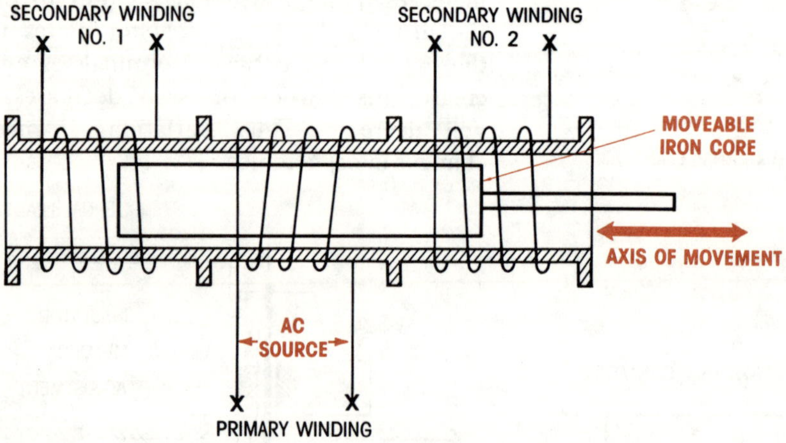

(B) Linear variable-differential transformer (LVDT).

Fig. 6-38. Inductive transducers.

winding (primary) is connected to an ac source. The two outer windings have voltage induced from the primary winding. When the moveable core is placed in the center of the enclosure, the voltages induced in the two outer windings are equal. Any movement of the core in either direction will cause one induced voltage to increase and the other induced voltage to decrease. It is possible to measure the difference in voltage induced into the two outer windings in terms of the amount of movement of the core. The variation in flux linkage due to the movement of the metal core is responsible for the change in induced voltages. Thus, a linear movement (physical quantity) can be converted to an electrical quantity by this type of inductive transducer.

Piezoelectric Transducers

When certain crystalline materials are subjected to a mechanical stress, an electrical potential is developed across the material. Crystals such as quartz and Rochelle salt exhibit this characteristic. An application of the piezoelectric principle is the cartridge/needle assembly of a phonograph (see Fig. 6-39A). The cartridge contains a crystalline material which

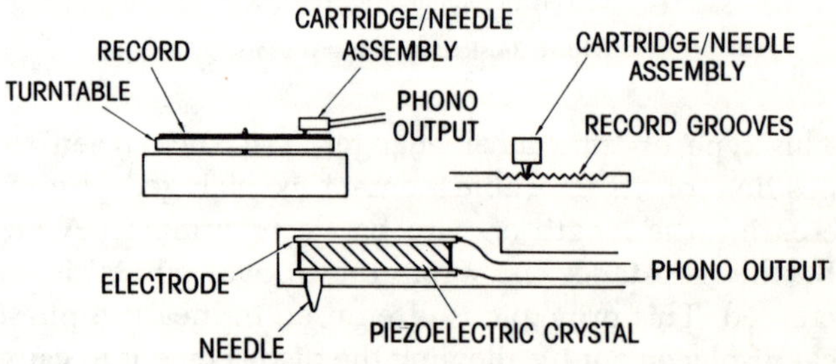

(A) Phonograph cartridge.

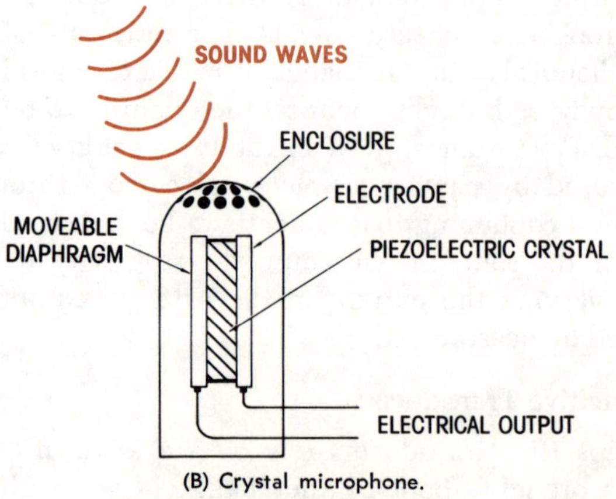

(B) Crystal microphone.

Fig. 6-39. Piezoelectric transducers.

vibrates in accordance with the variations in the grooves of a phonograph record as the needle (which is attached to the cartridge) rides through the record grooves. The crystalline material develops an electrical potential across its structure due to the mechanical vibrations. These small electrical variations are then amplified by the sound system. Thus, mechanical energy (vibrations) is converted to electrical energy by piezoelectric transducers.

It is also possible to convert sound energy to electrical energy with piezoelectric transducers. This is commonly done with crystal microphones in which sound waves cause vibration of a piezoelectric crystal. An electrical potential is then developed across the crystal and is amplified by the sound system.

Thermoelectric Transducers

Transducers that produce a change in electrical characteristics due to a change in temperature are referred to as *thermoelectric* transducers. Some transducers of this type, such as the resistance thermometer and the thermistor, change in resistance when the temperature changes. Others, such as the thermocouple, produce a voltage when heated.

The resistance thermometer is ordinarily made of a fine wire-mesh assembly of nickel, platinum, or copper wire as shown in Fig. 6-40A. The wire is insulated within a thin ceramic or plastic material. If a constant dc potential is applied to the resistive element, a constant current will be indicated on the ammeter. However, if the resistive element is heated, its resistance will increase. The increase in resistance will cause the current reading to decrease. Due to the small diameter wire element used, resistance thermometers respond rapidly to changes in temperature. They can also be used with bridge circuitry for making precise comparative measurements of temperature.

Another thermoresistive device is the thermistor. Thermistors are temperature-sensitive resistors whose resistance increases (typically) with a decrease in temperature. This is referred to as a negative temperature coefficient of resistance. Various metal-oxide-semiconductor materials are used to construct thermistors of various types (see Fig. 6-40B).

Thermistors are manufactured in a wide range of resistance characteristics and temperature coefficients. The use of thermistors as temperature-sensing elements for industrial applications has increased in recent years. They can be used with calibrated meters for direct measurement of temperature. They can also be used with comparative bridge circuitry for making temperature measurements. When a thermistor is used in a comparative bridge circuit, a change in temperature will upset the balance of the bridge. The bridge is then rebalanced, with the difference in resistance for rebalancing calibrated to indicate temperature.

Thermovoltaic transducers are referred to as *thermocouples*. A thermocouple, illustrated in Fig. 6-40C, consists of two dissimilar metals that are fused together at one end. The metals are usually combinations of iron-constantan, copper-constantan, platinum-rhodium or other metals that exhibit the thermocouple principle. When the fused end of the two dissimilar metals is heated, a voltage will be developed at the ends that are not connected. The potential difference exists due to the different coefficients of expansion of the metals. The voltage produced by thermocouples is usually in the millivolt range. Various wire combinations are used to respond to different ranges of temperature. Thermocouples are transducers since they convert heat energy into electrical energy.

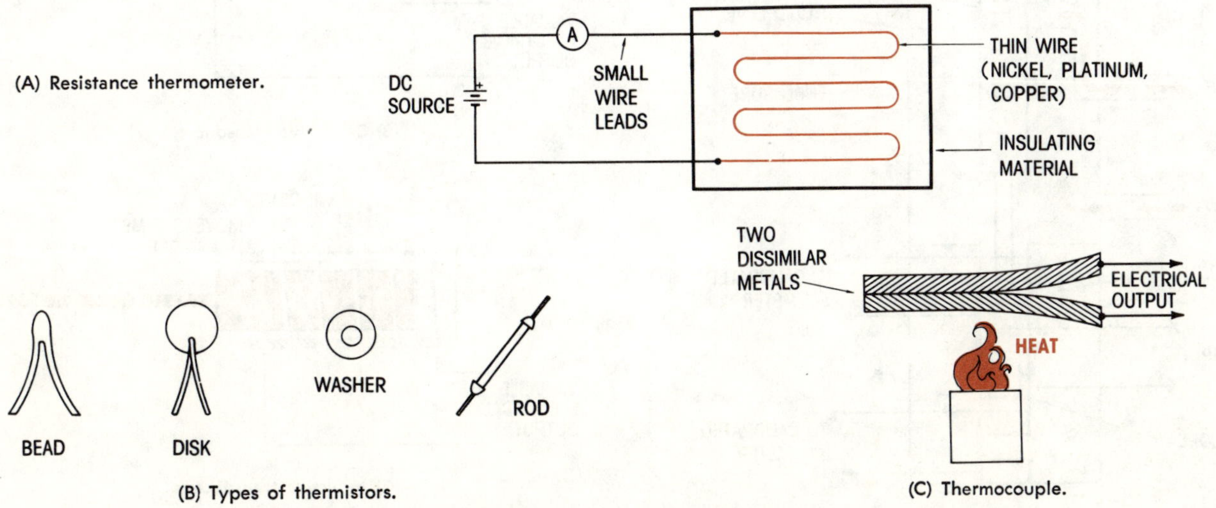

(A) Resistance thermometer.

(B) Types of thermistors.

(C) Thermocouple.

Fig. 6-40. Thermoelectric transducers.

Optoelectric Transducers

An additional type of transducer that has recently received emphasis for industrial applications is the optoelectric or photoelectric transducer. Such transducers are light-sensitive devices that convert changes of light energy into changes of electrical energy. As an example, when a photovoltaic cell is exposed to light, it will develop a voltage output. Other optoelectric transducers convert changes of light into changes of resistance. These are called *photoresistive* devices.

Measurement of Physical Quantities

The measurement of physical quantities usually is dependent upon some type of transducer that converts the physical quantity into an electrical quantity. The electrical quantity is then calibrated to correspond to a meter reading of some type. When the meter scale is calibrated properly, the physical characteristic can be measured directly by the indicator. The transducers previously described have applications with instruments that measure almost any physical quantity. In the following sections, we will discuss the measurement of specific physical quantities that are important for industrial processes.

Measurement of Density and Specific Gravity

The measurement of density is sometimes used in industry to control processes or to test products. Density is the mass per unit volume of a material, and is usually expressed in grams per cubic centimeter, pounds per cubic foot, or pounds per gallon. Water is ordinarily used as a basis of comparison. The ratio of the density of any fluid to the density of water is called its *specific gravity*.

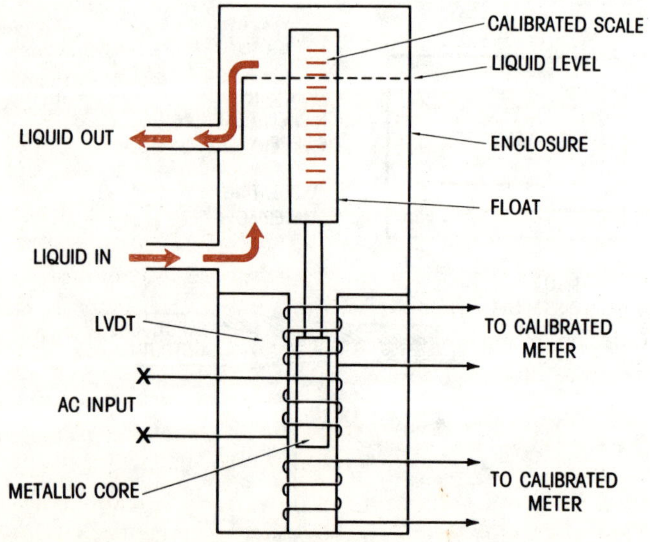

Fig. 6-41. A hydrometer for measuring specific gravity.

Fig. 6-41 shows a method that can be used to measure specific gravity. The instrument illustrated is referred to as a *hydrometer*. Note in Fig. 6-41 that a linear variable-differential transformer (LVDT) is used as the transducer. The float principle is used since the float will displace its own weight of liquid. For many years, this principle has been used for handheld float assemblies. When submerged in a liquid, the float will rise or fall according to the specific gravity of the liquid in which it is submerged. The float positions the LVDT, which produces an electrical signal in proportion to the displacement of its metallic core above or below its center. The output of the LVDT is connected to a calibrated meter, which displays the value of specific gravity of the fluid being sampled. The method shown is only one of several different methods that could be used to measure the specific gravity of a fluid. Other types of transducers could be used to convert the physical quantity of specific gravity into an electrical quantity.

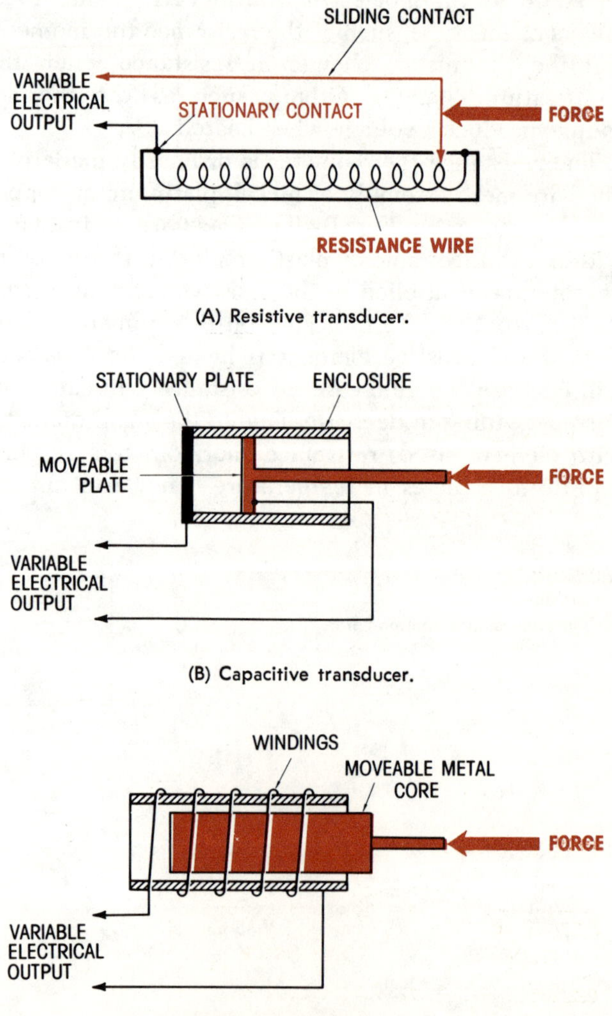

Fig. 6-42. Transducers for measuring displacement.

Measurement of Displacement

Displacement can be measured by means of various types of transducers. Fig. 6-42 illustrates how resistive, capacitive, and inductive transducers can be used with electrical indicators to show displacement. It is possible to measure displacements in the millimeter range or less. Displacement measurements are usually made with reference to a fixed position, or with reference to the force required to move from one position to another.

Measurement of Flow Rate

The measurement of the flow rates of liquids and gases is important for many industrial applications. Many processes rely upon specific flow rates for proper operations. As with most physical quantities, flow rates can be measured by utilizing various types of transducers. A common type of flowmeter is the magnetic flowmeter, illustrated in Fig. 6-43A, which is designed to measure the flow rate of electrically conductive fluids. The meter is placed into the fluid line so that all the fluid flowing in the line passes through it. Since the fluid line contains a conductive fluid, the fluid between the coil terminals produces a magnetic field which induces a voltage into the coil. The faster the rate of fluid flow, the higher in magnitude the voltage becomes. Thus the voltage developed in the coil is proportional to the fluid flow rate. The coil output is connected to a calibrated meter.

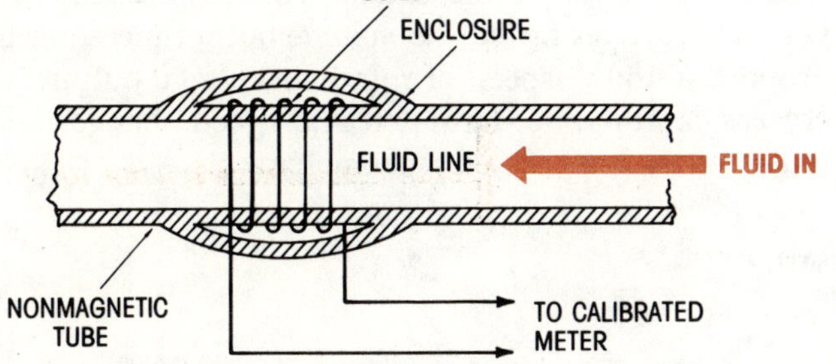

(A) A magnetic flowmeter.

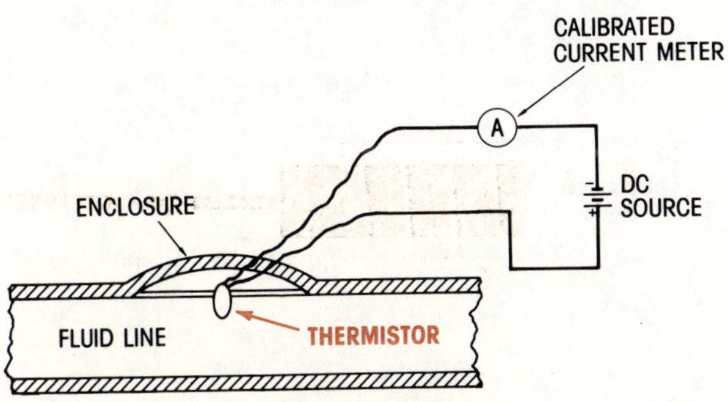

(B) A thermistor flowmeter.

Fig. 6-43. Methods of measuring flow rate.

Another method of measuring flow rate utilizes a thermistor as the transducer. The thermistor, as shown in Fig. 6-43B, is placed within the fluid line so that the fluid passes by it. A dc potential is applied to the thermistor to cause it to be heated. If there is no fluid flow, the thermistor is cooled at a uniform rate by the fluid, causing a constant reading on the current meter connected in series with the thermistor and voltage source. As the flow rate increases, the thermistor will decrease in temperature in proportion to the rate of fluid flow. As the temperature of the thermistor decreases, its resistance increases. The reading on the current meter will then decrease due to the increased resistance. The meter scale is calibrated to measure flow rate. A commercial flowmeter is shown in Fig. 6-44.

Courtesy The Foxboro Co.

Fig. 6-44. A commercial flowmeter.

Measurement of Pressure

The measurement of pressure can be accomplished by several different methods using a variety of transducers. Industrial applications of pressure measurement are very numerous. Pressure is defined as a force per unit area.

A simple way to measure pressure is with a U-tube manometer, illustrated in Fig. 6-45. The pressure created by a fluid forced into the manometer enclosure is used to balance with an unknown pressure. If only air pressure is applied to the manometer, the two columns of liquid will be at the same height above an established reference (points A and B). However, if a fluid is directed into the left column of the instrument, it exerts a certain pressure (P_1). The level of P_2 will then rise until an equilibrium condition is established (points C and D). The difference between points B and D is then a measure of the pressure directed into

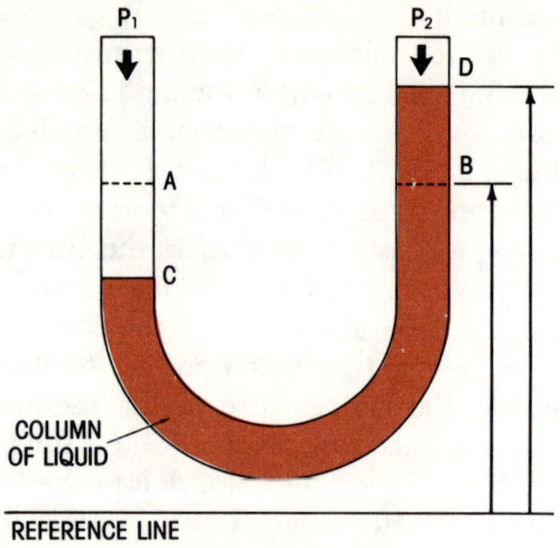

Fig. 6-45. A U-tube manometer.

the left column. If a heavy liquid such as mercury is used, very high pressures can be measured. There are several variations of the basic U-tube manometer for measuring different types and ranges of pressure.

Another method of measuring pressure relies on a capacitive transducer, such as the one illustrated in Fig. 6-37. An increase in pressure of a fluid within an enclosure will cause the moveable plate of the capacitor to come closer to the stationary plate. The capacitance will then increase. The scale of the indicator must be calibrated to measure the change in capacitance in terms of the fluid pressure.

A strain-gauge transducer can also be used to measure pressure, as illustrated in Fig. 6-46A. When increased pressure is exerted on the flexible steel from the pressure source, the strain-gauge wire will change dimensions. The change in resistance due to the change of dimension of the strain gauge is sensed by a comparative bridge network. The output of the bridge supplies current to cause deflection on a meter calibrated to measure changes of pressure exerted on the strain gauge.

Inductive transducers, discussed previously and illustrated in Fig. 6-38, or Bourdon tube spiral spring arrangements, such as the one illustrated in Fig. 6-46B, can also be used to measure pressure. The Bourdon tube is a highly elastic metallic element. Its elastic deformation is proportional to the applied pressure. In the arrangement shown, as the pressure increases, the Bourdon tube spring tends to straighten out, causing movement of a sliding contact connected to a fixed amount of resistance wire. The movement of the sliding contact will affect the resistance between points 1 and 2. These two points are connected to one leg of a comparative bridge. The difference in resistance caused by the increased pressure can then be sensed by the calibrated scale of the indicator. A commercial pressure indicator is shown in Fig. 6-47.

Measurement of Speed

A common type of industrial measurement is the measurement of speed. Usually, speed is measured as a rotary movement due to the nature of industrial equipment that has shafts, gears, pulleys, etc. Several different principles can be used for measuring speed. One method is referred to as a dc tachometer system and is illustrated in Fig. 6-48. This tachometer is connected directly to a rotating machine or piece of equipment. The principle involved in this system is that as the shaft of the small permanent-magnet dc generator rotates faster, the voltage output increases in proportion to the speed of rotation. Voltage output increases can then be translated into speed changes and indicated on a calibrated tachometer scale.

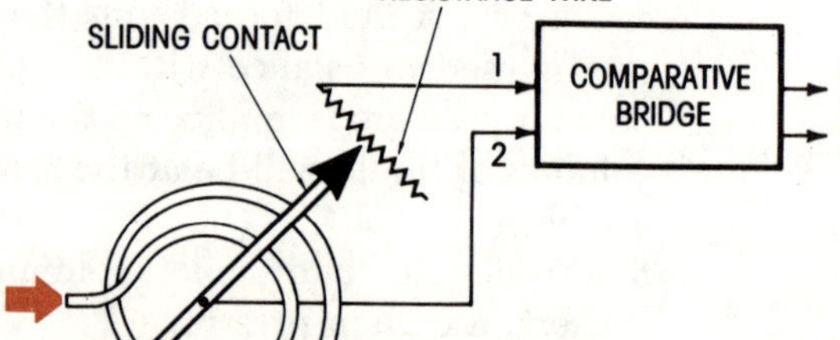

(A) A strain gauge.

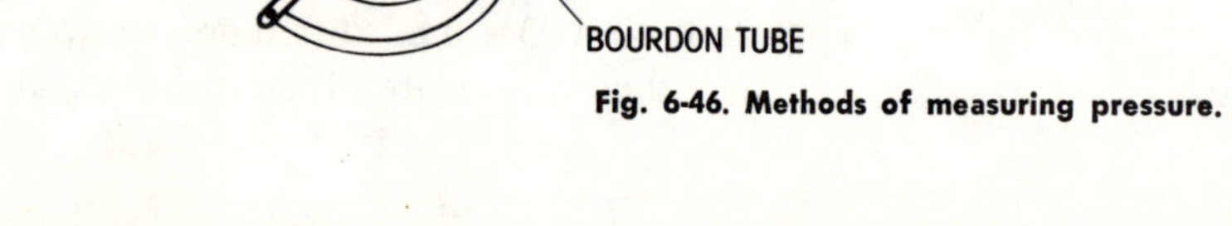

(B) A Bourdon tube.

Fig. 6-46. Methods of measuring pressure.

Courtesy The Foxboro Co.

Fig. 6-47. A commercial pressure indicator.

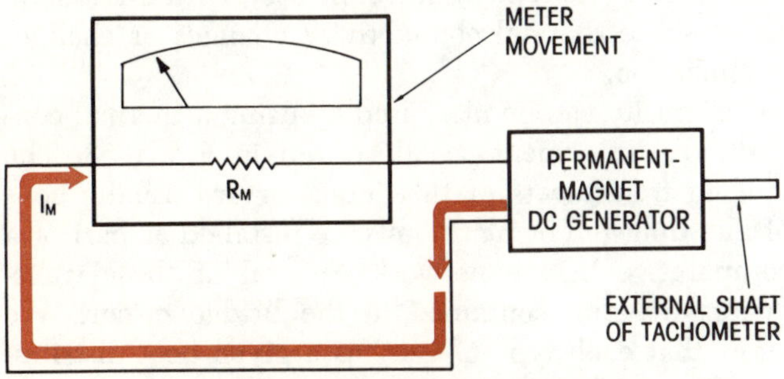

Fig. 6-48. A direct-current tachometer.

Electronic tachometers, such as the photoelectric tachometer shown in Fig. 6-49, are now used extensively due to their increased precision and ease of usage. In the photoelectric tachometer, movement is measured by providing a reflective material on the surface of the equipment or machine subject to measurement. The tachometer has a light source that is interrupted by the passage of the reflective material. A photocell converts the changes of light energy into electrical impulses. The electrical impulses control the movement of an indicator calibrated in revolutions per minute (r/min), or the pulse rate of a digital counter that provides a direct numerical readout of revolutions per minute.

Stroboscopic tachometers are also used in industry. This type of tachometer will create an illusion of motionlessness when the flash rate of its light source is the same as the speed of movement of the object being measured. The stroboscopic tachometer is basically a light source with electronic circuits to turn it on and off at a predetermined rate. A dial calibrated in revolutions per minute is adjusted until the flash rate equals the speed of movement of the measured object. At this time, an illusion of motionlessness of the object exists. The dial position will indicate the speed of movement.

Measurement of Temperature

Various direct and indirect methods are used for measuring temperature. The temperature of solids, liquids, or gases can be measured by direct contact by resistive or thermoelectric transducers. Simple thermometers are commonly used to measure temperature.

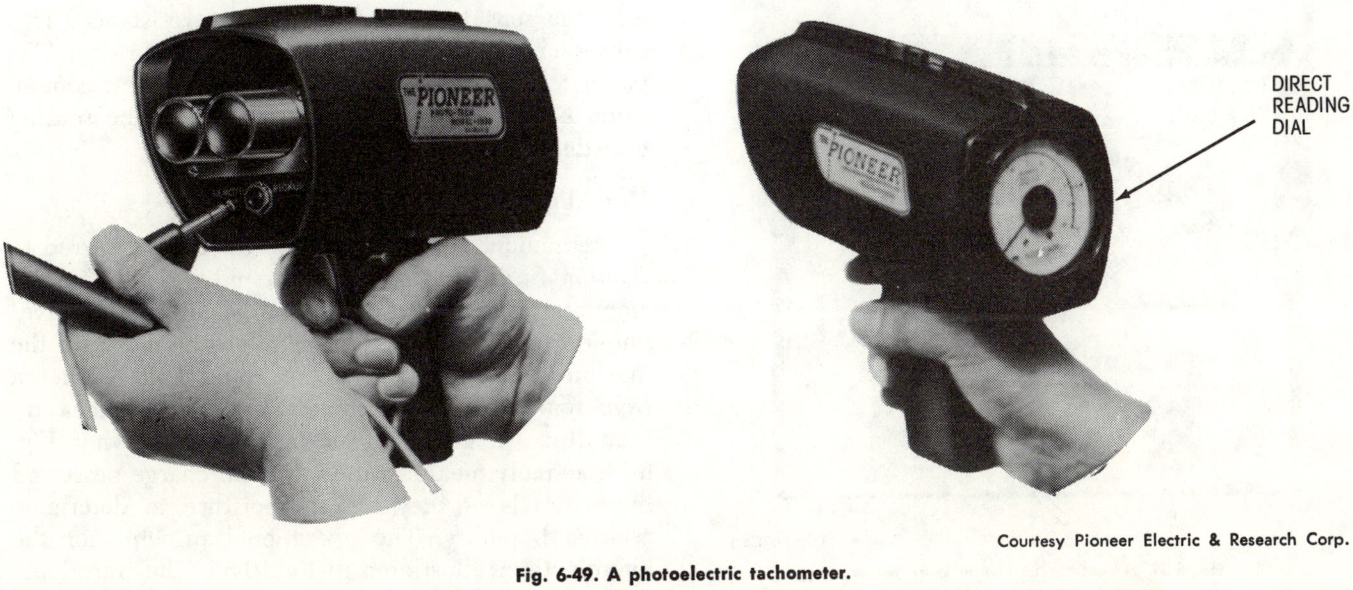

Courtesy Pioneer Electric & Research Corp.

Fig. 6-49. A photoelectric tachometer.

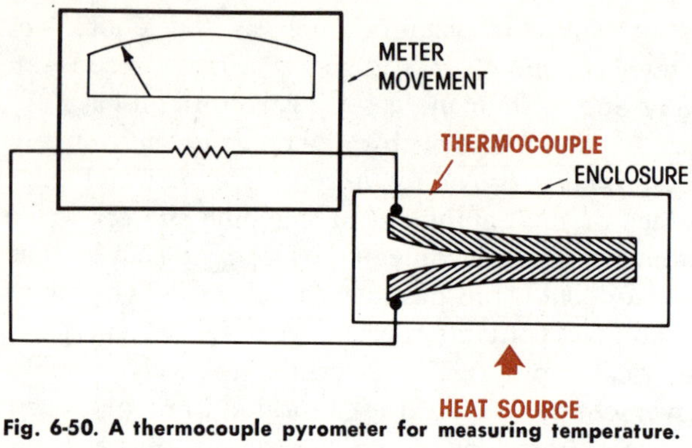

Fig. 6-50. A thermocouple pyrometer for measuring temperature.

In our discussion of temperature measurement, we will stress electrical methods.

Thermoelectric transducers, such as thermocouples and thermistors, are commonly used to measure temperature. The thermocouple pyrometer method of measuring temperature is illustrated in Fig. 6-50. As the thermocouple is heated to a higher temperature, its electrical potential becomes greater. Thus, the moving-coil indicator is deflected further to the right. The meter scale must be calibrated in terms of temperature. A commercial temperature tester is shown in Fig. 6-51.

Thermoresistive systems are also used for temperature measurements. A simple resistance thermometer element was illustrated previously in Fig. 6-40A. Fine insulated wire made of copper, nickel, or platinum is used as the resistance element. Insulating material is wrapped around the wire. Since the metal element has a positive temperature coefficient, its resistance will increase in temperature. The change in resistance is proportional to the change in temperature. The resistance thermometer is commonly used as one leg of a comparative resistance bridge.

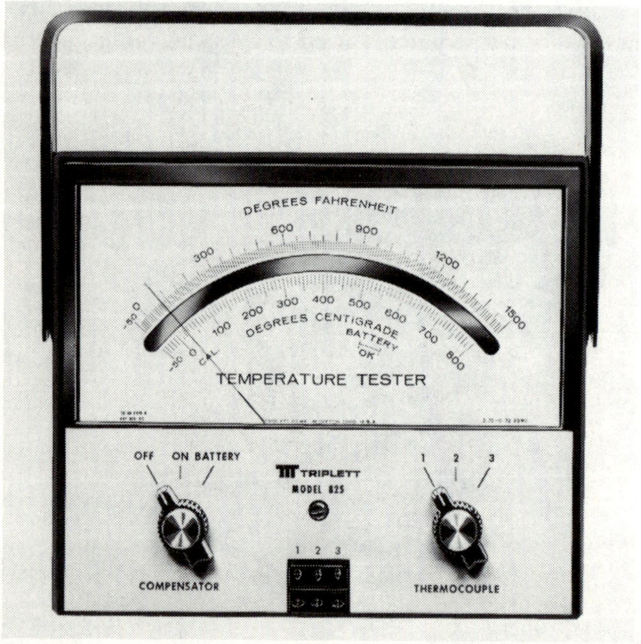

Courtesy Triplett Corp.
Fig. 6-51. A commercial temperature tester.

Thermistor temperature measurements are performed in essentially the same way as with the resistance thermometer. The thermistor, however, has a negative temperature coefficient.

Gas-Analyzing Measurements

Some commercial gas-analyzing instruments are shown in Fig. 6-52. A common method of gas analysis is called *gas chromatography*. This method allows the measurement of each compound that makes up a vapor or gas mixture. This principle of gas analysis involves the passage of a sample of gas in a tube along with an inert gas. The individual gas compounds will travel along a tube at a different rate of speed. An absorbent material, such as silica gel, is placed along the inside of the tube. Each compound absorbs into the silica gel, or similar material, at a different rate. The gas chromatograph can be used to analyze petroleum products and various other mixtures commonly used in industry. The type of detectors used with this system either sense thermal conductivity changes or changes in ionization.

Gas purity can be measured by using a thermal conductivity instrument, as illustrated in Fig. 6-53. The sensing transducers of this indicator are usually coils of fine tungsten or platinum wire installed as part of a comparative bridge network. Several of these transducers may be contained in the bridge circuit. We know that each type of gas has a given heat distribution factor. Thus, a particular gas affects the temperature of the sensing transducers. When a gas of a different composition is introduced into the measuring system, the sensing transducers change resistance. This causes an unbalanced condition in the bridge. The meter scale can be calibrated to show the gas composition based on the resistance change of the sensing transducers.

Humidity Measurements

The amount of water vapor in the air is referred to as *humidity*. The "absolute" humidity is the actual amount of water vapor present in the air. The "relative" humidity is the ratio of the absolute humidity to the maximum water vapor the air could hold. Either a psychrometer or a hygrometer can be used to measure humidity. A hygrometer, such as the one shown in Fig. 6-54, actually measures the electrical charge possessed by materials as they absorb moisture to determine relative humidity. The operational principle of the hygrometer is illustrated in Fig. 6-55. The transducer

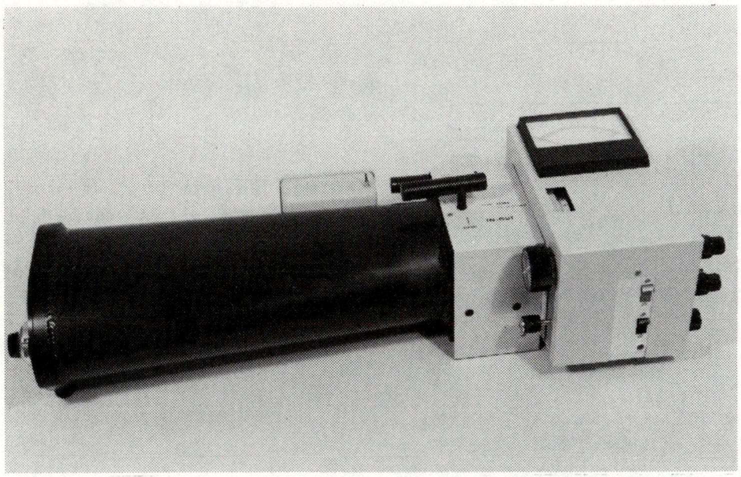

Courtesy Wilks Scientific Corp.

(A) A portable gas analyzer.

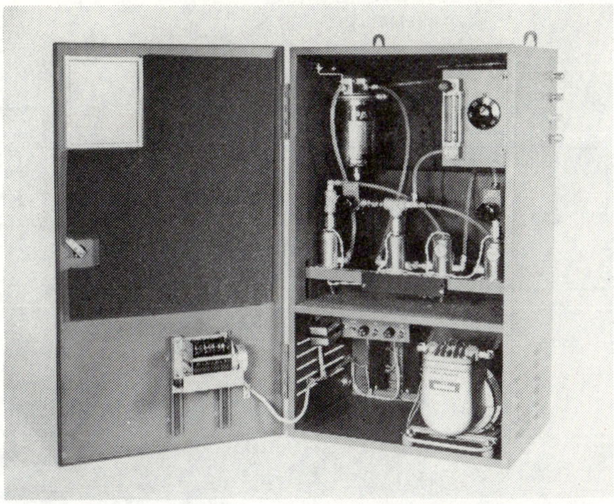

Courtesy Leeds & Northrup Co.

(B) A gas sample conditioner.

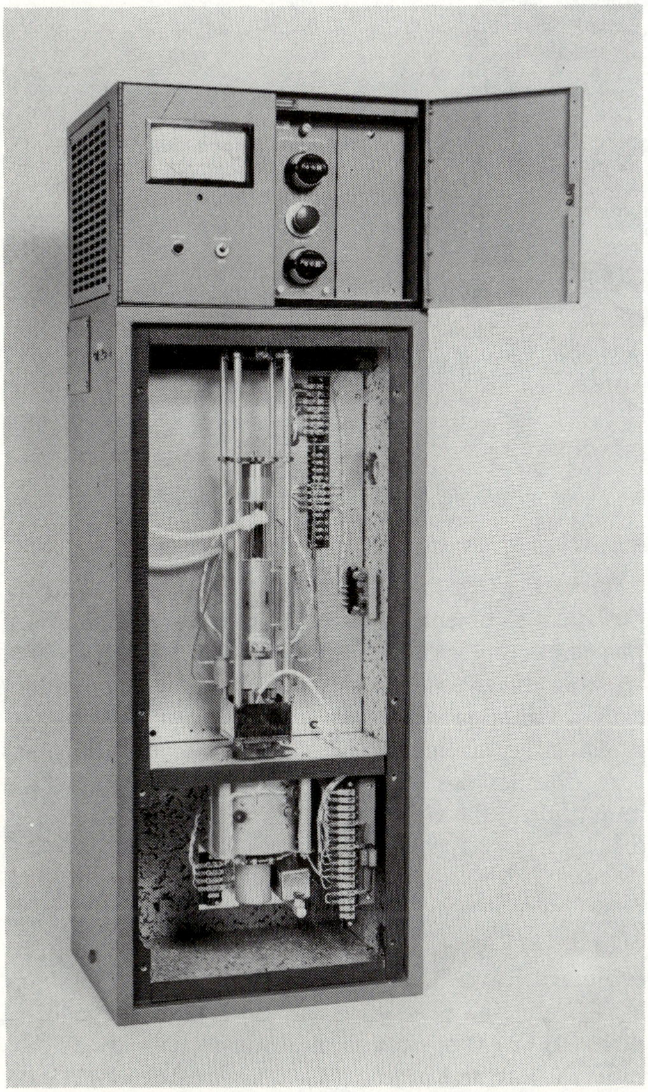

Courtesy Leeds & Northrup Co.

(C) An infrared gas analyzer.

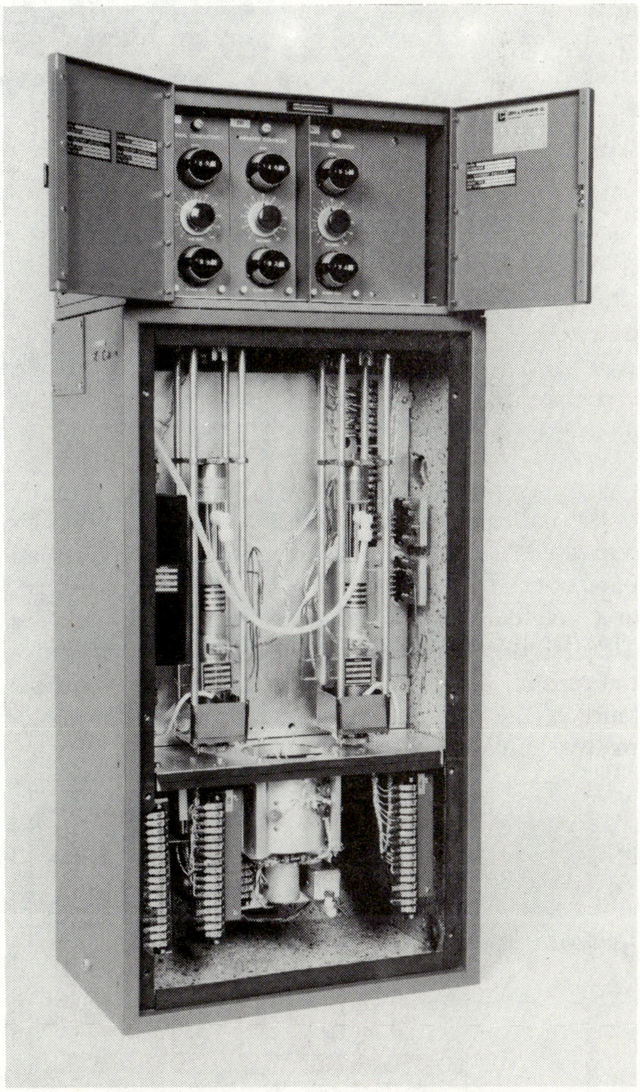

Courtesy Leeds & Northrup Co.

(D) A thermal conductivity gas analyzer.

Fig. 6-52. Commercial gas-analyzing instruments.

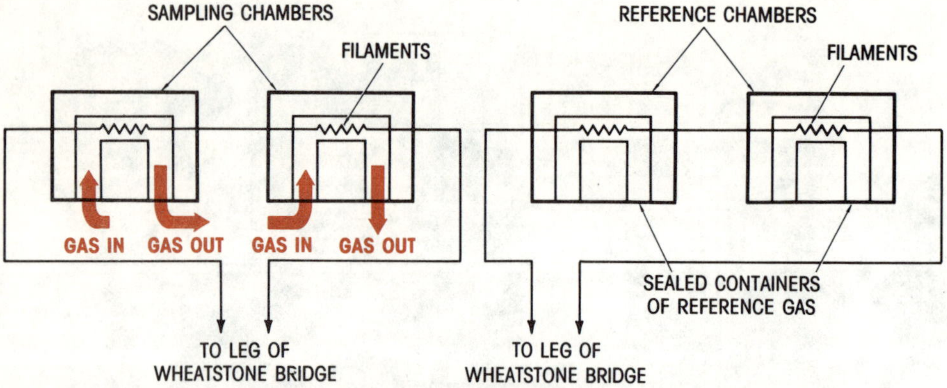

Fig. 6-53. Illustration of gas analysis by measurement of thermal conductivity.

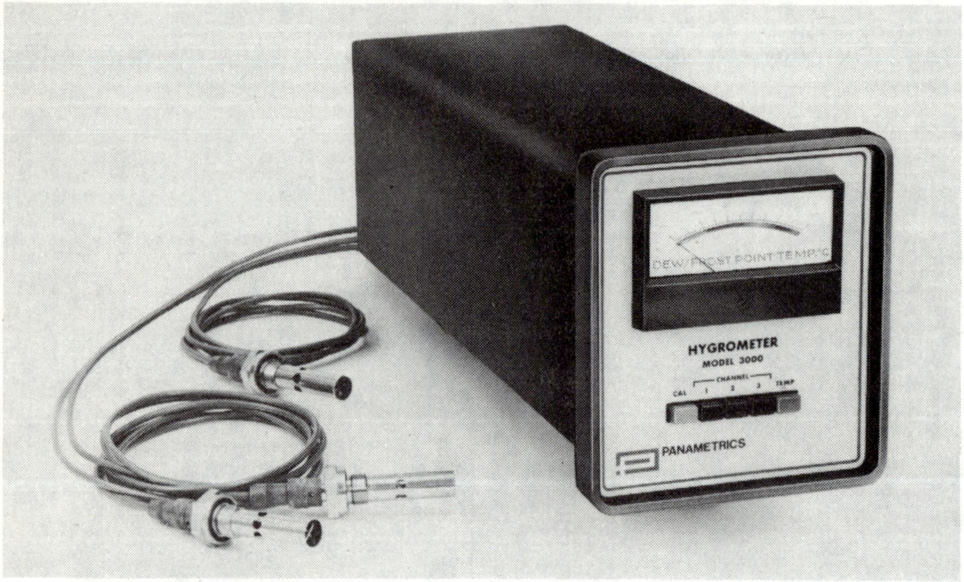

Courtesy Panametrics, Inc.

Fig. 6-54. A commercial hygrometer for measuring relative humidity.

of this indicator is a wire grid assembly. The wires are separated by a specific distance and a voltage is applied to each side of the grid. The grids are constructed of a metallic coated material that will absorb moisture. Thus, as the moisture contained in the air increases, the resistance between the grids decreases. The resistance of the transducer changes in proportion to the relative humidity.

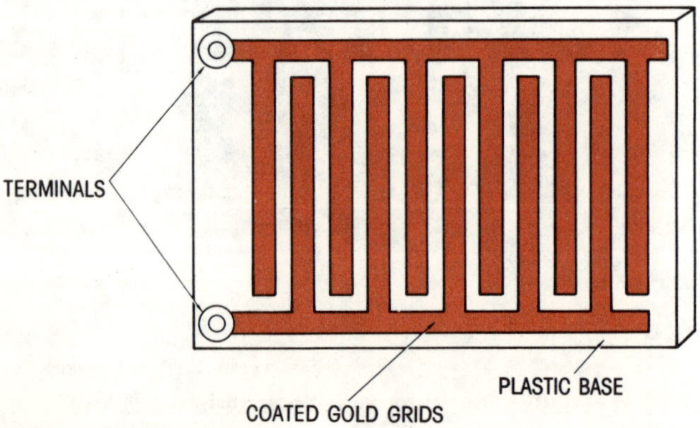

Fig. 6-55. Illustration of the hygrometer principle of operation.

Vibration Measurement

Vibration is a factor that must be controlled in industrial equipment. In order to measure vibration in a piece of equipment, a vibration meter with a piezoelectric transducer can be used (see Fig. 6-56). The reduction of vibration will improve the operating characteristics of the machinery. The transducer used to measure vibration has an output that is proportional to the amplitude of the vibrations.

SUMMARY

In this chapter, we have discussed several aspects of industrial indicating systems. Indicating systems can be generally classified as: (1) hand-deflection indicators, (2) comparative indicators, (3) cathode-ray-tube display indicators, (4) numerical-readout indicators, and (5) chart-recording indicators. Each of these types of indicators are used extensively in industry.

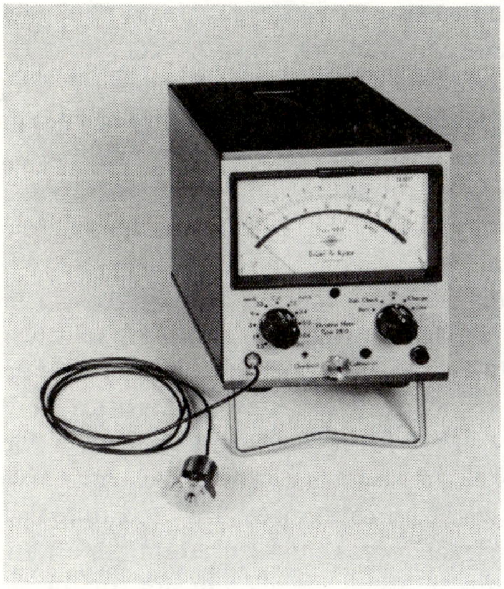

(A) A commercial vibration meter.

(B) Vibration meter with transducer attached to electric motor.

(C) Measuring vibration in a piece of industrial equipment.

Courtesy B&K Instruments, Inc.

Fig. 6-56. Vibration measurement.

Hand-deflection indicators, such as single-function voltage and current meters and vom's, are commonly used measuring devices for electrical quantities. It is also possible to monitor physical quantities, such as fluid-flow rate or pressure, with hand-deflection meters calibrated for such measurement. The majority of hand-deflection meters use the D'Arsonval, or moving-coil, meter movement. This movement can be modified by external circuitry to measure many electrical and physical quantities.

Direct current can be measured by using a moving-coil meter movement in combination with a resistance in parallel with it. This parallel resistance, which makes possible the measurement of higher ranges of current, is known as a *shunt*.

Direct-current voltage can be measured by using a meter movement with resistance of a specific value placed in series with it. This resistance is referred to as a *multiplier*. An important characteristic of any meter is its sensitivity. Sensitivity is determined by the ohms-per-volt rating of the meter movement.

Resistance can also be measured by using a moving-coil meter movement. In order to make resistance measurements, an internal power source must be used. The resistance scale used on a meter must be calibrated to correspond to the amount of deflection of the moving-coil meter movement.

Alternating current and voltage can also be measured with a moving-coil meter movement. Since the meter movement is a direct-current instrument, rectifiers must be used to convert the ac to dc. In general, ac meters of this type are less sensitive than their dc counterparts.

Electrical power is a product of the voltage and current of a circuit. It is possible to measure electrical power by using a dynamometer meter movement.

Comparative indicators, such as Wheatstone bridges and potentiometric devices, measure unknown quantities based on some standard or known value. Ordinarily, these methods are very precise.

The oscilloscope is a commonly used laboratory instrument. It is generally classified as a crt display indicator. Another type of indicator is the numerical-readout instrument. Numerical displays provide ease of reading and eliminate the need to interpret a meter scale. Another type of indicator is the chart-recording type. This type provides a permanent record of some quantity that is monitored either periodically or continuously.

Most industries monitor quantities that are power-related. Since industries use large quantities of power, they must control and monitor its use throughout the plant. Power-related measurements are made by using watt-hour meters, three-phase analyzers, power-factor meters, power-demand meters, frequency meters, and other specialized equipment such as synchroscopes, ground-fault indicators, clamp-type meters, and telemetering systems.

Industrial processes are monitored by indicating instruments that measure physical quantities as well as electrical quantities. Ordinarily, physical quantities are converted into electrical quantities for measurement. Transducers are used to accomplish this conversion. The basic types of transducers are resistive, capacitive, inductive, piezoelectric, thermoelectric, and optoelectric. These basic transducers can be used for almost any type of measurement. Some specific physical quantities that are measured by utilizing various types of transducers are density, specific gravity, displacement, flow rate, pressure, speed, temperature, gas purity or composition, humidity, and vibration.

STUDY QUESTIONS

1. *What are five general classifications of indicators?*
2. *What are some types of hand-deflection indicators?*
3. *How do hand-deflection indicators differ from other kinds of indicators?*
4. *What is the difference between a single-function meter and a multifunction meter?*
5. *Describe the D'Arsonval or moving-coil type of meter movement.*
6. *How can the range of a meter movement be extended to measure higher values of direct current?*
7. *What is the formula used to determine the value of shunt resistance required to extend the current capacity of a meter movement?*
8. *Describe the method used for determining the value of multiplier resistor needed to extend the range of a meter movement.*
9. *What is meant by the sensitivity of a meter?*
10. *How can resistance be measured using a moving-coil meter movement?*
11. *How is a resistance scale calibrated?*
12. *How can ac current and voltage be measured using a moving-coil meter movement?*
13. *How is electrical power measured? What is meant by true power? Apparent power? Reactive power?*
14. *What is a comparative indicator? What are some types?*
15. *How can an unknown resistance be measured using a Wheatstone bridge?*
16. *Discuss the basic operation of an oscilloscope.*
17. *What are some advantages of numerical-readout indicators?*
18. *How do chart-recording indicators differ from other types of indicators?*
19. *Discuss the operation of an X-Y recorder.*
20. *How are watt-hour meters used to measure electrical energy?*
21. *What are two methods for measuring three-phase power using single-phase wattmeters?*
22. *How is the power factor of a system measured?*
23. *How is the power-demand factor of an industry determined?*
24. *What are some methods used to measure frequency?*
25. *What is the purpose of using a synchroscope?*
26. *Why are ground-fault indicators commonly used in industry?*
27. *What are some applications of megohmmeters?*
28. *What is the major advantage of using a clamp-type current meter?*
29. *What are some applications of telemetering systems?*
30. *Discuss the following types of transducers:*
    a. *Resistive.*
    b. *Capacitive.*
    c. *Inductive.*
    d. *Piezoelectric.*
    e. *Thermoelectric.*
    f. *Optoelectric.*
31. *How are the following measurements accomplished?*
    a. *Density and specific gravity.*
    b. *Displacement.*
    c. *Flow rate.*
    d. *Pressure.*
    e. *Speed.*
    f. *Temperature.*

g. Gas purity.
h. Humidity.
i. Vibration.

PROBLEMS

1. Using a 1-mA, 20-ohm meter movement, calculate the shunt resistance required to extend its range to measure the following dc currents: (a) 5 mA, (b) 10 mA, (c) 100 mA, (d) 500 mA, and (e) 1 A.
2. Using a 500-μA, 500-ohm meter movement, calculate the multiplier resistance needed to extend its range to measure the following voltages: (a) 1 volt, (b) 10 volts, (c) 25 volts, (d) 50 volts, and (e) 100 volts.
3. What is the sensitivity of the meter movement of Problem 1? Problem 2?
4. A series ohmmeter is being designed using a 1-mA, 200-ohm meter movement. A 1.5-volt dry cell is used for the $\times 10$ range and the "ohms adjust" is a 100-ohm potentiometer. What value of series resistance is needed for the $\times 10$ range?
5. The voltage of a standard cell is measured with a potentiometer as 1.0189 volts. A 500,000-ohm resistor is then connected across the standard cell terminals, causing the reading to drop to 1.01879 volts. What is the internal resistance of the standard cell?
6. The four resistances of a wheatstone bridge are 500, 5000, 2000 and 200 ohms. An indicator having a 100-ohm resistance is placed from the junction of the 500- and 200-ohm resistors to the junction of the 5000- and 2000-ohm resistors. A 1.5-volt power source is connected across the other resistor junctions to complete the bridge. What is the current flow through the indicator?
7. Two wattmeters are connected to measure the power converted by a three-phase, wye-connected load. The readings of the meters are 5280 watts and 6590 watts. What is the total three-phase power of the system?
8. Three wattmeters are used to measure the power of a delta-connected system. The wattmeter readings are 610, 580, and 525 watts. What is the total three-phase power of the system?
9. The meters connected in a circuit have the following readings: voltmeter, 240 volts; ammeter, 10 amperes; wattmeter, 2100 watts. What reading would a power-factor meter have when connected to this circuit?

7

Mechanical Power Systems

Nearly all of the products manufactured by industry today are the end result of the application of some mechanical power system. Systems of this type are designed to transfer power from one point to another through a mechanical motion that can ultimately be used to do work. Punch presses that move up and down in a reciprocating motion, rotating machinery, and a number of linear changes can all be produced through the output of a mechanical power system. The value of this type of system to modern industry is almost beyond calculation.

MECHANICAL POWER SYSTEM BASICS

A mechanical power system must have a *primary energy source,* a *transmission path, control,* a *load* device, and possibly one or more *indicators* in order to function properly. These parts, you may recall, are the basic elements of all power systems. In order to distinguish this type of system from others, further classification is required. In this regard, a mechanical power system is defined as a system that produces some unique form of mechanical motion. As in other systems, the load is primarily responsible for producing this action.

The energy source of a mechanical power system also requires some degree of classification in order to distinguish it from other systems. Energy is present in three basic forms: heat, light, and mechanical motion. It is also available in the form of electricity, chemical action, and nuclear energy. These six energy forms are all related to some extent. This relationship is based on the fact that energy of one form can be readily changed or transferred into one of the other existing forms. The energy source of a mechanical power system operates by transferring energy from its basic form into something more useful. In this regard, either electricity or mechanical motion is used as the primary energy source of the system.

The transmission path of a mechanical power system is unique when compared with other systems. Electrical conductors, belts, rotating shafts, hollow pipes or tubes, and cables are often used to transfer power from the energy source to the remainder of the system. Electrons, moving in the form of an electric current, flow through conductors providing one distinct type of transmission path. In addition to this, hydraulic fluid or air can be used to pass mechanical energy through a system. Rotating machines are also used to develop mechanical energy and transfer it to other parts of the system by belts, chain drives, and shafts. The primary responsibility of the transmission path is to pass mechanical energy from the source to the load device.

The control function of a mechanical power system is designed to alter the flow of power through the system. Full and partial control are essential functions commonly associated with all mechanical power systems. Control of this type is represented by changes in hydraulic fluid and air flow, pressure, direction, force, and speed. Devices such as pressure regulators, valves, gears, pulleys, couplings, and clutches are commonly used throughout the system. The mechanical energy of an electrical system is also controlled through partial and full control devices after it has been developed by the source. Clutches, gears, brakes, and coupling elements are often used to control such things as force, speed, and direction.

The load of a mechanical power system is designed to produce some unique type of motion that can be harnessed to do useful work. Three distinct kinds of motion are important in industrial applications. These include rotary, linear, and reciprocating motion. As a general rule, these motions can be produced by either electricity or fluid power according to system needs. Industrial loads are primarily designed for continuous operation for long periods of time. Motors, relays, solenoids, actuators, and cylinders are typical devices used to achieve this system function.

In a mechanical power system, indicators are called upon to measure a number of physical quantities. This includes such things as pressure, flow rate, speed, direction, distance, force, torque, and electrical quantities. Many of these quantities must be monitored periodically to assure proper system operating efficiency. In addition to this, a number of indicators are used to test system conditions during maintenance procedures. As a general rule, mechanical power system indicators are not used as frequently as those of other systems. There are several unique indicators in operation today that have been designed particularly to measure the physical changes that take place in a mechanical power system.

MECHANICAL POWER SYSTEM TYPES

Applications of mechanical power systems are so numerous that it is difficult to place them into divisions of study without some degree of duplication. A large part of all mechanical power system applications used in industry today are achieved by electricity. There are, however, a few unique electrical devices that are used primarily to produce mechanical motion. Relays, contactors, solenoids, servomechanisms, and actuators fall into this category. Devices of this type are often classified as having both mechanical and electrical energy producing capabilities. We will concentrate on the mechanical aspects of these devices in this chapter.

Fluid power systems also represent a very important part of mechanical power system applications in industry today. The term *fluid*, in this case, refers to both hydraulic and pneumatic system applications. These systems are very similar in many respects. There are, however, many unique differences and applications of both system types found in industry. Fluid basics will be discussed pointing out some of the common principles that apply to both system types. Specific hydraulic and pneumatic system applications will then be investigated as they apply to unique industrial applications.

In the past, hydraulic and pneumatic systems were rarely ever included in a study of industrial electronics. Mechanical power system applications have primarily been limited to only those things produced by electricity. A number of sophisticated industrial systems have recently been developed that produce mechanical energy through a combination of both fluid and electrical system applications. Hybrid systems of this type are playing a very important role in the total industrial electronics field. A person working in this field must, therefore, be familiar with both fluid and electrical basics in order to truly understand the operation of a hybrid mechanical power system.

Fluid Power Systems

Fluid power systems represent that part of industry responsible for developing a useful form of mechanical power through the transmission of a liquid or gas in a pipe. Systems of this type are particularly useful in transferring power to unaccessible locations over moderate distances. A very fine degree of control can be achieved that has a wide variation of speed capabilities and a reversing characteristic. In many system applications, fluid power represents the only practical solution to mechanical power transmission.

Fluid power systems have a number of unique characteristics that distinguish them from other power systems. A very delicate force of a few ounces, for example, has the capability of controlling a larger force of 36,000 tons or more. Through the use of tape-controlled machines, fluid systems can provide changes that will move within tolerances of plus or minus one ten-thousandth of an inch. Fluid power systems can also provide rotary motion at speeds of nearly 100 miles per hour or can develop creeping speeds that only move a fraction of an inch per minute. Systems of this type are very compatible with other power systems. Electrical, electronic, and mechanical control of fluid power is very common today. Fluid systems are efficient, dependable, easy to maintain, and economical to operate for long periods of time. Approximately 90% of all the machine tools used in industry are controlled or operated by fluid power.

The Fluid Power Principle

Nearly everyone working in industry has had an opportunity to see a fluid power system in operation at one time or another. A classic example of this principle is commonly found in automobile service stations that have hoists to lift cars for servicing. A system of this type uses both air and oil to develop the power needed to lift an automobile.

An automobile hoist operates on the principle that air added to the top of a long cylinder confined in an oil-filled tube can be forced to move upward under pressure. The tube and cylinder are normally placed in the floor of the station in such a way that the entire unit will retract when air pressure is removed. The cylinder will rise out of the floor when air is forced into it through the action of a control valve. The air source is developed by an electric-motor-driven compressor. Full control of the hoist is achieved by manual manipulation of the air valve by the operator. Fig. 7-1 shows a simple fluid power system automobile hoist.

Pascal's Law

In 1653, a French scientist named Blaise Pascal discovered a simple physical law that states, "A pressure applied to a confined fluid is transmitted undiminished throughout the fluid. It acts on all surfaces in a direction at right angles to those surfaces." This statement eventually became known as *Pascal's law.* Air and hydraulic fluid both respond to the basic principles set forth by this law. Pascal's law, therefore, serves as the basic premise of all fluid power systems.

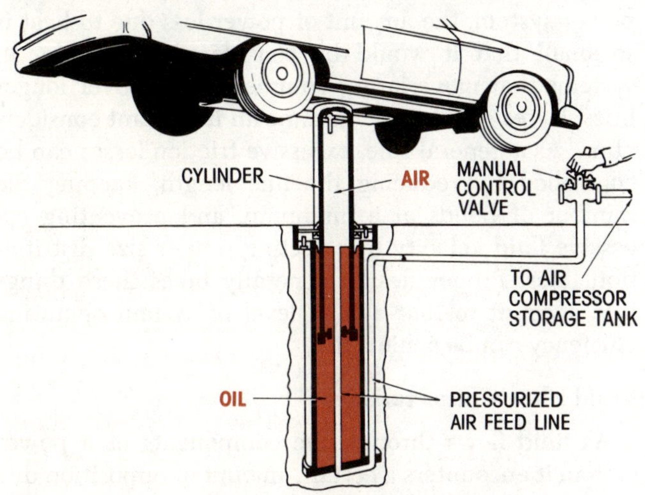

Fig. 7-1. A fluid power system automobile hoist.

Fig. 7-2 shows a graphic example of Pascal's law. The force applied to the fluid inside of the cylinder at piston A is instantly transferred to all parts of the cylinder. Piston B moves an amount that is equal to the force originating at A. This action occurs when both pistons

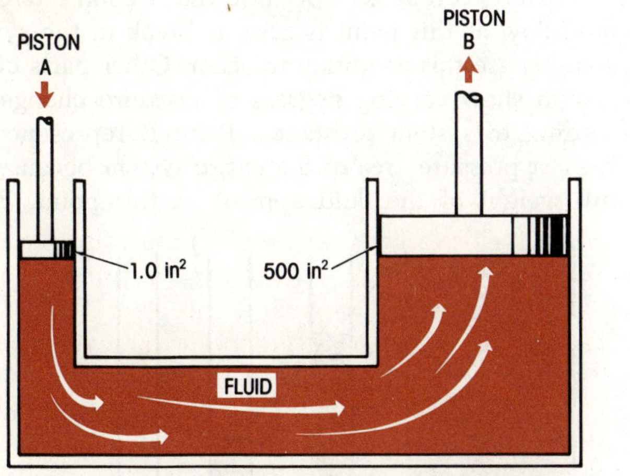

Fig. 7-2. A graphic illustration of Pascal's law.

are of the same physical size. The fluid of this system easily conforms to the inside shape of the cylinder. The same force acting on piston B is also applied to the inside walls of the cylinder. The strength of the cylinder walls must also be capable of withstanding this amount of pressure.

Force, Pressure, Work, and Power

The *force* applied to the piston of the system just described is defined as any cause which tends to produce or modify motion. To move a body or mass, an outside force must be applied to it. The amount of force needed to produce motion is primarily based on the inertia of the body. Force is normally expressed in units of weight. Weight is defined as the gravitational force exerted on a body (or mass) by the earth. Since the weight of a body is a force (not a mass), we must use units of force to express both weight and force. The basic unit of force in the present U.S. system is the pound (lb). In the international or metric system, the basic unit of force is the newton (N).

Pressure is a term used to describe the amount of force applied to a specific unit area, and is expressed in pounds per square inch (lb/in^2) in the U.S. system, or newtons per square meter (N/m^2) in the metric system. The name *pascal* (Pa) has been assigned as the basic unit of pressure in the metric system ($1 \text{ Pa} = 1 \text{ N/m}^2$). At sea level, the pressure of the atmosphere on the surface of the earth is 14.7 lb/in^2. In industrial applications, giant hydraulic presses are capable of squeezing metals with a pressure of 100,000,000 lb/in^2. Mathematically, pressure is expressed as:

$$P = \frac{F}{A}$$

where,

P is the pressure in pounds per square inch or pascals,
F is the force in pounds or newtons,
A is the area in square inches or square meters.

An interesting and important thing about force and pressure is that they only represent a measure of *effort.* A measure of what the system actually accomplishes is called *work.* In the fluid system just described, work is accomplished when the force applied to piston A causes it to move a certain distance. Work is commonly expressed in foot-pounds or newton-meters (joules). The mathematical formula for this relationship is:

$$W = F \times D$$

where,

W is work in foot-pounds or newton-meters (joules),
F is force in pounds or newtons,
D is distance in feet or meters.

A more realistic concept of work must take into account the length of time that it is being accomplished. The term *power* is commonly used to express this relationship. The term *horsepower* is commonly used in industry to express mechanical power. Moving 33,000 pounds a distance of one foot in one miunte, or 550 pounds a distance of one foot in one second is a physical expression of horsepower. Electric motors are rated in horsepower or fractional values of it.

A Simple Fluid Power System

A very simple static fluid power system is illustrated in Fig. 7-3. In this system, a 100-lb force is applied to piston A, which has an area of 1.0 in². A pressure of 100 lb/in² is developed by this action and is transferred through the fluid to piston B. The area of piston B is shown to be 100 in². Since the transfer of pressure through the fluid is equal to all parts of the cylinder, each square inch of piston B will receive 100 lb of force. As a result, 100 lb/in² × 100 in² equals 10,000 lb of force applied to piston B.

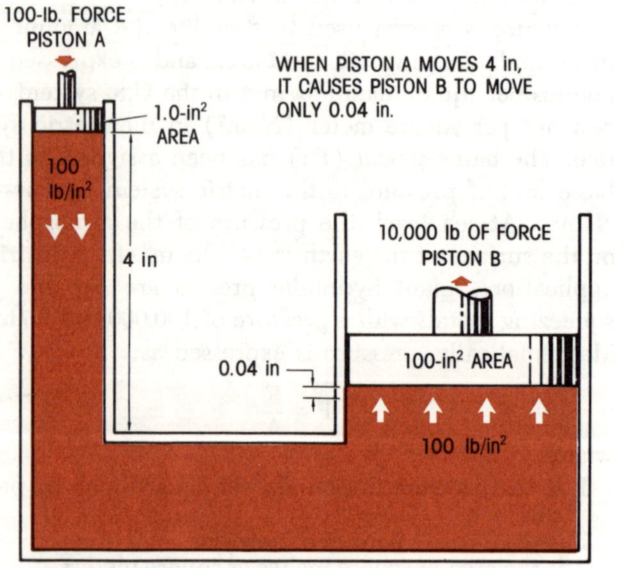

Fig. 7-3. Illustration of a simple static fluid power system.

The distance that piston B moves is directly proportional to the area ratio of the two pistons. Moving the one-inch piston four inches into the cylinder forces four cubic inches of fluid to be displaced. This displaced volume of fluid is based on the area of the piston times the distance it moves into the cylinder. One square inch times four inches, therefore, equals four cubic inches of fluid displacement. Spread over the 100 in² surface of piston B, this displacement causes piston B to move only 1/100 of the distance traveled by piston A. In this case, 1/100 of four inches only equals 0.04 inch of motion. Piston B, therefore, has more force applied because of its size, but travels only a small distance. The amount of work done by pistons of a static fluid power system shows an unusual relationship. The work done by piston A is 100 pounds times four inches or 400 inch-pounds. At piston B, the amount of work achieved is also 400 inch-pounds. This is determined by multiplying the applied piston force by the distance moved. Therefore, 10,000 pounds times 0.04 inch equals 400 inch-pounds of work.

In actual practice, fluid power systems do not show a 100% transfer of power from input to output. Fluid moving along cylinder walls, for example, encounters a form of surface friction. The amount of power loss developed by this friction appears primarily as heat developed on the walls of the cylinder. In a static power system, the amount of power loss due to heat is so small that it would be considered negligible. In systems that move larger volumes of fluid over longer lines, losses of this type become an important consideration. As a general rule, excessive friction losses can be controlled by reducing the line length, keeping the number of bends at a minimum, and preventing excessive fluid velocity by selecting proper size distribution lines. Proper design generally takes these things into account so that a high level of system operating efficiency can be achieved.

Fluid Flow Characteristics

As fluid flows through the components of a power system it encounters a certain amount of opposition due to friction. In a fluid system, this opposition is generally called *resistance*. System pressure is developed as a result of fluid being forced against the resistance of surface areas of system components. A direct relationship, therefore, exists between system pressure and component resistance.

Fig. 7-4 illustrates the friction/pressure relationship of a static fluid power system. The pressure at point F is referenced at zero because the friction offered to fluid flow at this point is zero. A break in the system could cause this condition to occur. Other parts of the system show varying degrees of pressure change according to system resistance. Point B represents the highest pressure area of the entire system because the full weight of the fluid appears at this point. Fluid

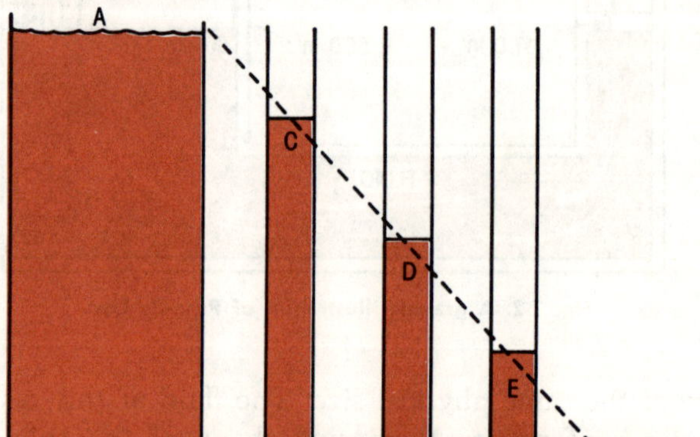

Fig. 7-4. Illustration of the friction/pressure relationship of a static fluid power system.

flowing from point B to point F must change all of its potential energy into heat energy. The moving fluid in this case also causes a drop in pressure as it passes from points B to F. This pressure drop increases at each location point beyond B. At the same time, the source pressure decreases at each point an equal amount. Pressure drops of this type are undesirable when power is being transferred through a system. In some applications, a drop in pressure is often used to trigger the starting of a second operation in a sequential system.

Pressure drop is also a very important characteristic in a flowing fluid power system. This condition is much more significant than that of the static system just discussed, because nearly all industrial systems respond to flowing air or fluid. Pressure drops of this type can be caused by flow turbulence created where abrupt changes in direction occur. Refer to the corners of the system in Fig. 7-5. System restrictions encountered by fluid flow are also a source of pressure drop. Control valves and tubing sizes are largely responsible for this type of drop. Small lines tend to increase the speed of fluid flow, which in turn causes an increase in surface friction. Notice the pressure drops near each of the restricted areas of Fig. 7-5. The length of a system line is another important pressure-drop factor that must be considered. In this case, fluid sees more resistance when it travels long distances. Proper system design usually minimizes this factor. A very interesting thing to note about pressure drop takes place when the flow ceases. Pressure drop, in this case, stops and the pressure reaches a stable value throughout the system. A faulty pump or loss of electricity could cause this condition to occur. A break in the system line, by comparison, normally causes a complete loss of pressure or a very pronounced change in pressure value. These two pressure conditions are very evident when compared with the characteristics of a normally operating system.

Compression of Fluids

Investigations of the fluid power system to this point have been of a rather general nature and could apply to both hydraulic and pneumatic systems. Compression of system fluid, however, represents a major difference between the two systems. In general, we can say that all gases and liquids are compressible under certain conditions. In a hydraulic system, oil is used to transfer power throughout the system. Ordinarily, a hydraulic fluid is not compressible except in extremely long transmission lines or under rather high pressures. Under normal operating conditions, compression of a hydraulic fluid is not considered to be a real problem. A volume reduction of approximately 0.5% for every 1000 lb/in² of pressure is typical. In most industrial applications, this amount of hydraulic fluid compression is not considered significant.

Fluid power systems of the pneumatic type are primarily designed to respond to fundamental laws that apply specifically to gaseous fluids. Industrial applications of this type nearly always use air as the operating medium. Air must first be compressed by the system source before it can be effectively used to transmit power. When air is reduced in volume by the compression process, its pressure increases. Compressed air is produced, stored in a tank, and then released into the system when it is needed. Pneumatic systems normally

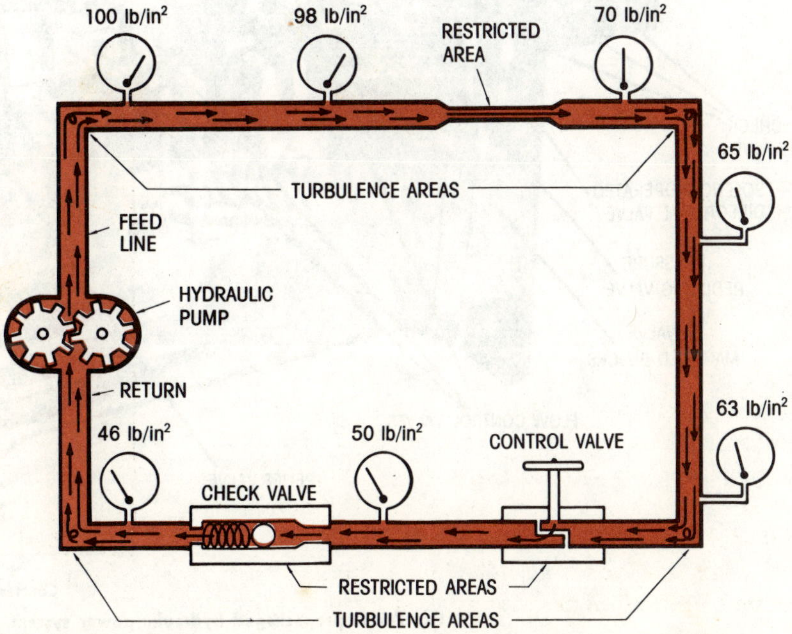

Fig. 7-5. Illustration of pressure drops around a flowing fluid power system.

release air into the atmosphere after it has been used by the system.

Fluid compression represents a major difference between hydraulic and pneumatic power systems. Hydraulic fluid is not compressed under normal operating conditions to any great extent. Pneumatic systems, by comparison, respond only to compressed gases, and primarily to air. Because of this basic difference, there are some unique physical features in system components. As a general rule, however, the basic function of each component is similar for both hydraulic and pneumatic systems.

Industrial Hydraulic Systems

Most of the hydraulic power systems used in industry today are of the circulating fluid type. This type of system employs a hydraulic pump as a power source to move the fluid. The pump is driven by mechanical energy that has been produced electrically. Hydraulic fluid under pressure is made to circulate through the transmission lines, or pipes, of the system in a ready state of operation. System control is achieved by such things as directional valves, metering devices, flow control valves, and regulators. The load of this system is primarily designed to do some type of useful work. Mechanical motion produced by the load is either rotary or linear. Cylinder actuators and hydraulic motors are typical components that achieve this function. A number of hydraulic indicators are also used to show different operating conditions and to test system components. Pressure, temperature, and flow are typical system values monitored by indicators.

Fig. 7-6 shows a prepackaged hydraulic power system that is often used in an industrial setting. The hydraulic pumping unit of this system is representative of those found in small to medium pressure applica-

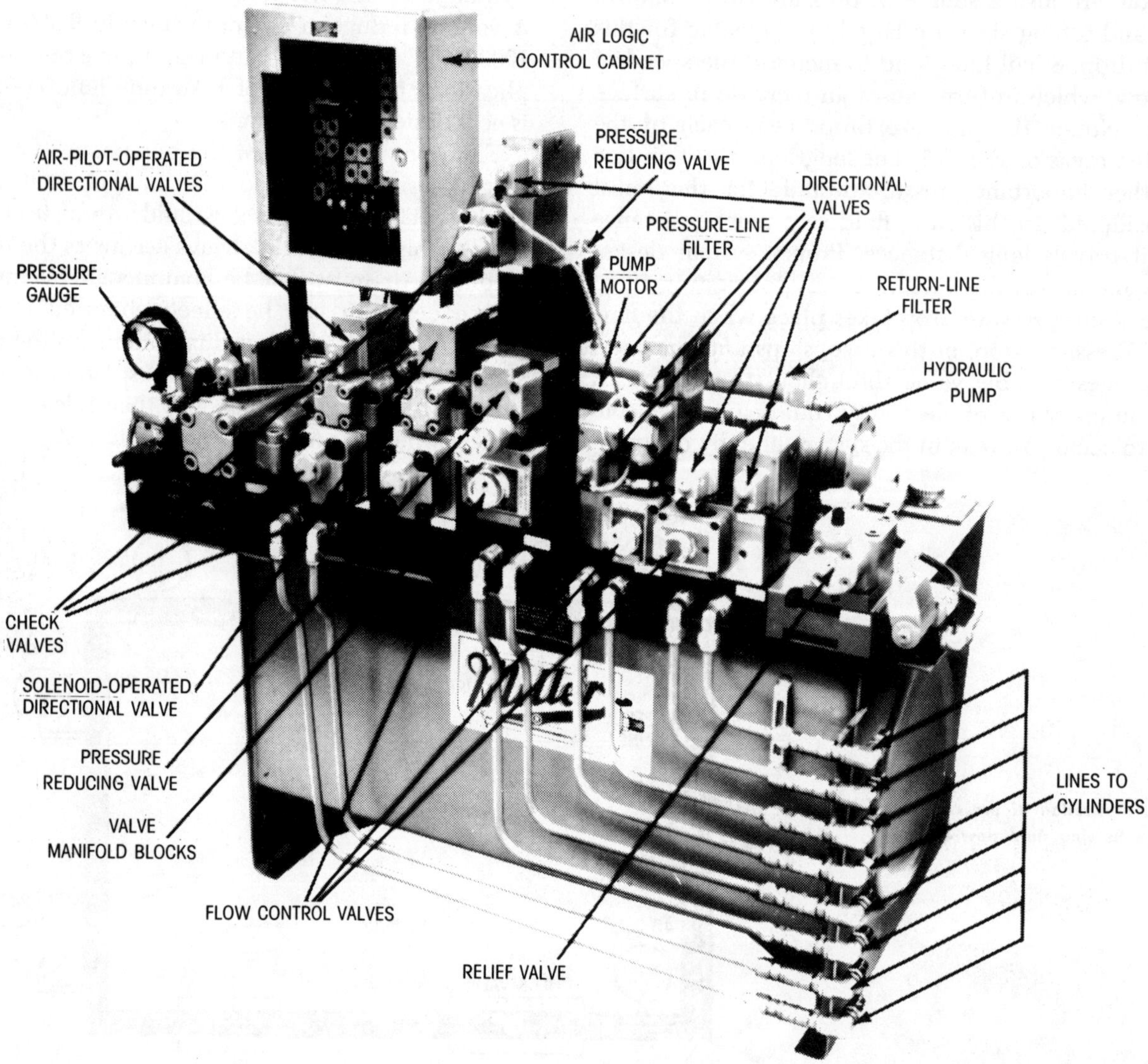

Courtesy Miller Fluid Power Div., Flick-Reedy Corp.

Fig. 7-6. A prepackaged hydraulic power system.

tions. A unit of this type often houses an electric motor, a hydraulic pump, a pressure relief valve, gauges, and a reservoir. The primary energy source of this unit is electricity. Electricity must first be changed to mechanical energy through rotation of the motor. The motor, in turn, is used to drive the pump. Hydraulic fluid from the pump then circulates through the system and builds up pressure as it encounters system resistance.

Most of the hydraulic power systems used in industrial applications today are equipped with a filtering device. In Fig. 7-6, the pressure-line filter is placed in the feed line. In this position, the hydraulic fluid is filtered before it has an opportunity to enter other system components. Any dirt particles coming from the pump or reservoir are removed from the fluid flow before entering the remainder of the system. As a general rule, filtering of this type reduces component breakdown a great deal.

Filtering devices can also be placed in the return line to remove dirt from the fluid after it passes through the entire system. One theory for placing the filter in this location states that fluid dirt is more prevalent in system components that move. In complex hydraulic systems with numerous components, filters may be placed in both the feed and return lines to reduce contamination problems.

Control of the hydraulic power system in Fig. 7-6 is achieved by a number of different valves. These include check valves, reducing valves, directional valves, relief valves, and flow control valves. Some of these are manipulated manually by an operator, electrically by solenoid devices, or automatically by pilot pressure changes. In addition to this, valves can be tripped by an outside air source, temperature changes, or through the mechanical action of devices at different system locations. The control function of a hydraulic system is undoubtedly the most complex function of the entire system.

The load of the hydraulic system of Fig. 7-6 could be double-acting cylinders connected to the cylinder lines. When high-pressure oil from the feed line is applied to each cylinder, it would cause the piston to be thrust forward. Any oil on the front side of the piston would be forced into the return line by this stroke of the piston. Reversing the oil flow from a directional control valve would cause the action of the piston to be reversed. Mechanical motion of this type can then be harnessed to do some type of useful work.

A pressure gauge is used as the indicator of the hydraulic system in Fig. 7-6. This device must be monitored periodically to assure proper operating conditions. As a general rule, hydraulic indicators are used rather sparingly in systems of this type.

Industrial Pneumatic Systems

The pneumatic system of Fig. 7-7 is often used in small industrial air installations. The air-compressor unit serves as the source for the entire system. Typically, this unit employs electrical energy to produce rotating mechanical energy through the action of the

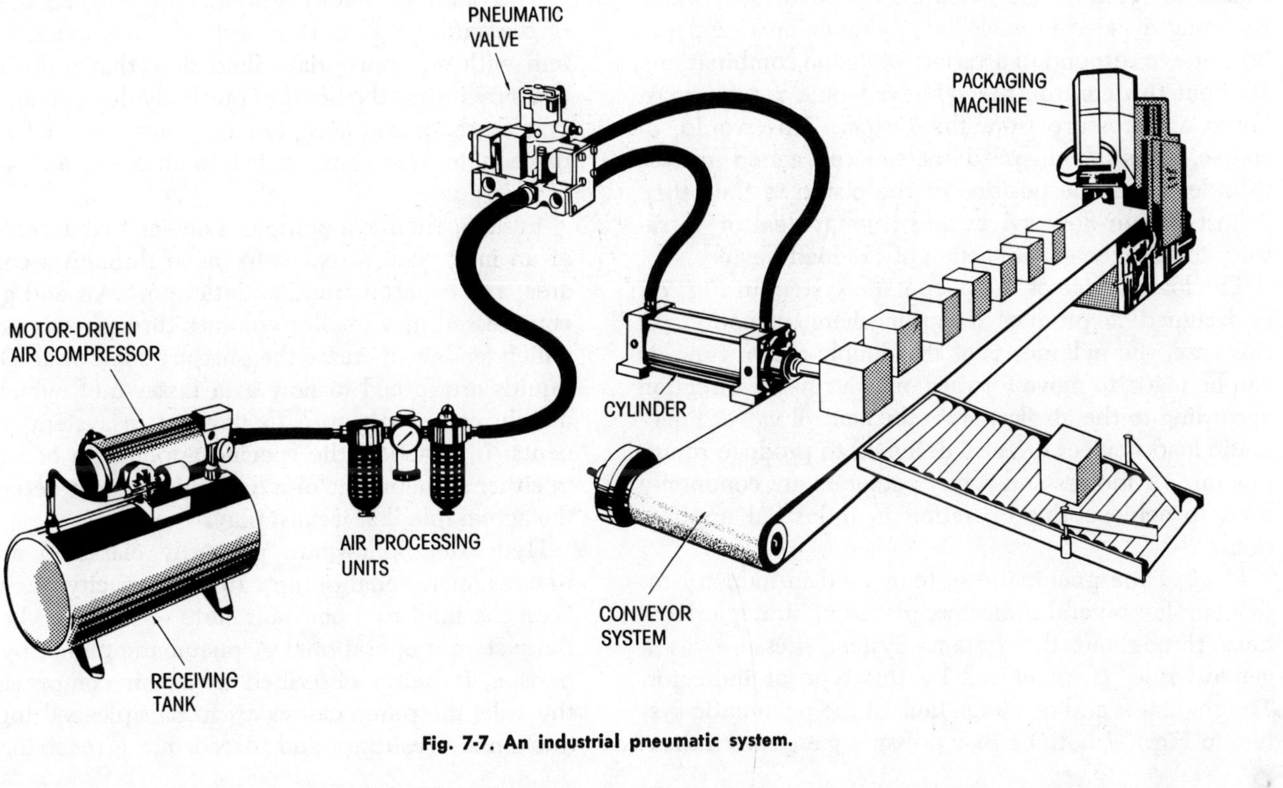

Fig. 7-7. An industrial pneumatic system.

motor. The compressor of this unit is driven by the motor. Compressed air is then stored in a receiving tank where it is eventually distributed to the system when needed. Receiver tank pressures range from 100 to 150 lb/in^2 as indicated on the tank pressure gauge. The feed line of the entire system is connected to the receiving tank for distribution throughout the plant.

Pneumatic systems must also employ some type of air-processing components in order to condition the air before it can be used. In systems with a small number of components, one conditioning unit is adequate. Systems with a large number of components often connect these units to each load device attached to the feed line. Conditioning the air involves filtering, pressure regulation, and lubrication. As a general rule, a conditioning unit is connected in the feed line between the compressor and the first control device.

Control of the pneumatic system of Fig. 7-7 is achieved by a manually operated, four-way control valve. Air flow to the cylinder can be directed to cause forward motion, reverse motion, or it can be turned off by the control valve. One line to the cylinder will receive air and the other line will exhaust air into the atmosphere through the control valve. Exhausting air instead of returning it to the system is a unique characteristic of the pneumatic system.

Most pneumatic systems used in industrial settings today employ adjustable air-flow valves to each load device. These devices are often placed in both lines as indicated in Fig. 7-7. Valves of this type are designed to regulate the actuating speed of a cylinder. By using a valve in each line, cylinder operation can be made to respond to a variety of action combinations. Without this control, the cylinder would receive maximum air pressure from the system. This would, of course, cause high-speed mechanical action of the cylinder from one position of the piston to the other. Adjustable air-flow valves add a great deal of versatility to the mechanical action of the load device.

The load device of the pneumatic system in Fig. 7-7 is designed to produce linear mechanical motion. In this case, the cylinder is of the double-action type. It can be made to move forward or in a reverse direction according to the air flow from the control valve. Pneumatic load devices are also designed to produce rotary motion. Air motors and rotary actuators are commonly used to achieve this operation in industrial applications.

Most of the pneumatic systems used in industry today employ several indicators placed at strategic locations throughout the system. System pressure, as a general rule, is monitored by this type of indicator. The regulator and receiving tank of the pneumatic system in Fig. 7-7 both employ pressure gauges as indicators. Pressure drops along the system feed line can also be monitored by observing pressure readings at key locations. System maintenance and troubleshooting generally rely heavily on the use of indicators to locate faulty components.

Fluid System Components

As discussed previously, the term *fluid system* is a generic term used to describe both hydraulics and pneumatics. In this section, we will discuss discrete components that apply to both systems. There is, of course, a great deal of similarity in these components, which justifies this approach. In some instances, these components can even be used interchangeably. As a general rule, however, hydraulic components are somewhat larger than their pneumatic counterparts. Hydraulic components also tend to be a bit more rugged in construction than those used in pneumatic systems. These differences can primarily be attributed to the fact that air is less dense than oil. Therefore, it takes larger volumes of air to transmit equal values of pressure and force. Hydraulic systems, by comparison, tend to be used more in applications that demand high pressure levels for heavy-duty machine operations. Except for the major differences in size, ruggedness, and pressure, hydraulic and pneumatic components operate on the same basic principles, and respond in a similar manner.

Fluid Pumps

The pump of a fluid system is often called the heart of the entire system. It is designed to provide the system with an appropriate fluid flow that will develop pressure just as the heart of our body does. In an actual system, the pump accepts mechanical power from the drive motor and converts it into an equivalent amount of fluid power.

Fundamentally, a pump is a device that accepts fluid at an inlet port, forces it to move through a confined area, and expels it from an outlet port. Air and gas are compressed into smaller volumes through this process, which tends to increase the pressure somewhat. Oil and liquids are forced to flow at a faster rate, which ultimately causes pressure to develop in system components. In practice, the specific application of a pump in either a pneumatic or a hydraulic system determines the actual role that it must play.

Hydraulic pumps are primarily classified as continuous operational pumps. In this capacity, they must keep the fluid in a constant state of motion whenever the system is operational. A pneumatic pump, by comparison, is better described as an air compressor. In this role, the pump causes air to be squeezed together into smaller volumes and forced into a receiving tank for storage. When the tank pressure builds up to a certain level, the compressor pump is turned off. Compressors are often used for short operational periods. These occur only when the demand for system air arises.

There are two general classifications of pumps used in fluid system applications today. The first classification is called a *nonpositive displacement* type of pump. This pump has no set amount of air or fluid that will be passed by the impeller blades during rotation. Flow is directly dependent upon the speed of the impeller blades. The second classification is called a *positive displacement* type of pump. This pump has a rather close clearance between the rotating member and the stationary components. As a result of this construction, a definite or *positive* amount of fluid will pass through the pump during each revolution. Fig. 7-8 shows a comparison of these two pump classifications.

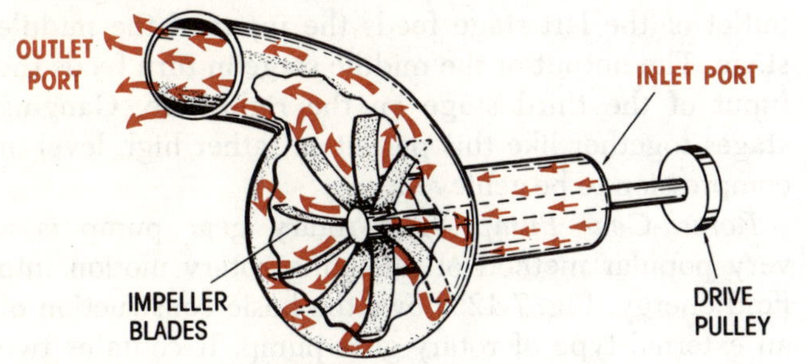

(A) Nonpositive displacement type.

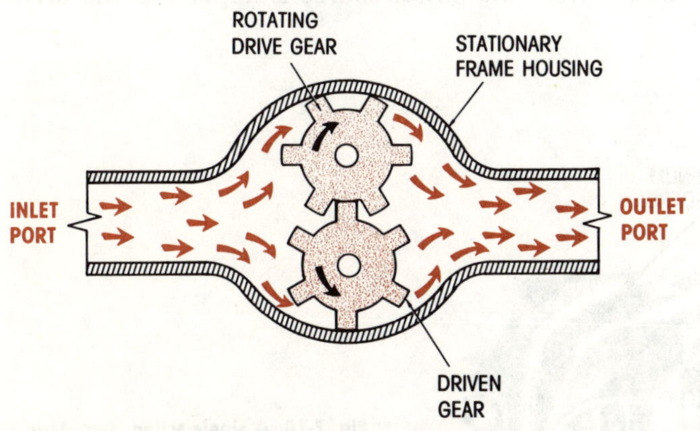

(B) Positive displacement type.

Fig. 7-8. Illustration of the two general classifications of pumps.

Basic Pump Types

The application of pumps in industry today is so varied that it is difficult to list all of the different types being used. A more realistic approach to this subject is to investigate the operation of a few basic pump types. With an understanding of these basics, a person should be able to apply them in a general way to practically all industrial pump applications encountered. This section will consider four basic pumping methods commonly found in industrial fluid power systems. Of these, the first three are of the positive displacement type and the fourth is of the nonpositive displacement type.

The reciprocating motion of a piston is one method often used in many high-volume and pressure pumping applications. Rotary motion is another common method used to produce pumping action. Gear and vane pumps operate on this principle. These pumps are available in a wide variety of designs and styles and represent the most common of all pump types. The final type of pump to be discussed here uses centrifugal

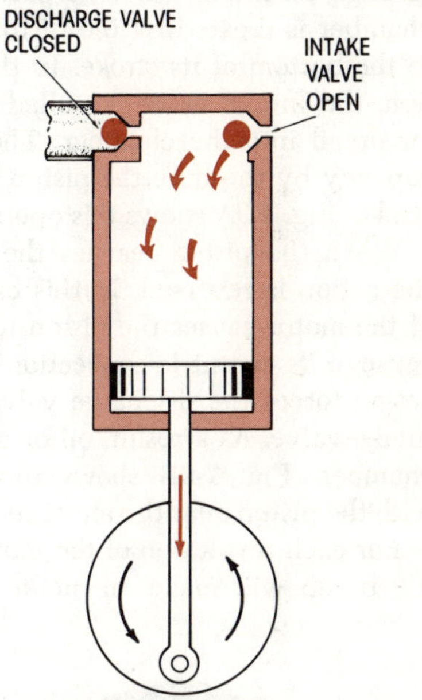

(A) Intake stroke.

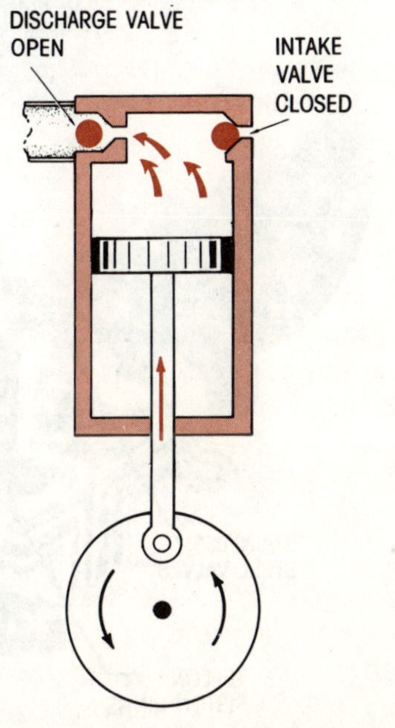

(B) Pumping stroke.

Fig. 7-9. Illustration of the operating principle of a reciprocating pump.

force to drive an impeller blade. This pump is more speed dependent than the other basic types. Each pumping method discussed is unique and has many industrial applications. This discussion involves only four distinct pumping methods being used in industry today. It by no means exhausts all of the possibilities that can be used.

Reciprocating Pumps—A very common type of industrial pump used in both hydraulic and air-compressor applications operates on the reciprocating principle. This pump forces either air or oil from a chamber by the reciprocating action of a moving piston. During the intake stroke, air or oil is drawn into the chamber through an intake valve. A partial vacuum inside the chamber is created by the piston as it is being pulled to the bottom of its stroke. In this condition of operation, the intake valve is pulled open, thus admitting air or oil into the chamber. The chamber is filled to capacity by the time the piston reaches the end of its stroke. Fig. 7-9A shows this operational step.

When the piston reaches the bottom of its stroke, the action is reversed. In this case, the rotary motion of the motor causes the piston to change direction because of its eccentric connection to the drive disk. This action forces the discharge valve open and closes the intake valve. As a result, oil or air is forced out of the chamber. Fig. 7-9B shows this operating condition with the piston near the top of its stroke.

For each revolution of the motor shaft, a reciprocating pump will make an intake and discharge stroke similar to the one just discussed. This would be classified as a single-stage, single-acting pump. Piston area and chamber volume are the key factors in determining the potential output of this pump. In some situations, two or more stages or cylinders may be driven by the same motor shaft.

Fig. 7-10 shows a typical single-acting, two-stage air compressor of the reciprocating type. A partial cutaway of one cylinder shows the piston and its connecting rod. The alternate cylinder appears on the opposite side and is in its housing. The compressor and motor of this unit are designed to occupy the same housing.

Fig. 7-11 shows a sectional view from the end of a V-type, water-cooled compressor. This unit is a multistage, single-acting compressor. Water is circulated around each stage head as a coolant by a water pump located at the lower right side. The intake and outlet valves of this unit are in the head of each stage. The outlet of the left stage feeds the input of the middle stage. The output of the middle stage in turn feeds the input of the third stage on the right side. Ganging stages together like this permits a rather high level of compression to be achieved.

Rotary-Gear Pumps—The rotary gear pump is a very popular method of changing rotary motion into fluid energy. Fig. 7-12 shows the basic construction of an external type of rotary-gear pump. It contains two gears enclosed in a precision machined housing. Rotary motion from the power source is applied to the drive

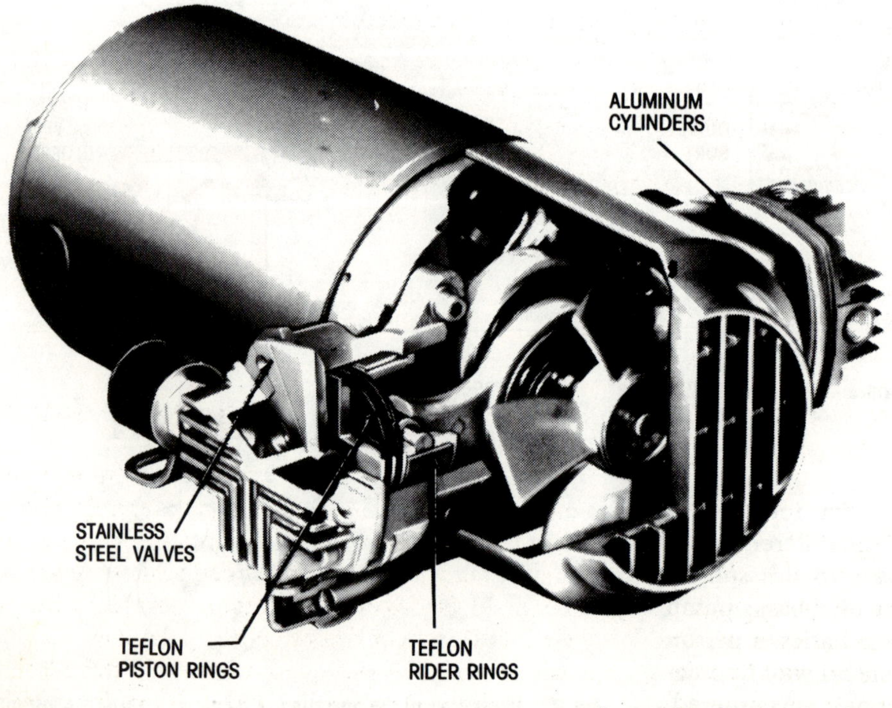

Fig. 7-10. A single-acting, two-stage reciprocating air compressor.

Courtesy Gast Manufacturing Corp.

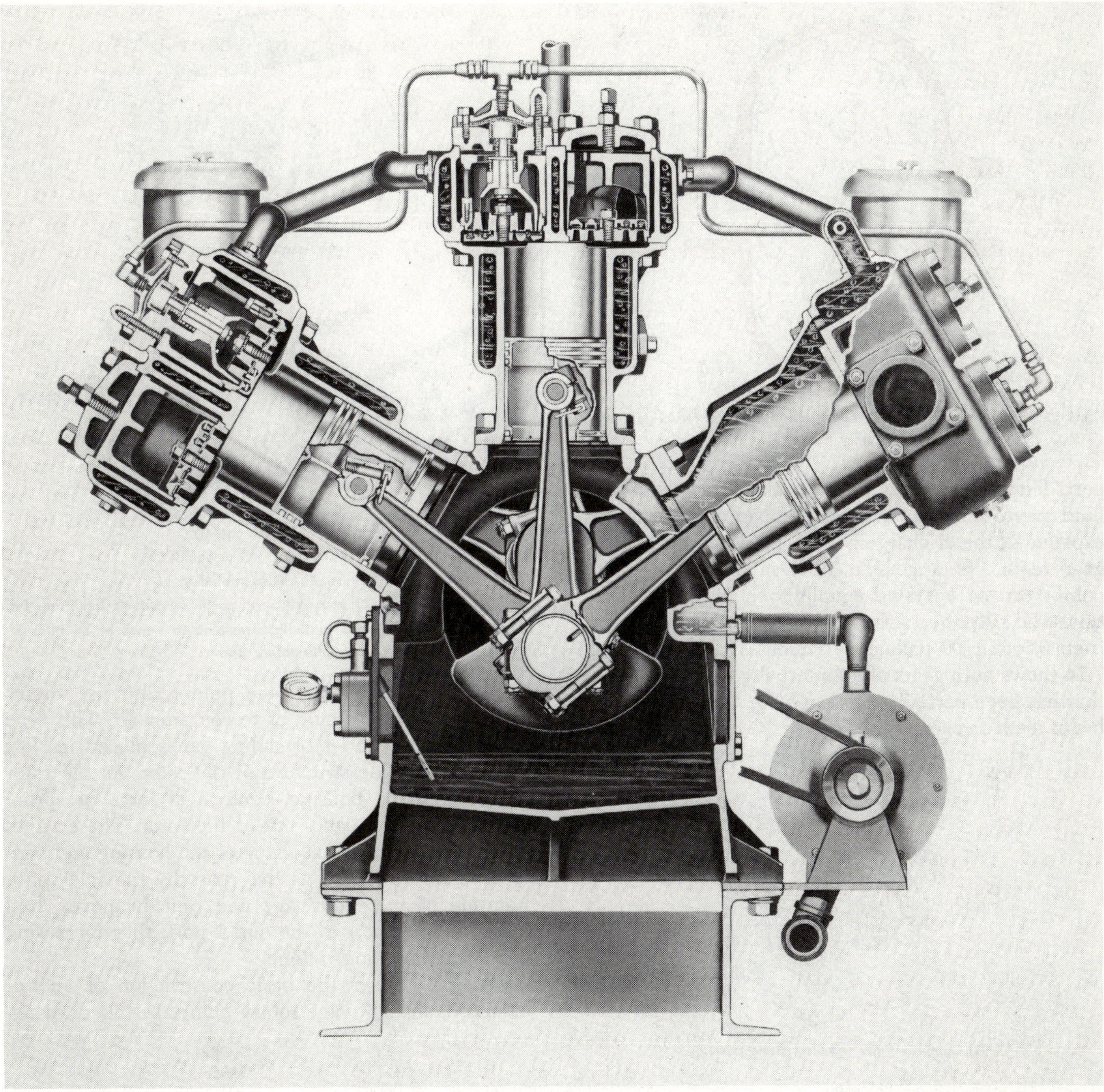

Courtesy Gardner-Denver Co.

Fig. 7-11. A multistage, single-acting air compressor.

gear. Rotation of this gear causes the second or driven gear to turn, with the teeth of the two gears meshing in the middle.

The basic operation of an external-gear rotary pump relys on unmeshed gears for carrying fluid away from the inlet side of the pump. Fluid trapped in these teeth is transferred around the periphery of both gears to the discharge side. When the gears remesh on the discharge side, fluid is forced to pass through the discharge port. Very little fluid is permitted to return to the inlet side of the pump due to the close mesh structure of the gears.

Another type of rotary-gear pump uses internal gears (see Fig. 7-13). In this type of pump, one gear rotates within another gear. The inner gear (idler gear) is designed so that it has one or more fewer teeth than the driven outer gear (rotor gear). As the idler gear rotates within the rotor gear, the gear teeth unmesh at the inlet port and remesh at the discharge port. As the gears unmesh, fluid is drawn into the inlet

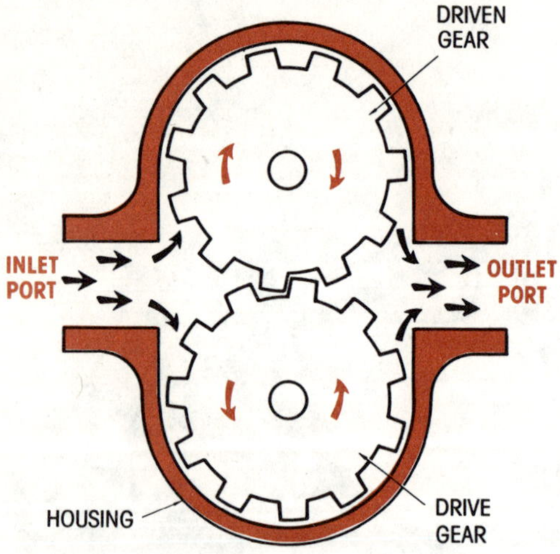

Fig. 7-12. Illustration of the basic construction of an external-gear type of rotary pump.

port, filling the spaces between the gear teeth. The fluid moves smoothly around the head crescent and is expelled at the discharge port by the remeshing of the gear teeth. As a general rule, internal-gear rotary pumps can be operated equally well in either direction. The output capacity of this type of pump ranges from 0.5 to 1100 gallons per minute (gal/min). Fig. 7-14 shows both gears of an internal-gear rotary pump that has been partially disassembled. Notice the number of teeth on each gear.

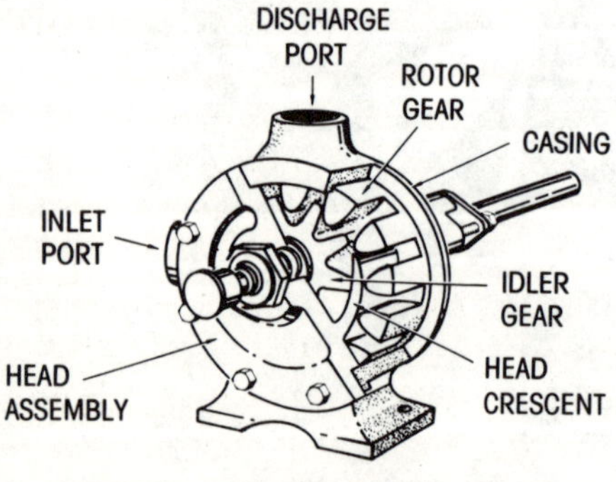

(A) Cutaway view showing basic parts.

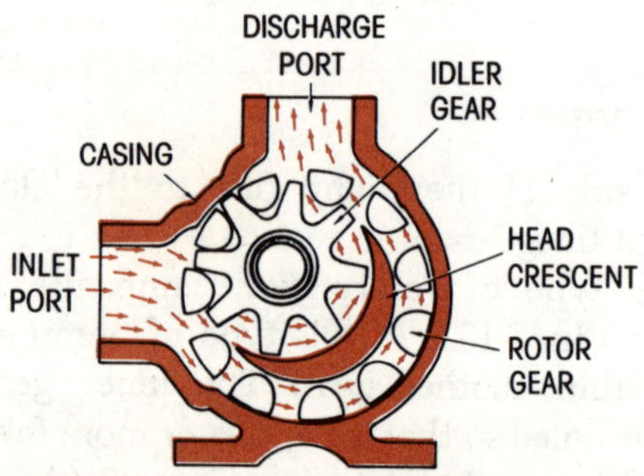

(B) Illustration of the principle of operation.

Courtesy Roper Pump Co.

Fig. 7-13. Basic parts and principle of operation of an internal-gear type of rotary pump.

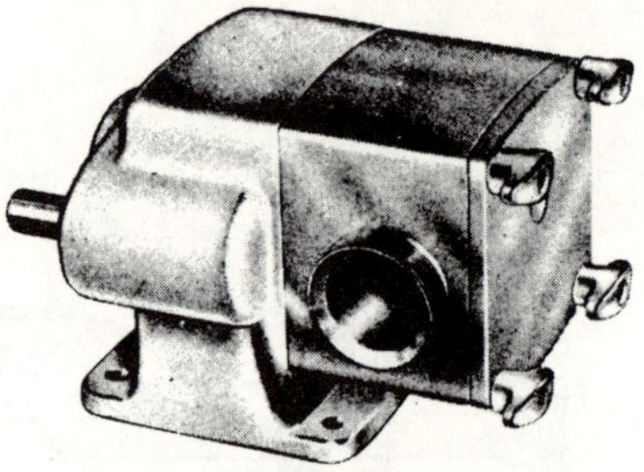

(A) Assembled view.

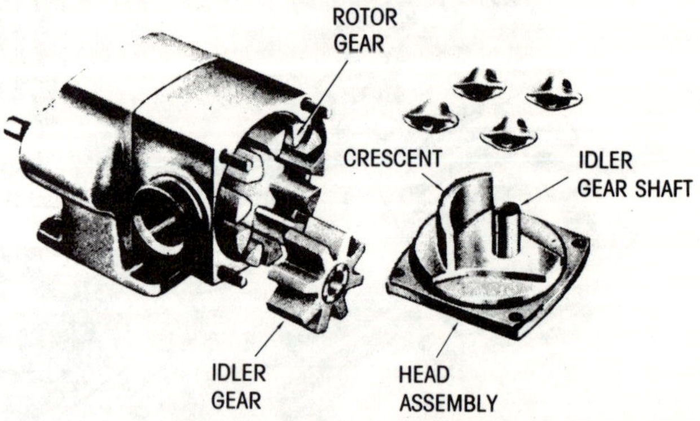

(B) Partially disassembled view.

Courtesy Viking Pump Div., Houdaille Industries, Inc.

Fig. 7-14. Assembled and partially disassembled views of an internal-gear rotary pump.

Rotary-Vane Pumps—Vane pumps also use rotary motion to circulate fluid or to compress air. This type of pump has a series of sliding vanes placed in slots around the inside structure of the rotor. As the rotor turns within its housing, centrifugal force or spring action forces the vanes out of the rotor. These vanes conform to the internal shape of the housing and capture volumes of fluid as they pass by the inlet port. Rotation of the rotor/vane unit quickly moves fluid from the inlet port to the outlet port, thus increasing flow or compressing volume.

Fig. 7-15 shows the basic construction of an unbalanced, straight-vane rotary pump. In this drawing,

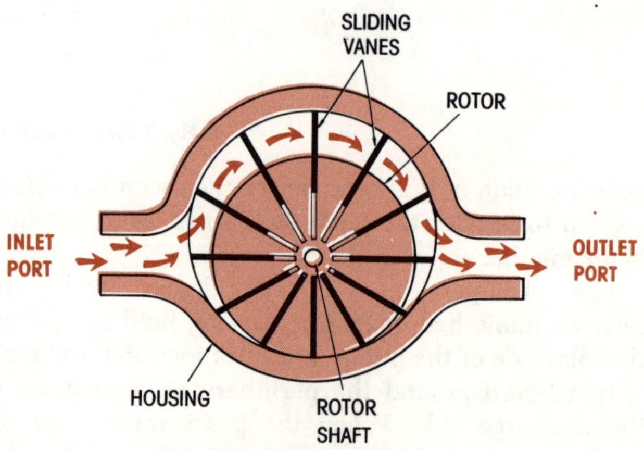

Fig. 7-15. Illustration of the basic construction of an unbalanced, straight-vane rotary pump.

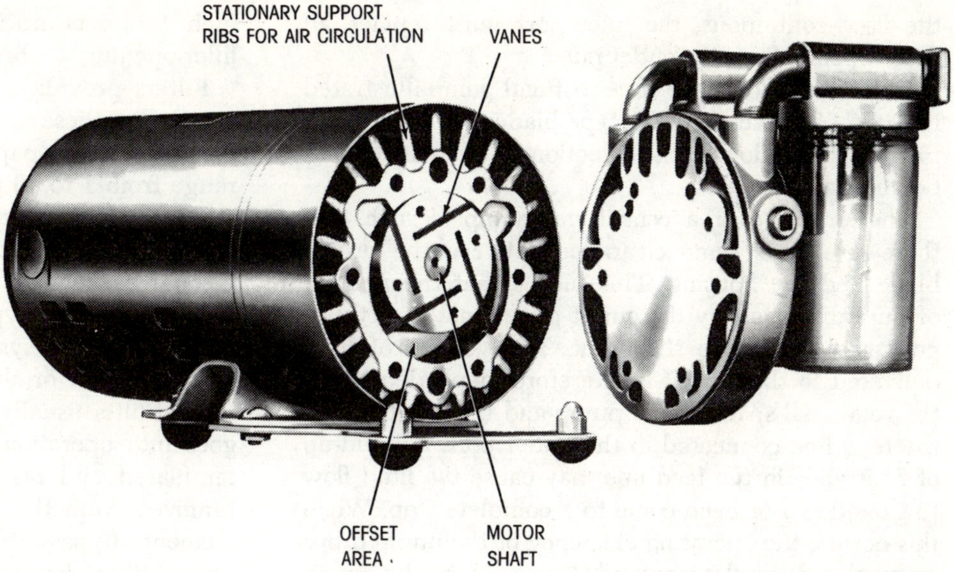

Fig. 7-16. A small air compressor that uses an unbalanced, straight-vane rotary pump.

Courtesy Gast Manufacturing Corp.

the rotor is offset toward the bottom of the housing. Through this type of construction, larger volume areas are present at the top of the rotor and small areas are present at the bottom. As a result, large volumes of oil or air can be made to move across the top with little or no return through the bottom. Offsetting the rotor in this way accounts for the unbalanced term commonly associated with this pump.

Balanced-vane pumps, by comparison, are designed so that the rotor is in the center of its housing. Large areas are made to appear at both the top and bottom of the housing. Separate inlet and outlet ports are developed by each side of the pump. Double pumping action of this type develops a smoother flow of air or oil than the unbalanced type of pump.

Fig. 7-16 shows a small air compressor that uses an unbalanced, straight-vane rotary pump. Only four vanes are used on the rotor of this pump. Notice the offset of the rotor and the large volume area on the bottom side of the pump. The supporting ribs around the inside of the housing provide air circulation from the motor for cooling. Approximately 50 lb/in² of compressed air can be developed by this pump. Rotary-vane pumps, as a general rule, are designed for single-direction operation only. The drive motor and pump of this unit are both housed in the same enclosure.

Centrifugal Pumps—A wide variety of centrifugal pumps are used today to circulate fluids. This type of pump is classified as having a nonpositive displacement of fluids. Pumps of this type are primarily used for low-pressure, high-volume flow applications in industry. Displacement of the fluid passing through a centrifugal pump is of an undeterminate amount compared with that of the other pumps discussed. The output of this pump is, therefore, dependent upon the rotational speed of its driving source.

In Fig. 7-17, two distinct types of centrifugal pumps are illustrated. The pump of Fig. 7-17A is commonly known as a *volute-type* pump because of the spiral (or volute) shape of its housing. When fluid enters the inlet port of the volute pump, it is set into rotation by the revolving blades of the impeller. This generates a centrifugal force which causes the fluid to move outward toward the inner wall of the housing. The volute-shaped housing then causes the fluid to circulate in a spiral-like path toward the outlet port. In order to keep

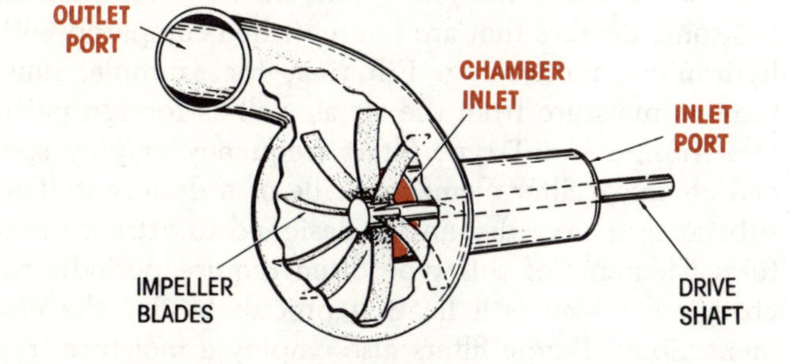

(A) Volute type.

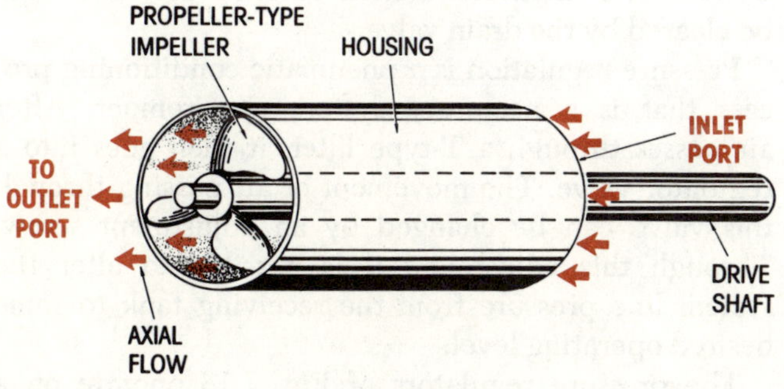

(B) Axial-flow type.

Fig. 7-17. Two types of centrifugal pumps.

the flow continuous, the inlet port must replace all fluid expelled from the outlet port.

In the axial-flow type of centrifugal pump illustrated in Fig. 7-17B, the propeller-type blades of the impeller maintain fluid flow in the direction of the axis of rotation of the drive shaft.

The structure of a centrifugal pump is such that there is always some clearance between the driven blade and the housing. This means that the amount of fluid displaced by the outlet port is not directly or positively related to the input. The volume of fluid delivered to the output is, therefore, dependent upon the rotational speed of the pump and the resistance of the feed line connected to the outlet port. A build-up of resistance in the feed line may cause the fluid flow to slow down or even come to a complete stop. When this occurs, the operating efficiency of the pump drops to zero and no flow occurs. Any fluid in the pump simply rotates inside without being expelled. An increase in pump speed could be used to solve this problem. As a general rule, pumps of this type are only used for transferring large amounts of fluid at low pressure.

Conditioning Components

Fluid power systems employ a number of devices to condition air or hydraulic fluid before it is processed through the outer system components. Typically, these devices are placed in the system to prolong component life by reducing the flow of foreign particles. The number of conditioning components employed by a system is largely dependent upon the type of system. This ranges from a simple hydraulic system with a line filter or strainer to a rather sophisticated system that employs filters, strainers, and heat exchangers. Pneumatic systems, by comparison, condition air by filtering it to remove dirt and water, regulating the pressure to the proper level, and adding oil as a lubricant.

Hydraulic Conditioning—The number of components, type of control devices, and condition of the operating environment are the major factors to consider in determining the amount of conditioning needed in a hydraulic system. For systems with manually operated control valves in a clean environment, a simple intake strainer may provide sufficient conditioning. For systems that operate for long hours in a dirty environment with precision control valves, micrometer filters and several strainers are considered a necessity.

Strainers are generally defined as coarse element devices placed in the pump inlet port or reservoir. These devices are usually of the stainless steel screen type with a wire mesh rating of 60 to 200 wires per square inch. Strainers are frequently placed in the reservoir filler opening, air breather, and pump inlet feed line.

Filters provide a finer grade of fluid conditioning. Typically, these devices are made of some porous medium such as paper, felt, or fine wire mesh. Ratings range from 1 to 40 micrometers. A micrometer is one-millionth of a meter or 0.00003937 inch. This rating refers to the particle size that is permitted to pass through the filter.

In-line and T-type filters are commonly used in many hydraulic systems. The T-type filter has a removable bowl or shell for element replacement. This type of filter usually employs a bypass relief valve that goes into operation when the element becomes contaminated and restricts flow. In-line filters must be removed from the line to clean or replace the filter element. Bypass relief valves are optional with this type of filter, depending on its application.

Some hydraulic systems employ heat-exchanger units to maintain the temperature of the fluid at a desired level. Machinery that is operating near a furnace, or that is used to control hot-metal rolling mills, often requires heat-exchanger units to cool the hydraulic fluid. Forced-air fan units, water-jacket coolers, and gaseous cooling devices are often used to achieve this function. As a general rule, applying heat is not a problem when a system becomes cold. It produces heat during its normal operation. Fluid heating would only be required in portable systems during a cold starting condition.

Pneumatic Conditioning—Pneumatic conditioning has some devices that are unique when compared with hydraulic conditioning. Filtering, for example, must remove moisture from the air as well as foreign particles. In-line and T-type filters frequency employ special chemical filter elements made of a desiccant. This substance is very dry and is designed to attract moisture. Elements of this type often require periodic recharging, which is a heating process to dry the element. Some T-type filters also employ a moisture trap at the bottom of the bowl. If inspection of the glass bowl shows an accumulation of moisture, it should be cleared by the drain valve.

Pressure regulation is a pneumatic conditioning process that is a necessary system requirement. After air passes through a T-type filter, it then goes into a regulator valve. The movement of air passing through this valve can be changed by an adjustment screw. Through this adjustment it is possible to alter the system line pressure from the receiving tank to some desired operating level.

The pressure regulators of Fig. 7-18 operate on a balance between atmospheric air pressure and system line pressure. Atmospheric pressure is applied to the

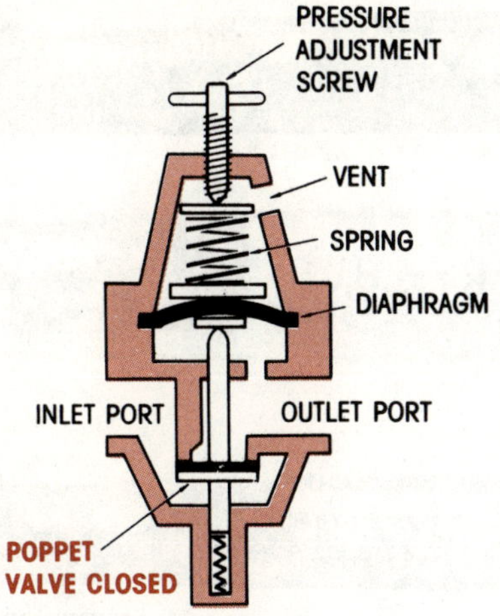

(A) System line pressure low.

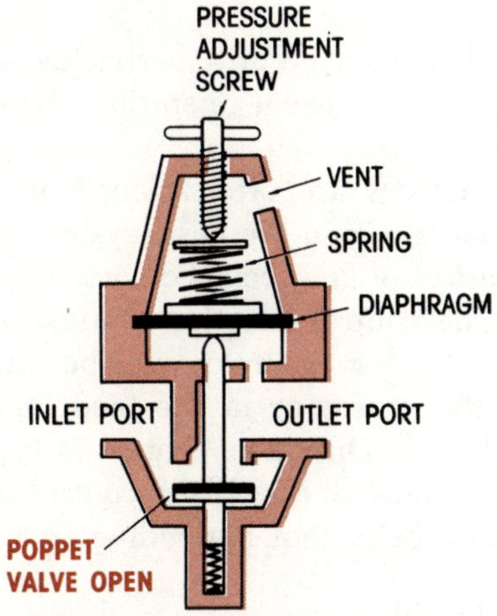

(B) System line pressure high.

Fig. 7-18. Illustration of the operating principle of an air-pressure regulator valve.

top of the diaphragm through a vent. System pressure is applied to the bottom of the diaphragm. Turning the adjusting screw adds mechanical pressure to the atmospheric pressure applied to the top of the diaphragm. When this pressure is greater than the system pressure, the diaphragm is forced down. This action opens the poppet valve, which in turn admits more air from the receiving tank (see Fig. 7-18A).

When the system line pressure becomes greater than the adjustment-screw/atmospheric-pressure setting, the diaphragm is forced upward. This action closes the poppet valve, thus maintaining the pressure at a set level. Fig. 7-18B shows this condition of operation. Should the system line pressure drop, the process would repeat, thus maintaining an even level of line pressure.

A regulator is simply a pressure balancing device that maintains system line pressure at a preset level. In order for this device to function properly, pressure from the receiving tank must be greater than the pressure setting of the regulator valve. To increase the line pressure, the adjusting screw must increase the spring pressure on the diaphragm. Loosening the adjustment screw lowers the line pressure by reducing the diaphragm pressure. Regulators may appear at several places in a pneumatic system.

Lubricators are unique conditioning devices found in pneumatic systems. This type of device simply adds a small quantity of oil to the air after it leaves the regulator. Through this conditioning process, valves and cylinders tend to last longer and operate more efficiently.

Fig. 7-19 shows a typical lubricator that is used to supply a mist of oil to the transmission feed line of a

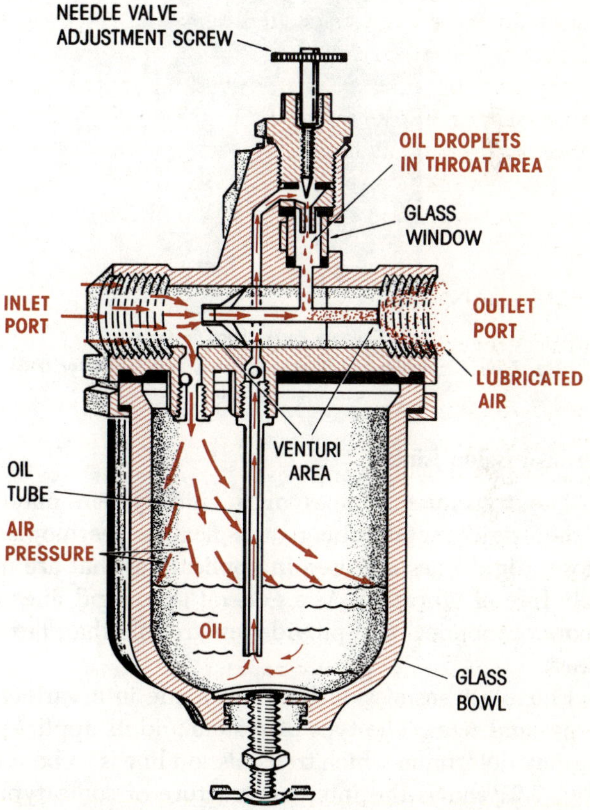

Fig. 7-19. Illustration of the operating principle of a pneumatic lubricator unit.

pneumatic system. When air enters at the inlet port, it is directed into a narrowed area called the *venturi.* While passing through this area, the flow increases in velocity. Pressure developed in the narrow venturi area is lower than that of the larger area. As a result of this condition, oil is forced from the glass bowl into the oil tube and transported to the top of the unit. The needle valve is then adjusted to regulate the oil flow so that small droplets fall into the throat area. The air velocity at the bottom of the throat causes these droplets to break into a fine mist of oil and mix with the passing air. Ultimately the lubricated air is forced to pass into the system through the outlet port.

In pneumatic systems, conditioning components such as the filter, regulator, and lubricator are often placed together in a combination unit. This combination is commonly called an FRL unit. Fig. 7-20 shows a typical FRL conditioning unit that achieves all three system functions.

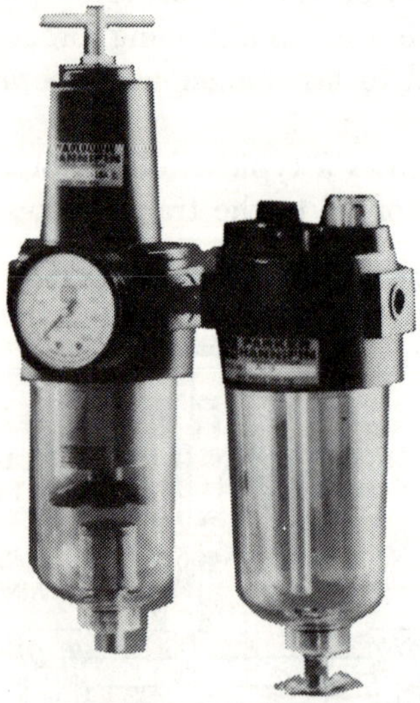

Courtesy Parker Hannifin Corp.

Fig. 7-20. A combination filter, regulator, and lubricator (FRL) conditioning unit.

Transmission Lines

The transmission lines of a fluid system may be either rigid metal tubing or a flexible thermoplastic hose. Rigid lines are used in applications that are usually free of vibration. As a general rule, rigid lines are more economical and provide less trouble than flexible lines.

Flexible transmission lines are made in a variety of types and sizes. The type of system and its application largely determine which transmission line is to be used. Fig. 7-21 shows the internal structure of some typical flexible hoses. Three basic elements of a flexible hose are considered when selecting it for a specific application. These include the tube or inner lining, the reinforcement material, and the outside cover material. These considerations determine pressure limits, temperature operating range, and outside exposure resistance.

System Control

Control is an essential operation that applies to all fluid power systems. Through this operation, fluid power can be made to do some type of useful mechanical work. Primarily, control is achieved by components that alter system pressure, direction, and volume of

Courtesy Parker Hannifin Corp.

Fig. 7-21. Some typical fluid system flexible hoses.

fluid flow. This may be of a partial nature, or it may completely stop a system operation, depending on its application.

Control devices are often placed at a number of different places within the basic system structure. The actual location of a specific device is largely determined by the control function it must achieve. Pressure regulators, for example, may be attached to the output of the source or in the feed line attached to the load device. Direction control is typically found near the load device. Fluid flow control may be found in any part of the system, depending on a specific need.

Pressure Control

Pressure control refers to those operations that alter the pressure level of an operating system. These include relief valves, reducing, by-passing, sequencing, and counter-balance. As a general rule, control devices of this type are named according to the operation they achieve.

Fig. 7-22 shows the internal workings of an orifice-disk relief valve. When the line pressure attached to the right side of the valve exceeds the pressure setting of the adjustment screw, a flexible diaphragm is forced up from the bottom. This action forces the connector rod to move, which in turn causes the orifice disk to shift to the position as shown in Fig. 7-22B. The flow path is completed through the holes of the two disks. Flow takes place from right to left. At pressure levels less than the adjustment screw setting, the path is obstructed as shown by the position of the orifice disks of Fig. 7-22C.

In hydraulic systems, pressure-relief valves are used to dump the output of a positive displacement pump back into the reservoir when the pressure rises to a dangerous level. In this pressure control application, the relief valve serves as a safety control device.

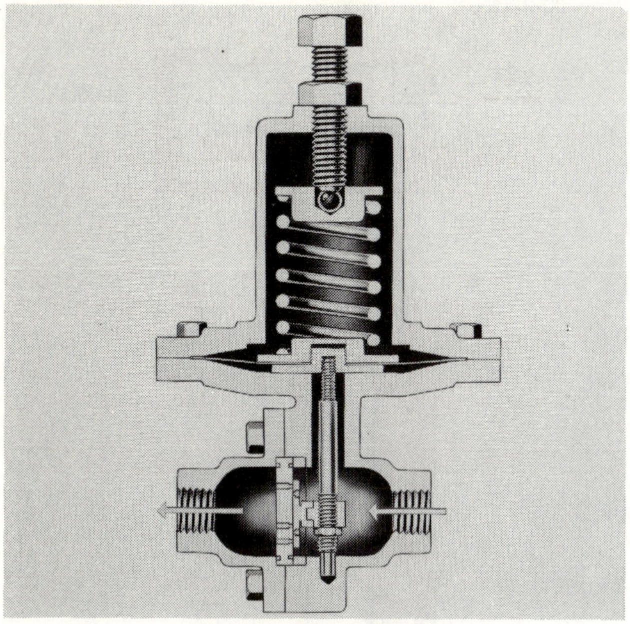

(A) Cutaway view of valve.

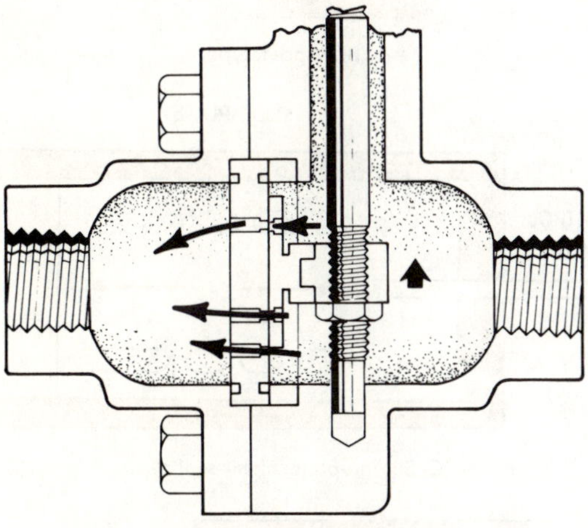

(B) Area of orifice disk showing flow path open.

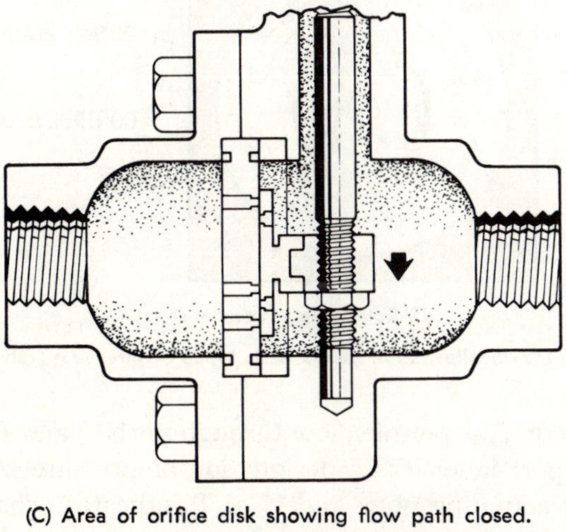

(C) Area of orifice disk showing flow path closed.

Courtesy Jordan Valve Div., Richards Industries, Inc.

Fig. 7-22. An orifice-disk pressure-relief valve.

In pneumatic systems, pressure-relief valves are used to meter smaller amounts of air or to bypass specific parts of the system when activated. In this case, excess air is not released into the atmosphere. The output port of a reducing valve may be altered in size to achieve metering control.

Pressure control may also be used to establish specific operating sequences. Valves that achieve this type of control simply direct pressure in some predetermined sequence at certain pressure levels. Control of this type is achieved by a relief valve type of construction. When the main system pressure overcomes the valve setting, it shifts to a different port. Sequencing valves are usually included in automated machining operations.

Direction Control

Directional control is achieved by devices designed to start, stop, or reverse system flow without causing an appreciable change in pressure or flow rate. One-, two-, three-, and four-way valves are some of the more common directional control devices. These devices may be actuated by pressure, mechanical energy, electricity, or by a manual operation.

The control action of a directional valve can be achieved in a variety of different ways. One-way valves, for example, may operate on the seated-ball principle of the relief valve. One-way valves are commonly called *check* valves in most applications. This valve will permit flow in only one direction.

Fig. 7-23 shows the operation of a check valve. When pressure is applied to the inlet port, it drives the ball away from its seat. When this occurs, the flow path becomes open and fluid will pass through the valve. In the reverse direction, pressure will force the ball into its seat. This action will prevent flow in this direction.

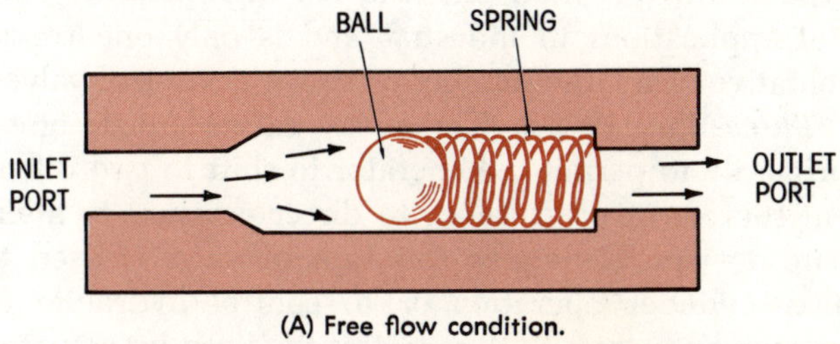

(A) Free flow condition.

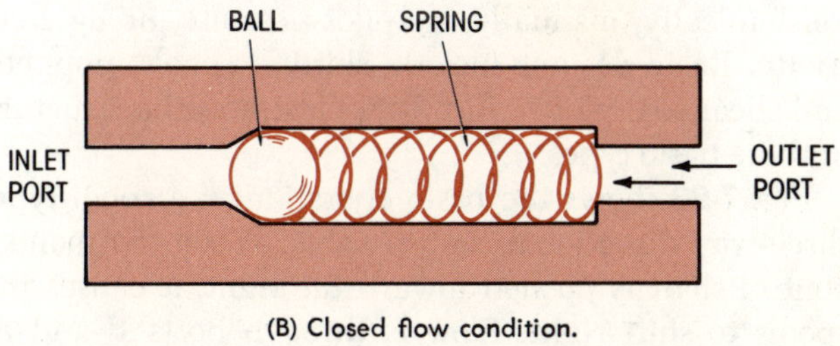

(B) Closed flow condition.

Fig. 7-23. Illustration of the operating principle of a check valve.

Valves of this type are often used to permit free flow around some controls when the flow direction is reversed.

Two-Way Valves—Two-way valves are simple directional control devices that have input and output connections placed in series with the transmission line. Valves of this type are designed primarily to achieve full control. They simply permit flow or shut it off. This type of control is achieved by placing gates, plugs, discs, spools, and numerous other objects in the line in such a manner that normal flow is obstructed.

Fig. 7-24 shows the internal workings of a ball-type directional control valve. This type of valve has a ball

Courtesy Jamesbury Corp.

Fig. 7-24. Cutaway view of a ball type of two-way valve.

that can be rotated manually by the outside control handle. When the handle is in line with the feed-line connections, flow will occur. Turning the handle to a right angle positions the ball to stop the flow. This type of valve is used primarily for high-pressure control applications in industry, and is only one representative of a large number of two-way control valves.

Three-Way Valves—Three-way valves are designed primarily to permit the operator to shift to two different sources of pressure, or to direct pressure to alternate devices. Generally, this type of valve is used to alter cylinder operation or to control hydraulic or pneumatic motors. Valves of this type can be actuated mechanically, manually, by pilot pressure, or by electricity. Basic designs include shifting spools, poppets, and shear seal plates. Fig. 7-25 illustrates the structure of these basic types.

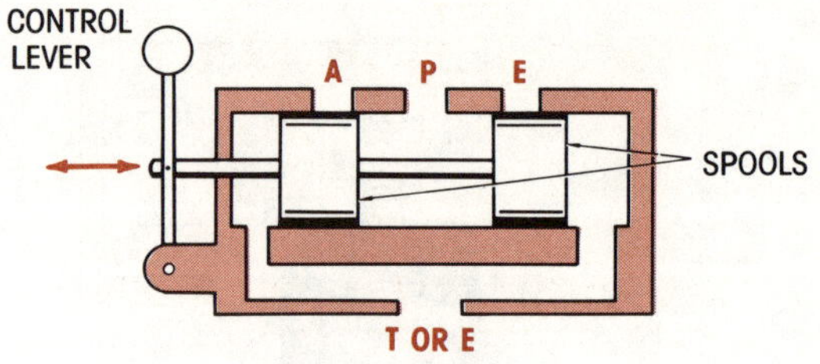

(A) Spool type.

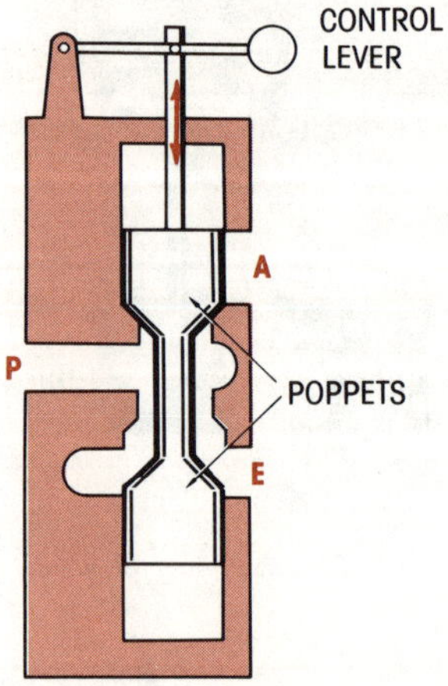

(B) Poppet type.

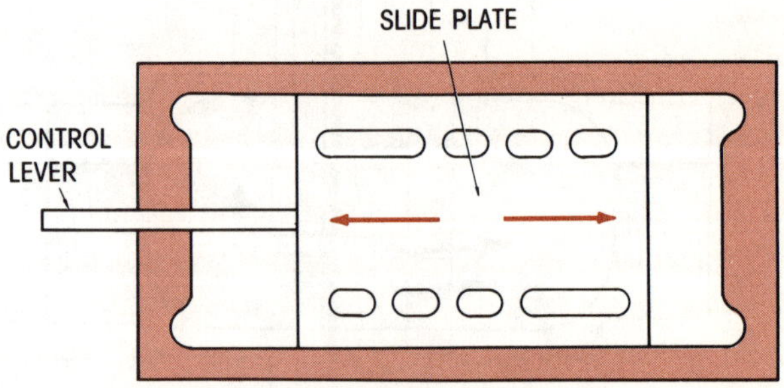

(C) Sliding-plate, shear-seal type.

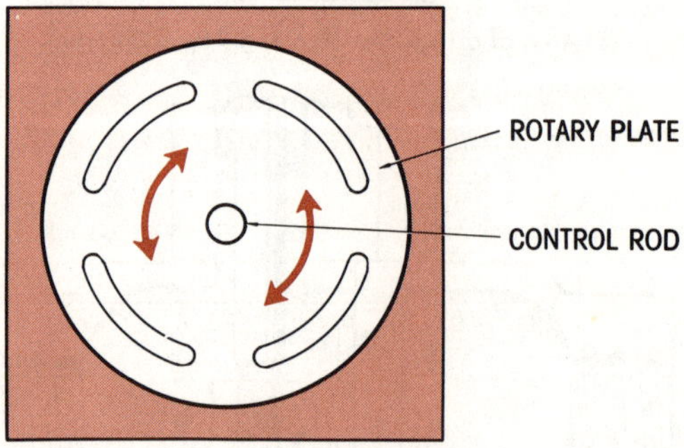

(D) Rotary-plate, shear-seal type.

Fig. 7-25. Illustration of the basic types of three-way valves.

Fig. 7-26 shows the basic operation of a spool-type, three-way directional control valve. When the manual control shaft is pushed toward the right, it causes the spools to shift right. Flow is through ports P and A. When the shaft is pulled to the left, the spools shift to the left. This permits flow through ports P and E or T. The port label P is a designation of pressure; A indicates actuating port; and E or T indicates exhaust or tank. These letter designations are assigned by the American National Standards Institute (ANSI).

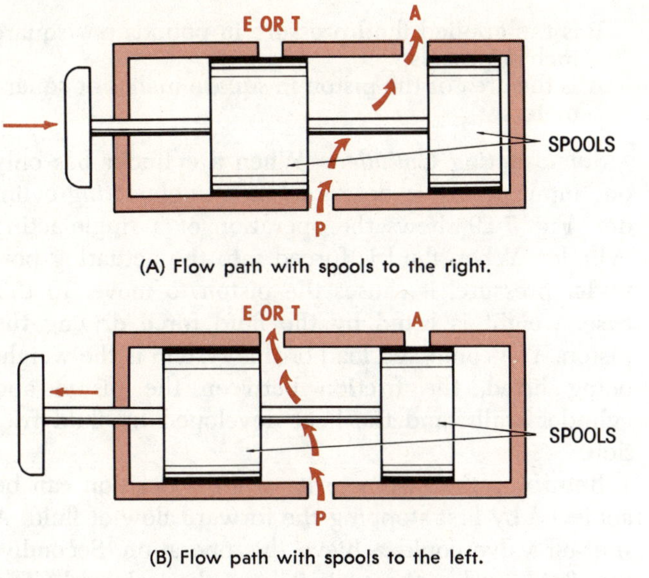

(A) Flow path with spools to the right.

(B) Flow path with spools to the left.

Fig. 7-26. Illustration of the operating principle of a spool-type, three-way valve.

As a general rule, three-way valves are designed for only two-position operation. Some valves do, however, have a neutral or off position. This additional position simply increases the control capabilities of the valve.

Four-Way Valves—Four-way valves are used in control applications to start, stop, or reverse the direction of flow. In its simplest form, this type of valve has four working connections. It is also manufactured in two, three, four, five, and six different position combinations. Four-way valves are commonly used to control forward and reverse actuation of a double-acting cylinder, or to reverse the rotational direction of a fluid motor.

Fig. 7-27 shows the basic operation of a four-way, spool-type valve. In position 1, the pressure feed line is neutralized, or in the off position. Position 2 shows the flow direction from P to A with an exhaust from B to E_2. Position 3 shows flow from P to B with the exhaust from A to E_1. With this type of control, flow direction can easily be reversed by manually shifting spool location. Actuation of this valve can be achieved mechanically, manually, electrically, or with a pilot pressure.

Flow Control

Flow control refers to those components designed to alter the volume or flow rate of hydraulic or pneumatic systems. The rate at which air or hydraulic fluid is delivered to the load of a system determines its operational speed. Motor speed, for example, is directly dependent upon the flow rate of the applied fluid. By altering this rate, a motor can be made to operate at a wide variety of speeds.

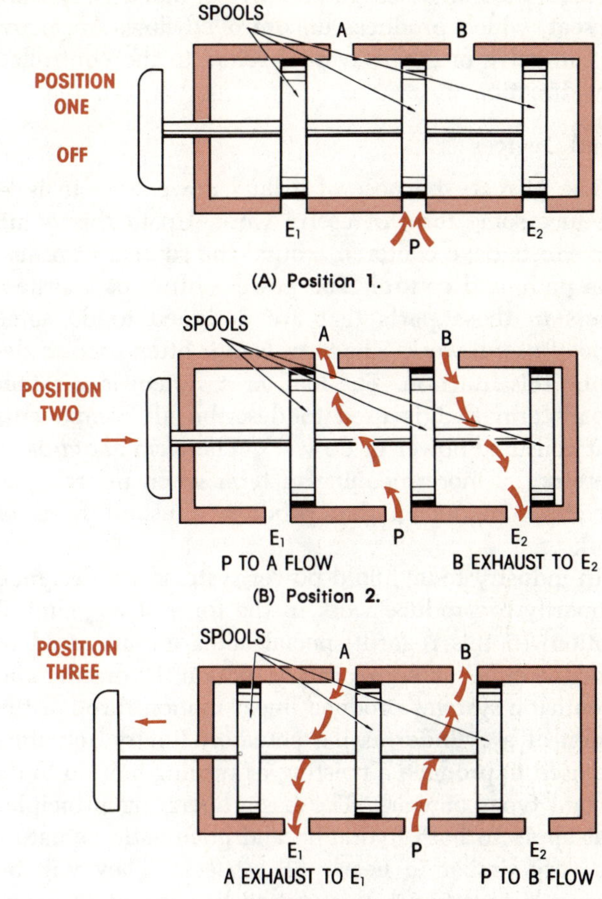

(A) Position 1.

(B) Position 2.

(C) Position 3.

Fig. 7-27. Illustration of the operating principle of a spool-type, four-way valve.

Cylinder actuating speed is also controlled by fluid flow devices. To alter the linear motion of a cylinder, fluid may be controlled at the input feed line, the return line, or a combination of both. The term *metering* is often used to describe this function.

Fig. 7-25 shows the operation of a flow control valve. Construction of this valve permits controlled flow from P to F. Flow level in this direction is adjusted by the needle valve. The letters P and F refer to the pressure and free-flow connections of the valve. Flow from F to

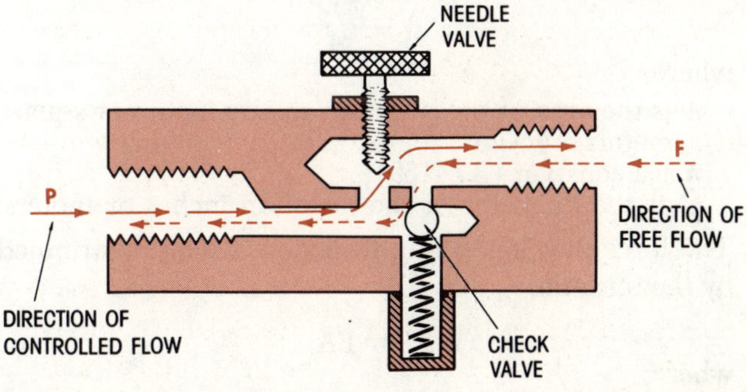

Fig. 7-28. Illustration of the operating principle of a flow control valve.

P forces the ball of the check valve to move away from its seat, which produces uncontrolled flow. An arrow on the valve or ANSI symbol refers to the controlled flow direction.

Load Devices

The primary purpose of a fluid power system is to produce some form of useful work. Up to this point, our efforts have centered around the source, transmission path, and control. The power output of a system refers to those parts that are designed to do some type of useful work. The term *load* is often used to describe this function. The load of a system is a rather broad term that is used to describe all components that consume power or do work. The term *actuator* is, therefore, a more meaningful term when referring to specific components that produce a useful form of work.

In industry today, fluid power systems are designed primarily to produce work in the form of mechanical motion. In this regard, special actuators are used to produce either linear or rotary motion. Hydraulic and pneumatic systems produce linear motion through the action of a cylinder as the actuator. Rotary actuators are used to produce a twisting or turning motion to do certain types of work. The basic operating principles that apply to both hydraulic and pneumatic actuators are very similar in nearly all respects. They will be discussed together in this section because of this similarity.

Linear Actuators/Cylinders

In industry today, cylinders are used to develop the force needed to lift, compress, hold, or position objects during different manufacturing processes. In order to produce linear motion, hydraulic fluid or air is forced into a cylindrical chamber under pressure. A piston placed within the chamber is free to move from the pressure applied to it. The area of the piston determines the amount of force it will develop from an applied pressure. The area of a round piston can be determined by the formula:

$$A = \frac{\pi D^2}{4}$$

where,

A is the area of the piston in square inches or square meters,

π is a constant (3.14159),

D is the diameter of the piston in inches or meters.

The force developed by this piston is then determined by the formula:

$$F = PA$$

where,

F is the force developed in pounds or newtons,

P is the applied fluid pressure in pounds per square inch or pascals,

A is the area of the piston in square inches or square meters.

Single-Acting Cylinders—When a cylinder has only one input port, it is described as a *single-acting* cylinder. Fig. 7-29 shows the operation of a single-acting cylinder. When fluid is forced into the actuating port under pressure, it causes the piston to move. In this case, weight is lifted by the fluid force driving the piston. The combined load of this system is the weight being lifted, the friction between the piston and cylinder walls, and the heat developed by fluid friction.

Returning the piston to its original position can be achieved by first stopping the forward flow of fluid. A shut-off valve could achieve this operation. Secondly, the fluid under the piston must be released. The weight of the load would then cause the piston to retract when the fluid is released. Hydraulic jacks and automobile lifts typically employ single-acting cylinders of this type.

Double-Acting Cylinders—Double-acting cylinders, as the name implies, have power action in two directions. Two ports are needed with this type of cylinder. Fluid is applied to one port and expelled from the other port during its extending operation. Retracting the piston is achieved by reversing the fluid flow. A four-way valve is commonly used to control the motion of this cylinder. Industrial applications are found in such things as punch presses, rolling mills, machine-tool clamps, paper cutters, and production-line actuators.

Fig. 7-30 shows the basic operation of a double-acting cylinder. To initially extend the piston rod, fluid

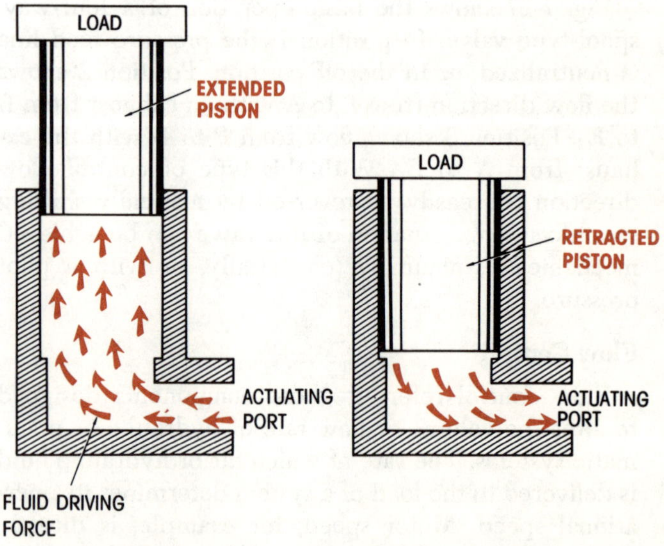

(A) Piston extended by fluid driving force.

(B) Piston retracted by release of fluid driving force.

Fig. 7-29. Illustration of the operation of a single-acting cylinder.

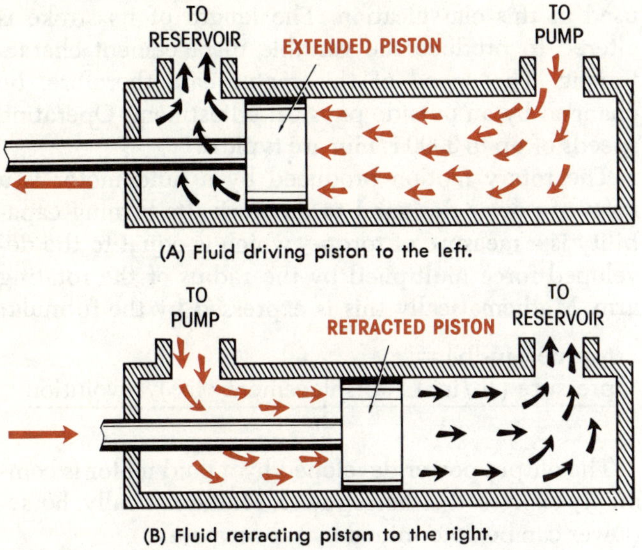

Fig. 7-30. Illustration of the operation of a double-acting cylinder.

must be applied to the right side of the piston and removed from the left side. This action would force the piston to move to the left. Switching the fluid flow would cause the piston to move to the right. As a general rule, the retracting force is somewhat less than the extending force. The area of the piston in this case is somewhat smaller due to the connection of the piston rod.

The double-acting cylinder just discussed is of the differential type. The linear action of the piston is determined by the pressure difference on each side of the piston. The piston rod extends only from one side of the cylinder. A nondifferential type of cylinder would have rods extending from both ends of the piston. Cylinders of this type can provide an equal force in either direction.

Fig. 7-31 shows a cutaway view of a double-acting pneumatic cylinder, with separate ports, a built-in directional control valve, and speed regulating valves combined in a single unit. The manufacturer describes this unit as an *air motor*. It produces continuous reciprocating motion when air is applied to the single inlet port. Cylinders of this type are becoming very popular today.

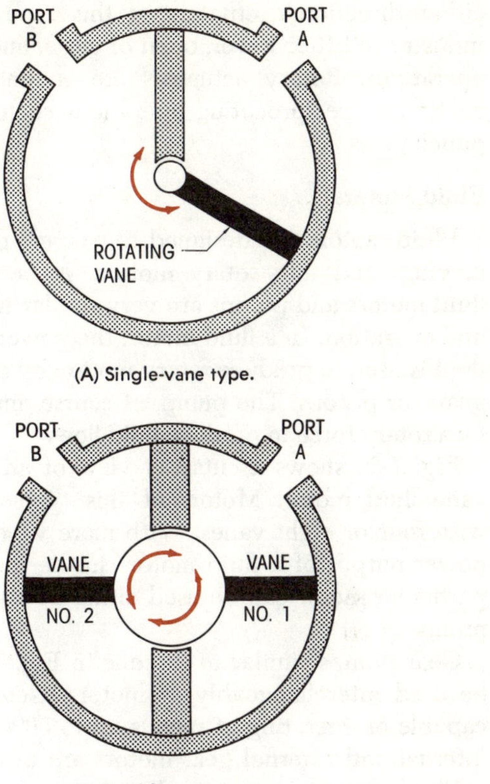

Fig. 7-32. Rotary actuators.

Rotary Actuators

Rotary actuators are fluid power devices designed to produce a limited amount of rotary motion in either direction. Fig. 7-32 illustrates two of the most common types. Fluid applied to port A causes the rotor to move in a clockwise direction a certain distance. Counterclockwise rotation is achieved by applying fluid to port B and expelling fluid from port A. The single-vane rotor can be made to turn approximately 280° in either direction. The double-vane rotor has twice the turning power, but can only turn approximately 100° in

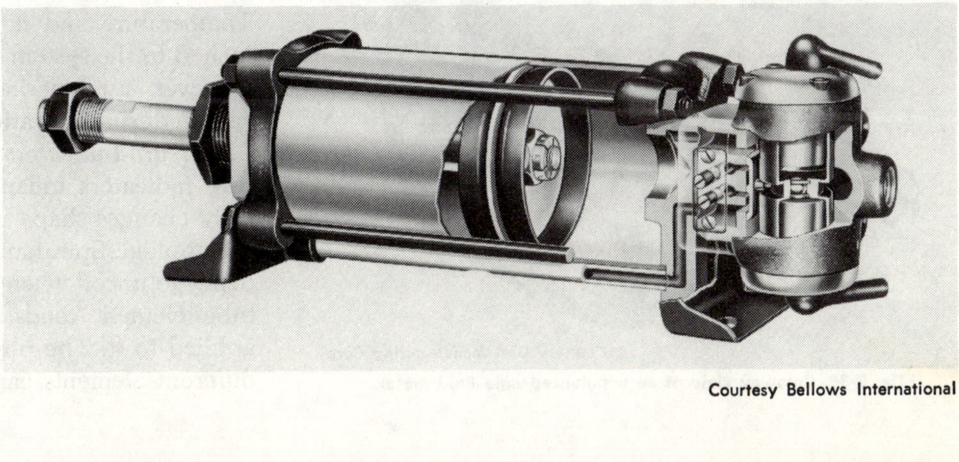

Fig. 7-31. Cutaway view of a double-acting pneumatic cylinder.

Courtesy Bellows International

either direction. Actuators of this type are used in industry to lift or lower, open or close, and in indexing operations. Rotary actuators are commonly used in continuous reciprocating operations such as that of a punch press.

Fluid Motors

Fluid motors are designed to convert the force of a moving fluid into rotary motion. As a general rule, fluid motors and pumps are very similar in appearance and operation. In a fluid motor, the power of a moving fluid is used to produce rotary motion by driving vanes, gears, or pistons. The pump, of course, must be driven by a rotary force to produce fluid flow.

Fig. 7-33 shows a cutaway view of an unbalanced-vane fluid motor. Motors of this type are designed with four or eight vanes. With more vanes, the rotary power output of a fluid motor can be increased. This particular motor can be used either as a motor or as a pump.

Gear pumps similar to the one in Fig. 7-13 can also be used interchangeably as motors. Gear motors are capable of operating at speeds up to 5000 r/min. Both internal and external gear motors are available today.

Fluid motors are generally classified according to the type of fluid displacement they possess. Gear, vane, and piston motors usually have a fixed displacement characteristic. This type of motor accepts a certain amount of fluid and moves it with each revolution. The operating speed of the motor depends entirely on the amount of fluid supplied by the source. Variable displacement motors, by comparison, are designed so that the amount of fluid circulated during each revolution can be changed. The piston type of motor is commonly used in this classification. The length of its stroke is altered to produce the variable displacement characteristic. The speed of this motor can, therefore, be changed by an outside physical adjustment. Operating speeds of up to 3000 r/min are typical.

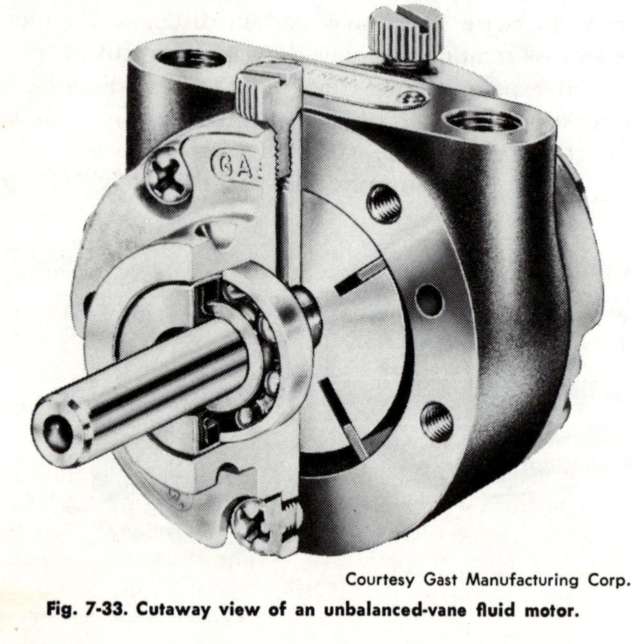

Courtesy Gast Manufacturing Corp.

Fig. 7-33. Cutaway view of an unbalanced-vane fluid motor.

The rotary motion produced by a fluid motor is a form of power designed to do work. Its turning capability is a measure of torque, which is equal to the developed force multiplied by the radius of the rotating arm. Mathematically this is expressed by the formula:

$$\text{torque (in-lb)} = \frac{\text{pressure (lb/in}^2) \times \text{displacement (in}^3\text{)/revolution}}{2\pi}$$

The output power developed by a fluid motor is commonly expressed as horsepower. Mathematically, horsepower can be determined by the formula:

$$\text{horsepower (hp)} = \frac{\text{torque (in-lb)} \times \text{speed (r/min)} \times 2\pi}{33{,}000 \text{ ft-lb/min}}$$

Motor performance information supplied by the manufacturer can normally be obtained to make these calculations.

Indicators

Undoubtedly the most significant measurement to be made in a fluid system is pressure. Indicators that measure pressure play a very important role in the overall evaluation of a system. Regulators and pneumatic receiver tanks often employ pressure indicators or gauges as permanent fixtures. A wide range of pressures must be measured today. Negative pressures (vacuums) as low as 0.00002 lb/in² up to positive pressures as high as 1,000,000 lb/in² must be measured in industrial equipment. This wide range of measurement requires a number of different indicating devices.

In addition to pressure, fluid systems often require measurements of flow and temperature. Indicators that measure and display these values are very valuable tools used to analyze system operating efficiency. Temperature and flow indicators are rarely ever attached to the system as a permanent fixture. They are, however, used periodically to test and evaluate the system during operational periods.

Pressure Indicators—A very large portion of all pressure indicators today employ an element that physically changes shape when different values of pressure are applied. Spiral and helix coil elements have a tendency to uncoil when pressure is applied. The Bourdon tube element tends to straighten when pressure is applied to it. The physical change produced by these different elements can then be used to move an indicator hand on a scale or a stylus on a paper chart. Fig. 7-34 shows these three basic pressure-sensing elements.

Fig. 7-35 shows a cutaway view of a typical pressure recorder with a spiral element. Pressures up to 80,000 lb/in^2 can be measured and recorded with this type of

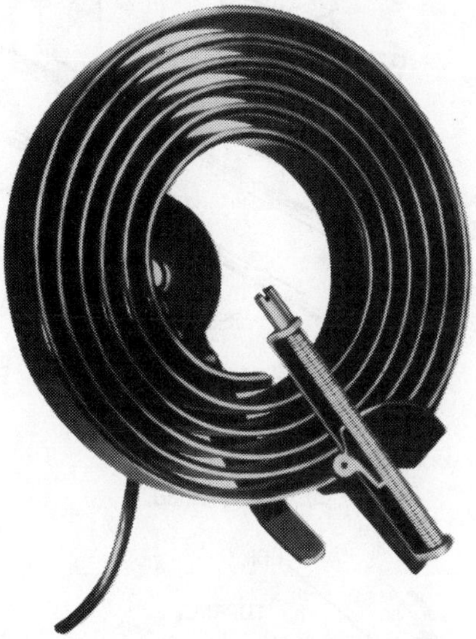

(A) Spiral element.

(B) Helix element.

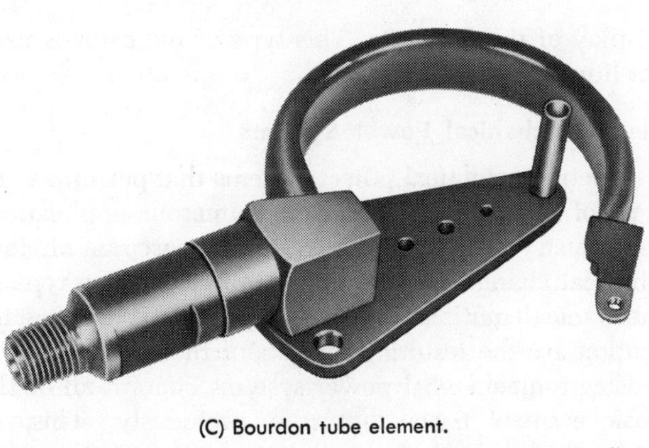

(C) Bourdon tube element.

Courtesy Honeywell, Inc.

Fig. 7-34. Three types of pressure-sensing elements.

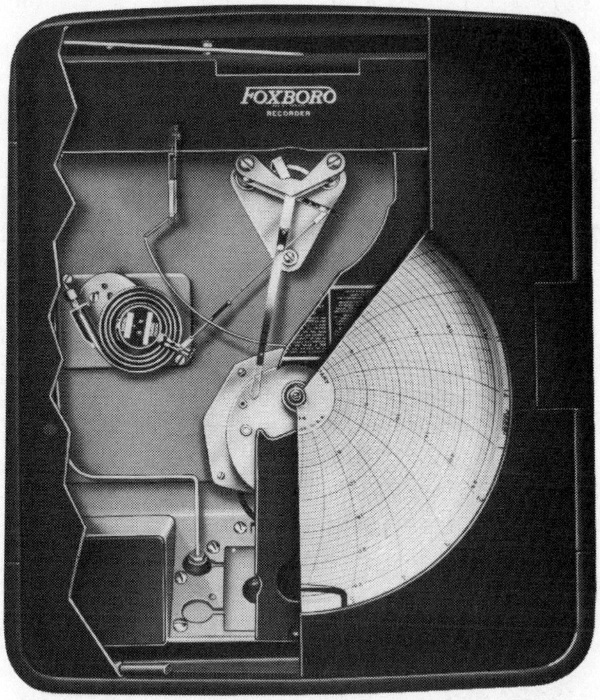

Courtesy The Foxboro Co.

Fig. 7-35. Cutaway view of a typical pressure recorder.

indicator. Instruments of this type are direct-reading pressure indicators. This means that the output is taken directly from the pressure element. Indicators of this type are rather delicate and are commonly mounted at a fixed location.

Flow Indicators—Flow indicators are primarily used to test flow rates from pumps, at key system locations, and at the inlet and outlet ports of actuators. By monitoring flow rates, it is possible to determine system efficiency and to minimize troubleshooting procedures. Measurements of this type are conveniently divided into three general catagories. These are the differential pressure method, the force method, and the velocity flow method.

Fig. 7-36 shows a simplification of three in-pipe methods of measuring flow. The differential pressure method is achieved by forcing fluid to pass through an orifice plate (see Fig. 7-36A). The pressure difference before and after the orifice plate is measured and recorded. For liquid flow measurements only, the force method can be used. This is achieved by placing a flow tube in an electromagnetic field (see Fig. 7-36B). Electrodes on each side of the tube measure a minute amount of voltage. This voltage increases in proportion to the flow. The velocity flow method employs a turbine blade or target that moves or turns with the flow (see Fig. 7-36C). The resulting physical change can then be used to move a gauge hand or the stylus

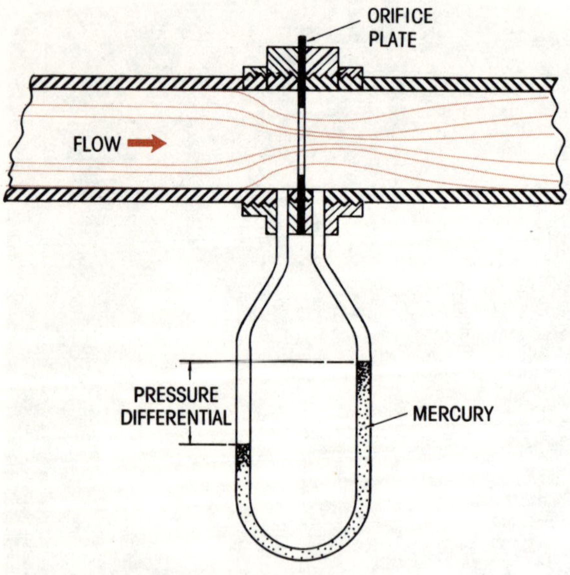

(A) Differential pressure method.

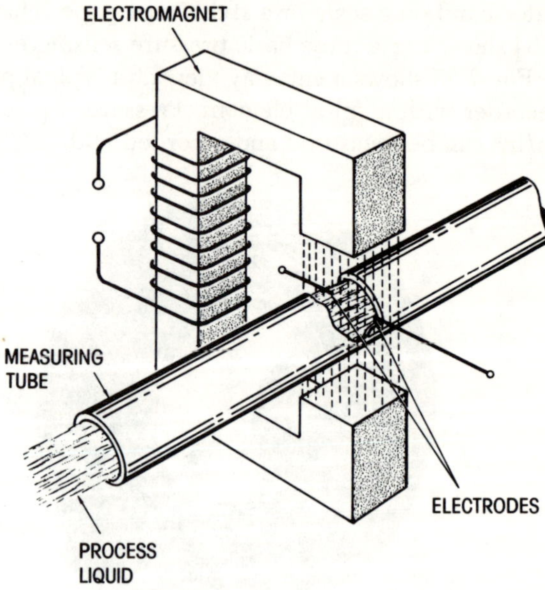

(B) Force method.

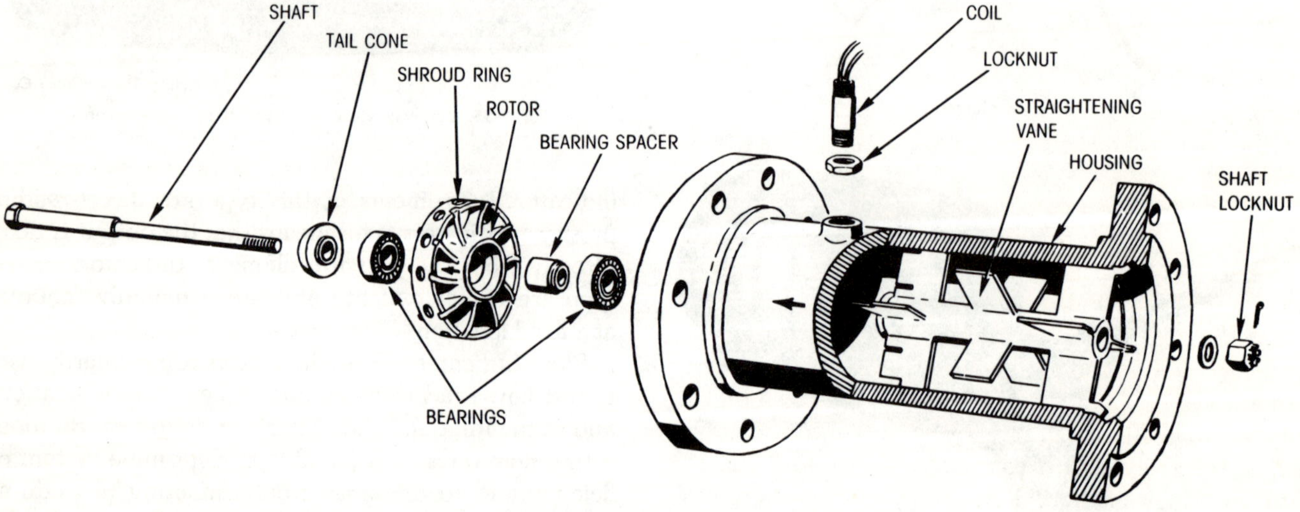

(C) Velocity flow method.

Courtesy The Foxboro Co.

Fig. 7-36. Illustration of three methods of measuring system flow.

of a chart recorder. Indicators of this type are portable and somewhat fragile.

A rather new type of flow indicator is shown in Fig. 7-37. This device has a clamp-on attachment that is placed around a pipe. Upon command from the control unit, an ultrasonic signal is sent out from the transmitter side of the clamp. A beam of ultrasonic waves travels through the pipe and its moving fluid. The receiver side of the clamp picks up the signal and forwards it back to the control unit. Fluid flow through the pipe causes the signal to change in proportion to the velocity of the flow. The control unit will analyze the received data and compute the flow in gallons per minute or hour. The control unit produces a digital display of the flow rate. This type of indicator is used for liquid flow only.

Electromechanical Power Systems

Electromechanical power systems that perform some type of mechanical work have numerous applications in industry today. These systems cause an obvious physical change that generally produces some type of mechanical motion. Linear, reciprocating, and rotary motion are the resulting outputs of these systems.

Electromechanical power systems contain all of the basic system parts discussed previously. This, of course, refers to the energy source, transmission path, control, load device, and indicator. The unique part

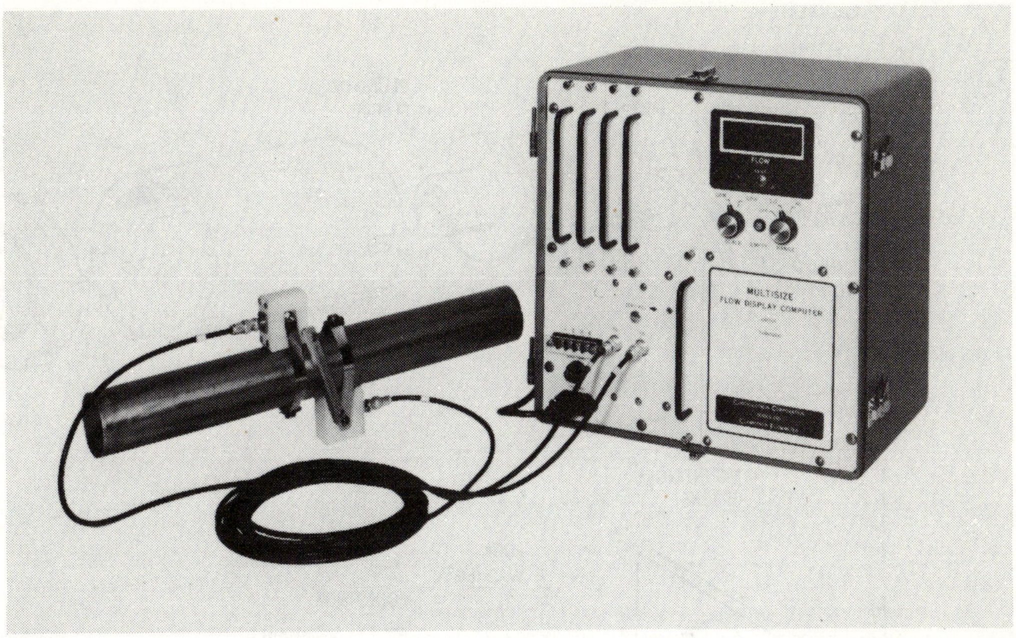

Fig. 7-37. A clamp-on type of flow indicator.

Courtesy Controlotron Corp.

of the electrical power system that we are concerned with in this chapter is the load device. You may recall that the load represents that part of the system that does work.

In some electrical power systems the load can be quite complex, since it represents all of the system parts that consume electrical power. In many cases, some of these parts do not do useful work. Heat, for example, is not necessarily helpful in the operation of an electric motor. It does, however, represent a part of the total system load. In practice, the motor must be designed to dissipate heat very effectively. The term *load* is, therefore, much too broad to be very meaningful in this discussion. A more appropriate term for the components to be discussed here is the *actuator*. Actuators, as a general rule, represent only part of the total system load.

Electrical actuators are primarily mechanisms designed to move or control some physical operation through an indirect means. Automation has greatly increased the importance of this type of component in industrial processing today.

Linear Actuator/Solenoids

In its simplest form, linear motion can be achieved by employing a solenoid as a linear actuator. This device employs an electromagnetic coil and a plunger core. When the coil is energized, the core is attracted by the magnetic field and pulled into it. Something attached to the core would then be subjected to the same mechanical movement. Solenoids are typically small in size and respond very quickly when energized. Fig. 7-38 shows an exploded view of a tubular type of solenoid.

Fig. 7-39 shows a variety of solenoids used in typical industrial applications. These devices are energized by a wide range of ac or dc voltage and current values. Solenoids are used in industry to physically change switches, to turn on hydraulic and pneumatic valves, to engage clutches of drive systems, and to latch locks. As a general rule, solenoids are inexpensive and have a rather long operational life expectancy.

Driven Linear Actuators

In industry today there is a need for linear actuators that provide reliable physical changes that can be accurately controlled at different speeds, be made to stop instantly, and to hold a position without further demand for power. Driven linear actuators are designed to achieve these operating conditions. The term *driven*, in this case, refers to electric-motor-operated mechanisms that produce linear motion. Fig. 7-40 shows a cutaway view of the internal parts of a driven linear actuator.

Fig. 7-41 shows a simplified version of a driven linear actuator. When the drive motor is activated, it rotates the threaded drive screw. A mating nut on the drive screw, which is attached to a drive rod, is thereby made to move axially. This action causes the drive rod to either extend or retract according to the direction of rotation of the drive motor.

When the drive motor of the actuator is set into motion, completion of the stroke or some type of overload will activate the overload cam. This action can

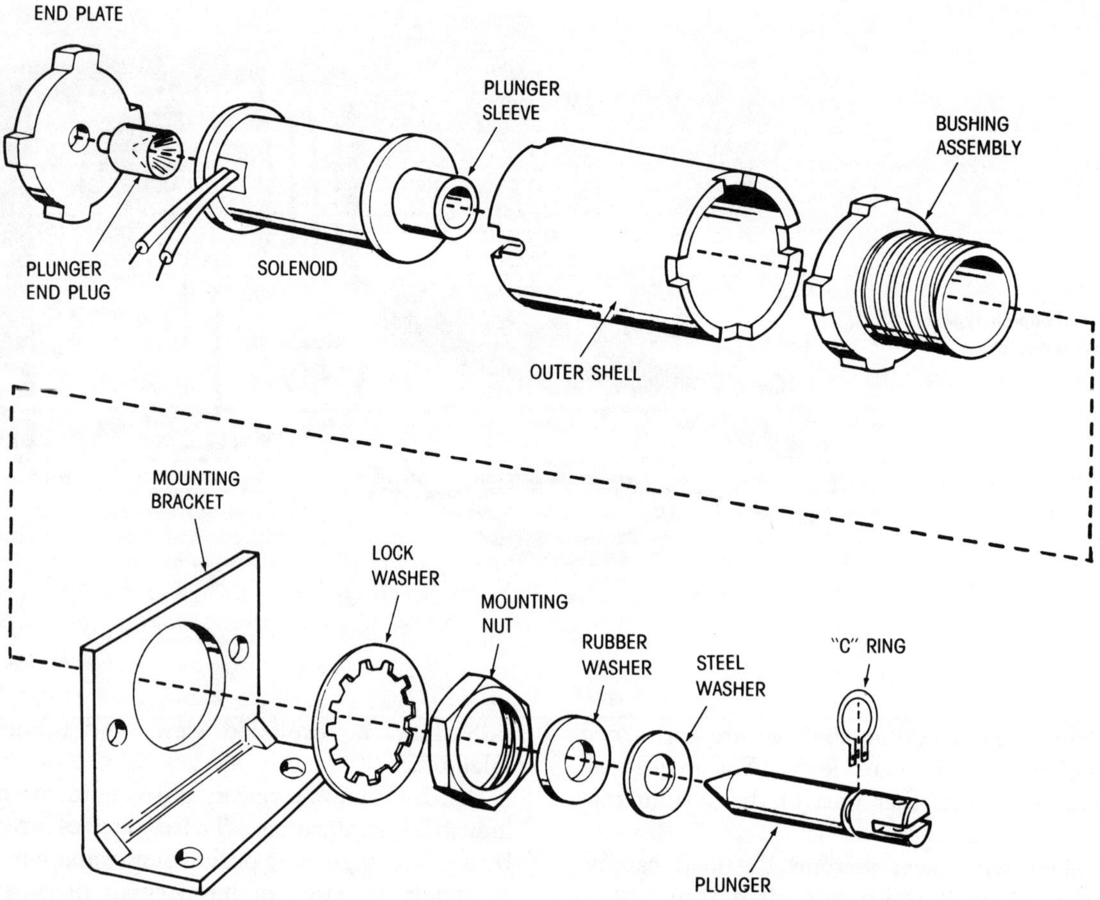

Courtesy Guardian Electric Manufacturing Co.

Fig. 7-38. Exploded view of a tubular solenoid.

be used to shut off the motor or reverse its rotation, according to the design of the actuator. This mechanism can also be triggered by an outside control unit. It will permit operation for certain time periods, or it can be triggered by a series of events. Feedback signals can also be generated to provide accurate positioning control in closed-loop applications. Driven actuators are used wherever electric linear motion is desired. This includes lever arms, cranks, slides, conveyor-belt positioning, and diverter-gate control.

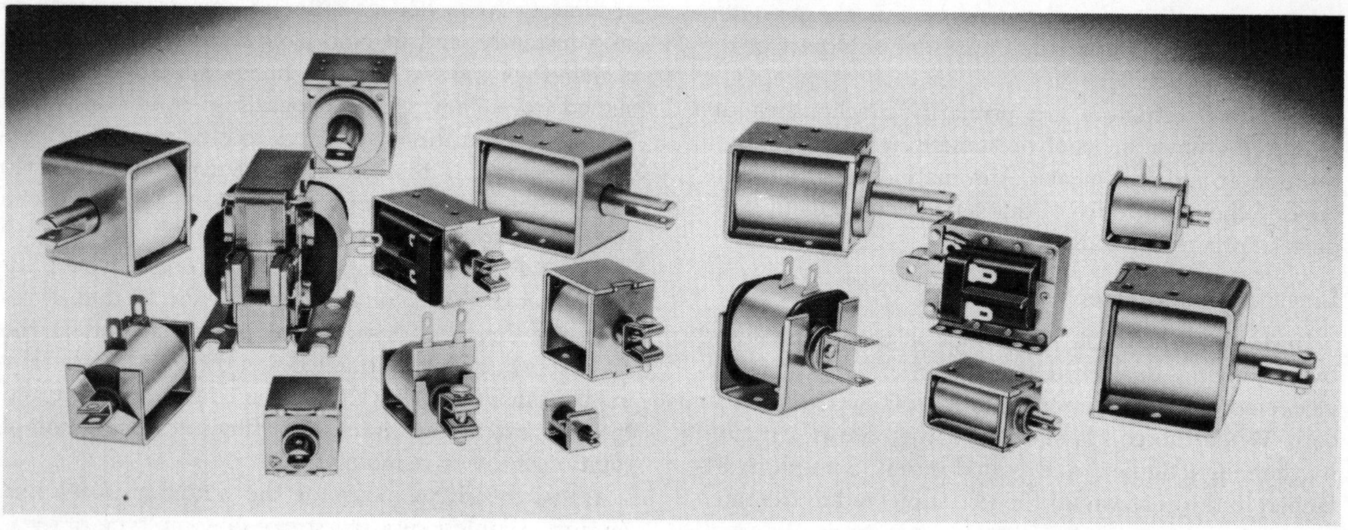

Courtesy Guardian Electric Manufacturing Co.

Fig. 7-39. An assortment of electromechanical solenoids.

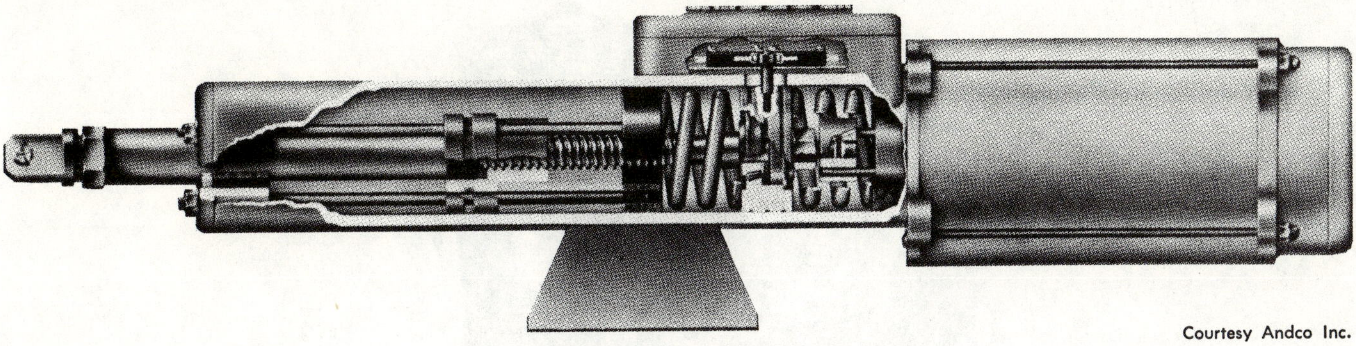

Courtesy Andco Inc.

Fig. 7-40. Cutaway view of a driven linear actuator.

The function of a linear driven actuator is fundamentally the same as that of a hydraulic or pneumatic cylinder. Electromagnetic actuators of this type have a number of unique advantages over their fluid counterparts. This includes such things as variable drive velocity, installation at remote locations, unlimited control capabilities, and overload protection. Devices of this type play a very important role in industry today.

Electromagnetic Relays

Electromagnetic relays represent one of the most widely used electromechanical devices in industry. Unlike the solenoid, the electromagnet of a relay contains a stationary core. Mounted close to one end of the core is a movable piece of magnetic material called the *armature*. When the coil is activated electrically, it produces a magnetic field in the metal core. The armature is then attracted to the core, which in turn produces mechanical motion. When the coil is de-energized, the armature returns to its original position by spring action.

The armature of a relay is generally designed so that electrical contact points respond to its movement. Activation of the relay coil will cause the contact points to "make" or "break" according to the design of the relay. In a strict sense, a relay could be described as an electromagnetic switching mechanism. There is an almost endless number of special-purpose relays and switch combinations in operation today. Fig. 7-42 shows a typical assortment of industrial relays.

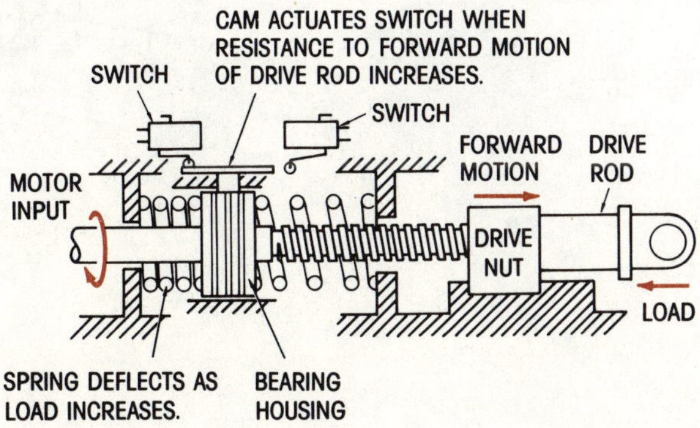

Courtesy Andco Inc.

Fig. 7-41. Simplified drawing of a driven linear actuator.

Motor Contactors

One very important industrial application of a relay type of device is the electric motor contactor circuit of Fig. 7-43. This device is a control element that actuates a motor by electromagnetic relay action. Pressing the "start" push-button switch completes a low-current path to the contactor coil. The contactor coil in turn produces a magnetic field that attracts the armature. Mechanical movement of the armature then completes an electrical path between the power-line and motor-load contact points. When this action takes place, the motor starts its operational cycle.

The coil contact points on the left side of the armature are pulled together by the same armature action just described. As a result, release of the "start" button at this time will not de-energize the contactor coil. A path to the coil source is now completed through the "stop" push-button switch and the coil contact points on the armature. The motor, therefore, continues to operate as long as electric power is applied.

Stopping the operation of a contactor-controlled motor is achieved by pushing the "stop" button. This action breaks the contactor coil source line, which de-energizes the coil and causes the armature to drop out of position. The line contactor points to the motor also break contact, which in turn de-energizes the power line to the motor. This path then becomes incomplete and motor operation stops immediately.

The relay action of a motor contactor is designed to have a latching characteristic that holds it into operation once it is energized. This condition of operation is a necessity in motor control applications. In addition to this, contactors are also called upon to actuate a series of operations in a particular sequence. This involves different logic combinations in order to provide an appropriate alternative.

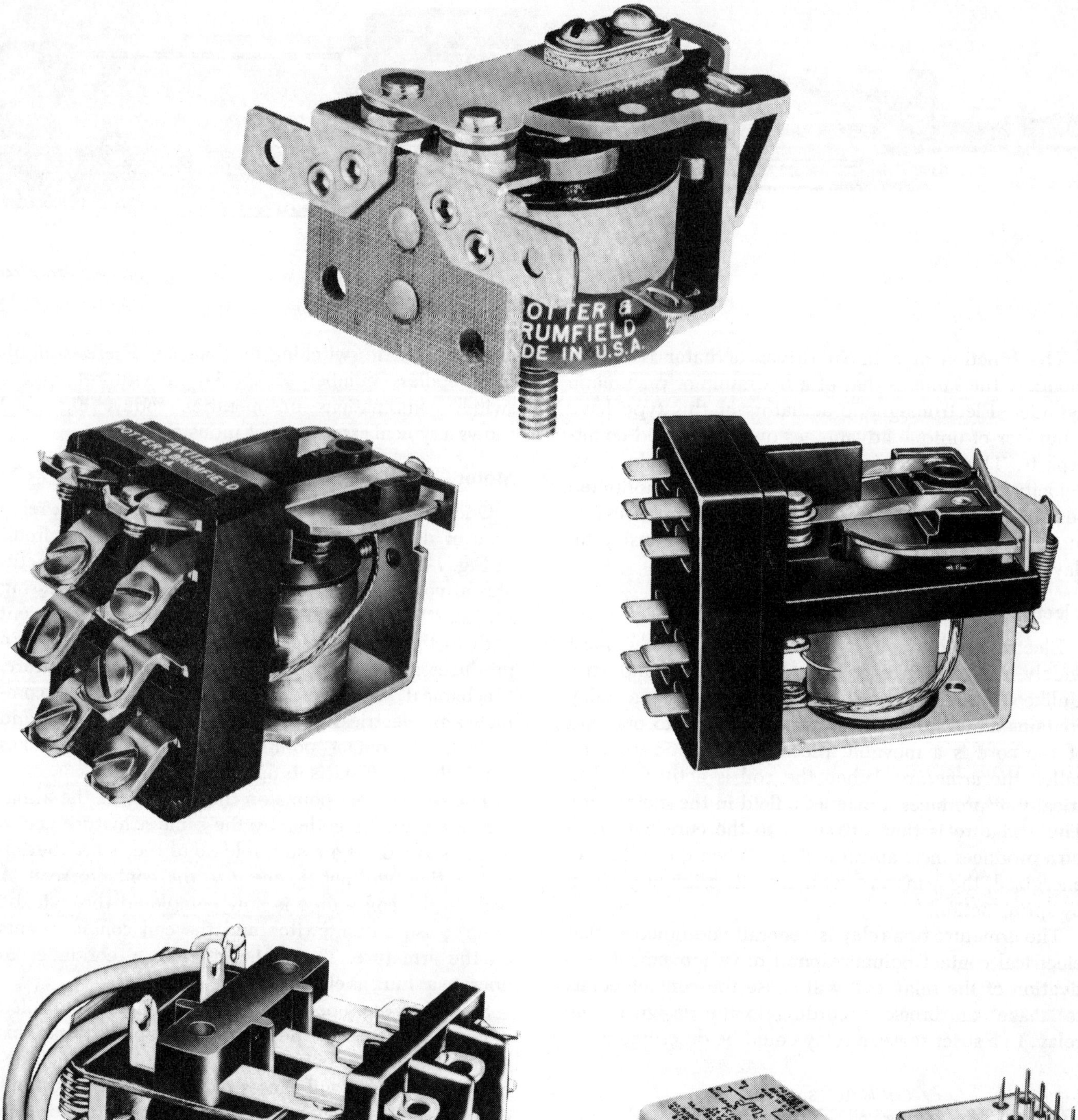

Courtesy Potter & Brumfield Div., AMF, Inc.

Fig. 7-42. A typical assortment of industrial relays.

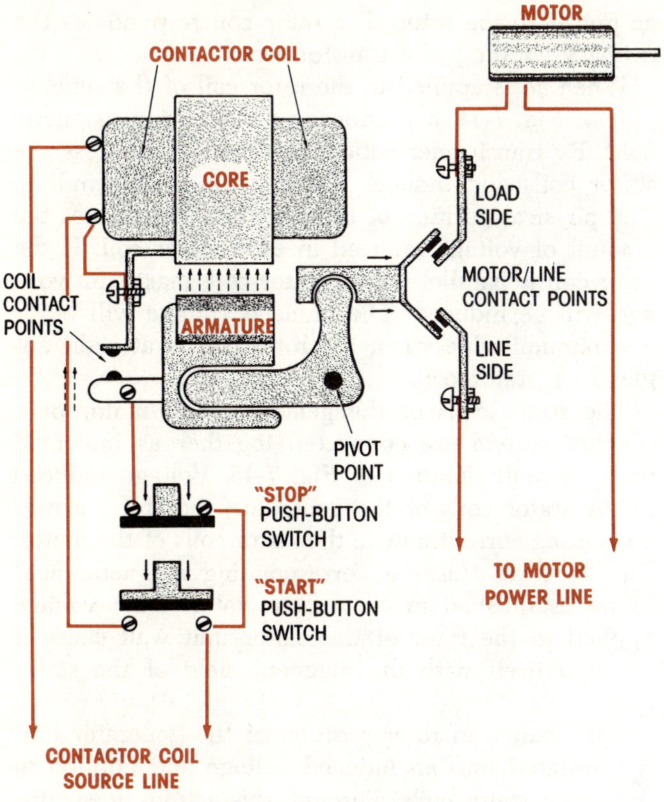

Fig. 7-43. A motor contactor circuit.

Contactors are also designed to provide delay of the starting and stopping operation of a motor after the appropriate button has been pushed. This type of operation only provides more alternatives for the vast number of industrial relay applications being used today.

Rotary Electrical Actuators

In industry today there is a need for devices that produce a type of rotary motion that is somewhat different from that produced by an electric motor. This type of actuator employs rotary motion to control the angular position of one shaft by positioning the shaft of a second device. Synchro systems and servomechanisms are used in industry to achieve this basic operation. Through these devices, it becomes possible to transmit rotary motion between locations without direct mechanical linkage.

Synchro systems are classified as two or more motor/generator units connected together in a way that will permit the transmission of angular shaft positions by electromagnetic field changes. When an operator turns the generator shaft of the unit to a certain position, it automatically rotates the motor shaft to an equivalent position at a remote location. Through this type of system, it is possible to achieve accurate control over great distances. Computers often employ these units to determine physical changes that take place in automation.

In synchro systems that require increased rotational torque or precise movements of a control device, servomechanisms are employed. A servomechanism is ordinarily a special type of ac or dc motor that drives a precision piece of equipment in specific increments. Systems that include servomechanisms generally require amplifiers and error-detecting devices to control angular displacement of a shaft.

Synchro Systems—A synchro system contains two or more electromagnetic devices that are similar in appearance to small electric motors. These devices are connected together in such a way that the angular position of the generator shaft can easily be transmitted to the motor or receiver unit. Fig. 7-44 shows

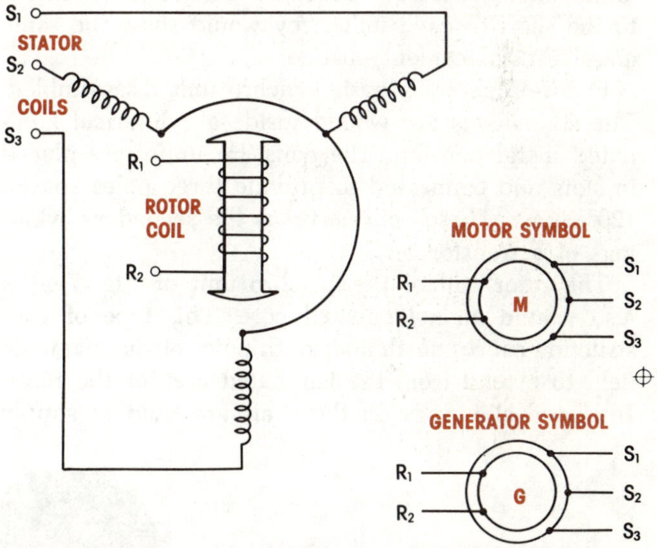

Fig. 7-44. Schematic diagram and symbols for a basic synchro unit.

a schematic diagram of a basic synchro unit. As a general rule, the generator and motor units are identical electrically. Physically, the motor unit has a metal flywheel attached to its shaft to prevent shaft oscillations or vibrations when it is powered. The letters G and M inside of the electrical symbol denote the generator or motor functions.

Fig. 7-45 shows the circuit diagram of a basic synchro system. Single-phase ac line voltage is used to power this particular system. Note that the line voltage is applied to the rotors of both the generator and motor. The stationary coils, or stator windings, are connected together as indicated.

When power is initially applied to the system, the motor will position itself according to the location of the generator shaft. No physical change will take place after the motor unit aligns itself with the generator position. Both units will remain in a stationary con-

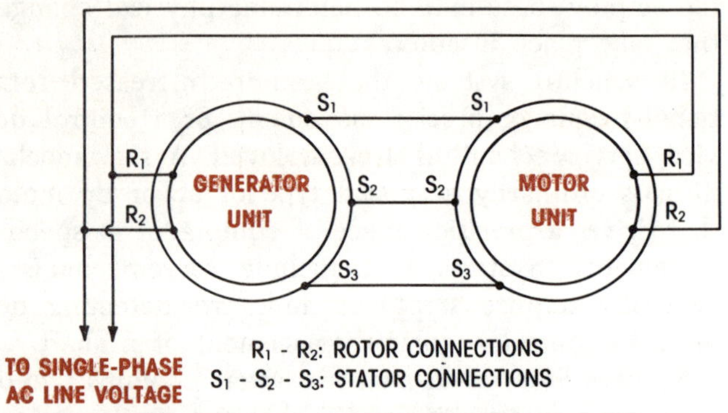

Fig. 7-45. Circuit diagram of a basic synchro system.

dition until some further action takes place. Turning the generator shaft a certain number of degrees in a clockwise direction will cause a corresponding change in the motor unit. If calibrated dials were attached to the shaft of each unit, they would show the same angular displacement change.

Fig. 7-46 shows a basic synchro unit disassembled. The stator coils are wound inside a cylindrical laminated metal housing. The coils are uniformly placed in slots and connected to provide three poles spaced 120° apart. These coils serve as the secondary windings of a transformer.

The rotor coil of the synchro unit of Fig. 7-46 is also wound on a laminated core. This type of construction causes north and south poles of the magnetic field to extend from the laminated area of the rotor. Insulated slip rings on the shaft are used to supply ac power to the rotor. The rotor coil responds as the primary winding of a transformer.

When ac is applied to the rotor coil of the synchro unit of Fig. 7-44, it produces an alternating magnetic field. By transformer action, this field cuts across the stator coils and induces a voltage in each winding. The physical position of the rotor coil determines the amount of voltage induced in each stator coil. If the rotor coil is parallel with a stator coil, maximum voltage will be induced. The induced voltage will be of a minimum value when the rotor coil is at right angles to a stator coil.

The stator coils of the generator and motor of a synchro system are connected together as indicated in the circuit diagram of Fig. 7-45. Voltage induced in the stator coils of the generator, therefore, causes a resulting current flow in the stator coils of the motor. This in turn causes a corresponding magnetic field to be established in the motor stator. Line voltage applied to the rotor of the motor unit wall cause it to align itself with the magnetic field of the stator coils.

Any change in rotor position of the generator unit is translated into an induced voltage and applied to the motor stator coils. Through this action, linear displacement changes can be effectively transmitted to the motor through three rather small stator coil wires. Systems of this type are becoming very important in industrial automatic process control applications today.

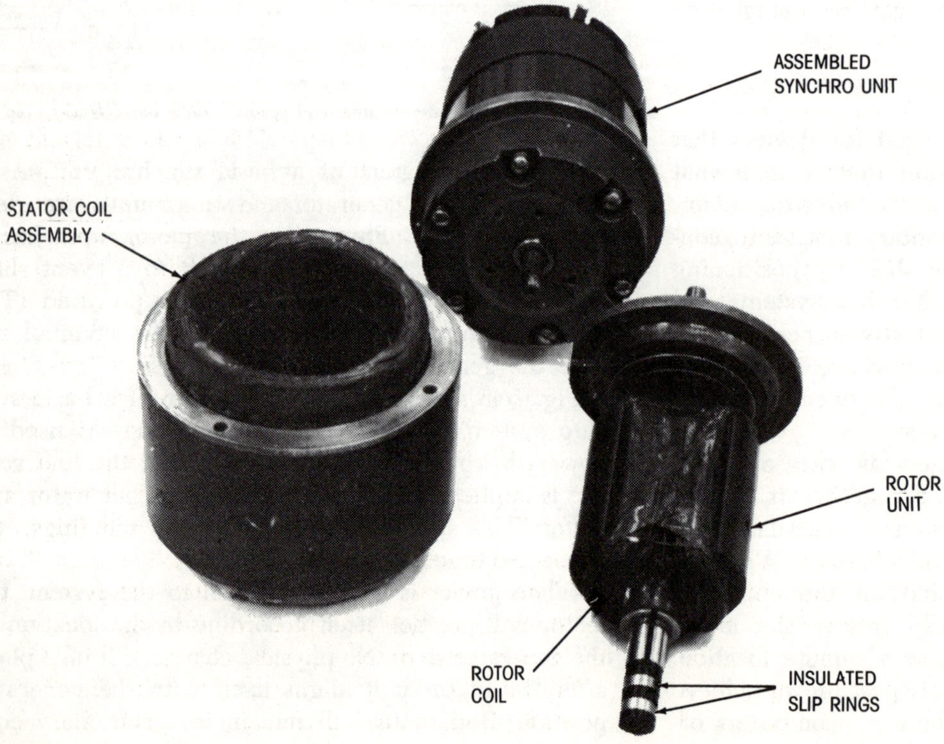

Fig. 7-46. Disassembled view of a synchro unit.

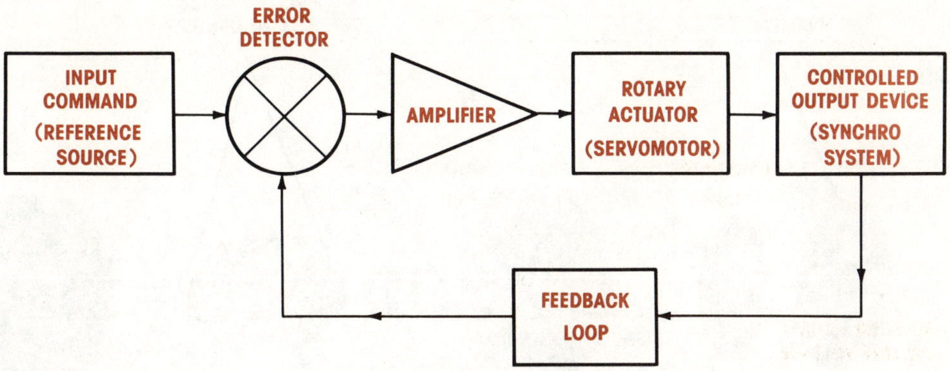

Fig. 7-47. Block diagram of a typical servo system.

Servo Systems—Servo systems are a specific type of rotating machine concerned with such things as changing the mechanical position or speed of an object. Mechanical position applications include numerical control machinery and process control indicating equipment. Speed applications are found in conveyor-belt control units, spindle speed control in machine tool operations, and disk or magnetic-tape drives for computers. As a general rule, a servo system is a rather complex unit that follows the commands of a closed-loop control path. Fig. 7-47 shows the components of a typical servo system.

The input of a servo system serves as the reference source or as a set point to which the load element responds. By changing the input in some way, a command is applied to the error detector. This device receives data from both the input source and from the controlled output device. If a correction is needed with reference to the input command, it is amplified and applied to the actuator. The actuator is normally a servomotor that produces controlled shaft displacements. The controlled output device is usually a synchro system that relays information back to the error detector for position comparison.

In the discussion that follows, we will center our attention on the servomotor part of the system. This component is primarily responsible for producing mechanical changes from an electromagnetic actuating device. A device of this type is normally coupled to the work load by a gear train or some mechanical linkage. Both ac and dc servomotors are used to achieve this operation.

Servomotors—A servomotor is a very special type of device that is used to achieve a precise degree of rotary motion. As a general rule, a servomotor is unique when compared with other electric motors. Servomotors, for example, are designed to do something other than change electrical energy into rotating mechanical energy. Motors of this type must first be able to respond accurately to signals developed by the system's amplifier. Secondly they must be capable of reversing direction quickly when a specific signal polarity is applied. Also, the amount of torque developed by a servomotor must be quite high. As a general rule, developed torque is a function of the voltage and current supply source.

Two distinct types of servomotors are used today to achieve these operating conditions. An ac type of motor, called a *synchronous* motor, is commonly used in low power applications. Excessive amounts of heat developed during starting conditions normally limit this motor to rather low output power applications. Dc stepping motors represent a unique application of the servomotor.

AC Synchronous Motors—The construction of an ac synchronous motor is quite simple. It contains no brushes, commutators, slip rings, or centrifugal-force switches. As shown in Fig. 7-48, it is simply made up of a rotor and a stator assembly. There is no direct physical contact between the rotor and stator. A carefully maintained air gap is always present between these two components. As a result of this construction, the motor has a long operating life and is highly reliable.

The speed of a synchronous motor is directly proportional to the frequency of the applied ac and the number of pairs of stator poles. Since the number of stator poles cannot be effectively altered after the motor has been manufactured, frequency is the most significant speed factor. Speeds of 28, 72, and 200 r/min are typical, with 72 r/min being a common industrial numerical control standard.

Fig. 7-49 shows the stator layout of a two-phase synchronous motor with four poles per phase. Poles N_1–S_3 and N_5–S_7 represent one phase while poles N_2–S_4 and N_6–S_8 represent the second phase. There are places for 48 teeth around the inside of the stator of Fig. 7-49. One tooth per pole, however, has been eliminated to provide a space for the windings. Five teeth per pole, or a total of 40 teeth, are formed on the stator. The four coils of each phase are connected in series to achieve the correct polarity.

The rotor of the synchronous motor in Fig. 7-48 is an axially magnetized permanent magnet. There are

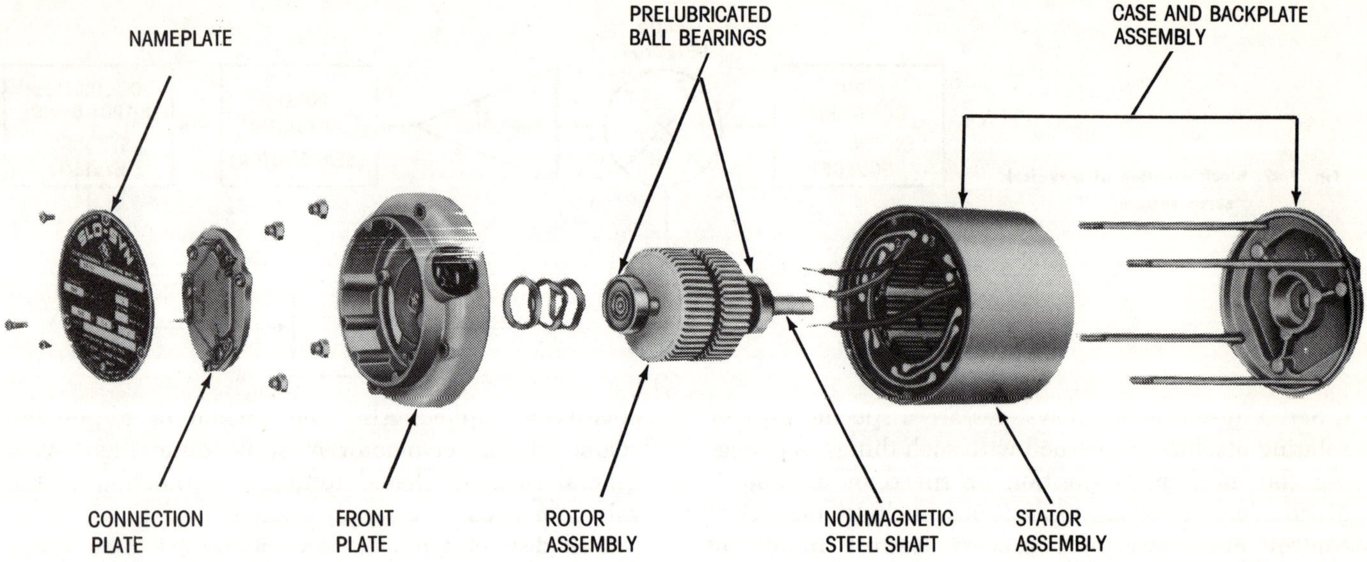

Courtesy The Superior Electric Co.

Fig. 7-48. Exploded view of an ac synchronous motor.

50 teeth cast into its form. The front section of the rotor has one polarity while the back section has the opposite polarity. The physical difference in the number of stator teeth (40) and rotor teeth (50) means that only two teeth of each part can be properly aligned simultaneously. With one section of the rotor being a north pole and the other section being a south pole, the rotor has the ability to stop very quickly. It can also produce complete direction reversals without hesitation because of this gear-like construction.

A circuit diagram of a single-phase synchronous motor is shown in Fig. 7-50. The resistor and capacitor of this circuit are used to produce a 90° phase shift in one winding. As a result, the two windings are always out of phase regardless of whether the switch is in the clockwise (cw) or counterclockwise (ccw) position. When power is applied, the four coils of one phase produce an electromagnetic field. The rotor is attracted and aligns itself to these stator coils. Then, 90° later, the four coils of the second phase produce a corresponding field. The stator is again attracted to this position. As a result of this action, the rotor sees a moving force across first one phase and then the other. This force gives the rotor the needed torque that causes it to start and continue rotation when power is applied.

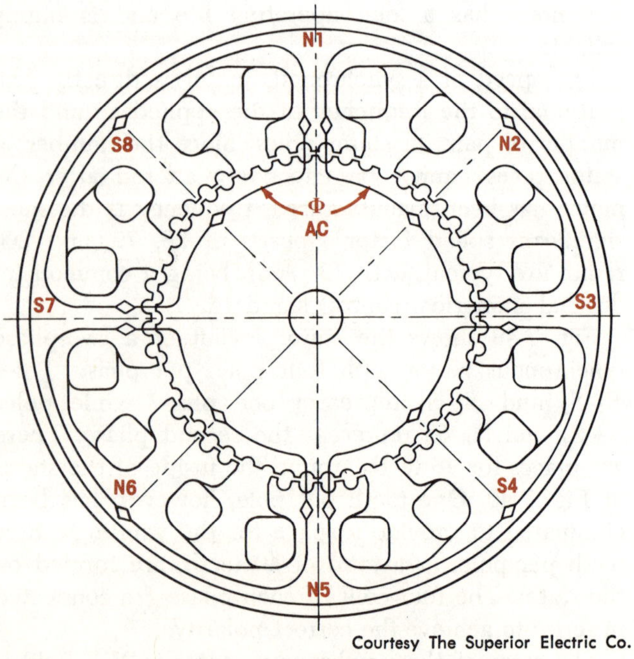

Courtesy The Superior Electric Co.

Fig. 7-49. The stator layout of a two-phase ac synchronous motor.

The synchronous motor just described has the capability of starting in one and one-half cycles of the applied line-voltage frequency. In addition to this, it can be stopped in five mechanical degrees of rotation. These two characteristics are primarily attributed to the geared rotor and stator construction. Synchronous motors of this type have one other important characteristic—they draw the same amount of line current when stalled that they do when operating. This characteristic is very important in automatic machine-tool applications where overloads occur frequently.

DC Stepping Motors—Dc stepping motors represent a unique rotary actuator that is used in the control of automatic industrial processing equipment. Dc motors of this type are found in nearly all high-power servomechanisms. They are more efficient and develop significantly more torque than the synchronous servomotor. The stepping motor is used primarily to change electrical pulses into rotary motion that can be used to produce mechanical movements.

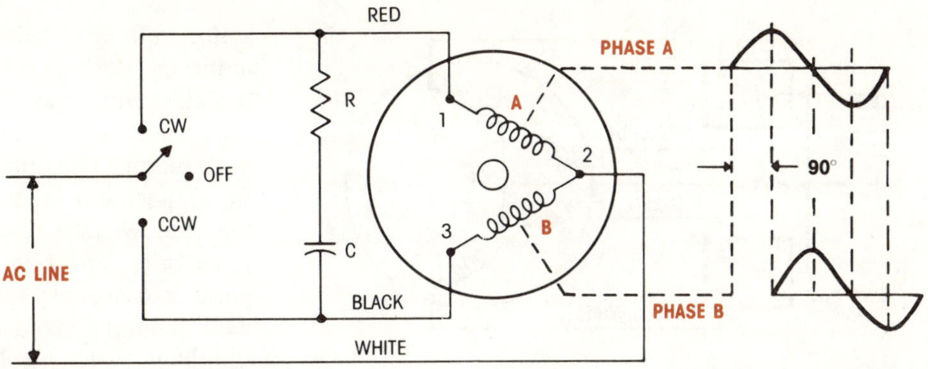

Fig. 7-50. Circuit diagram of a single-phase synchronous motor.

Courtesy The Superior Electric Co.

The shaft of a dc stepping motor rotates a specific number of mechanical degrees with each incoming pulse of electrical energy. The amount of rotary movement or angular displacement produced by each pulse can be repeated precisely with each succeeding pulse from the drive source. The resulting output of this device is used to accurately locate or position automatic process machinery.

The velocity, distance, and direction of a specific device can be moved or controlled by a dc stepping motor. The movement error of this device is generally less than 5% per step. This error factor is not cumulative regardless of the distance moved or the number of steps taken. Motors of this type are energized by a dc drive amplifier that is controlled by logic circuits. The drive amplifier circuitry is a key factor in the overall performance of this motor.

The basic construction of a dc stepping motor is very similar to that of the ac synchronous motor of Fig. 7-48. The stator construction and coil placement are also the same as that of the synchronous motor layout shown in Fig. 7-49. One manufacturer makes some servomotors that can be operated as either an ac synchronous motor or as a dc phase-switched stepping motor. In order to achieve this, the rotor must be of a permanent-magnet type of construction.

Stepping-Motor Operation—A very important principle that applies to the operation of a dc stepping motor states that," like magnetic poles repel and unlike magnetic poles attract." If a permanent-magnet rotor were placed between two series-connected stator coils, it would produce the situation of Fig. 7-51A. With power applied to the stator, the rotor would be repelled in either direction. The direction of rotation in this case is unpredictable. Adding two more stator coils to a simple motor, as indicated in Fig. 7-51B, would make the direction of rotation very predictable. With the stator polarities indicated, the rotor would align itself midway between the two pairs of stator coils. The direction of rotation can now be predicted, and is determined by the polarities of the

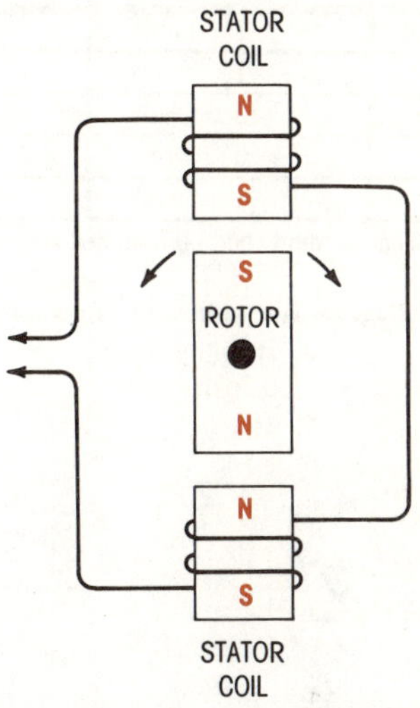

(A) With one set of stator coils, direction of rotation of rotor is unpredictable.

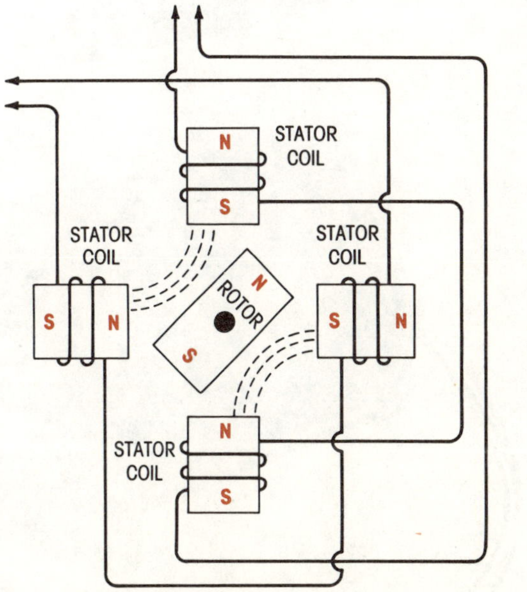

(B) With two sets of stator coils, and polarities as shown, rotor will align itself as shown.

Fig. 7-51. Illustration of the magnetic principle involved in the operation of a dc stepping motor.

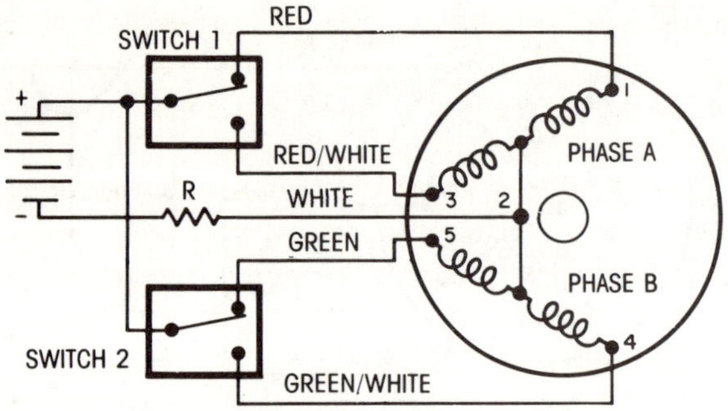

SWITCHING SEQUENCE *

STEP	SWITCH 1	SWITCH 2
1	1	5
2	1	4
3	3	4
4	3	5
1	1	5

* To reverse direction, read chart up from bottom.

Courtesy The Superior Electric Co.

Fig. 7-52. Circuit diagram and switching sequence of a bifilar-wound dc stepping motor.

stator coil sets. Adding more stator coil pairs to a motor of this type improves its rotation and makes the stepping action very accurate.

Fig. 7-52 shows an electrical diagram of a dc stepping motor. The stator coils of this motor are wound by a special type of construction called *bifilar*. Two separate wires are wound into the coil slots at the same time. The two wires are smaller in size, which permits twice as many turns as could be achieved with a larger sized wire. Construction of this type simplifies the control circuitry and dc energy-source requirements.

Operation of the stepping motor of Fig. 7-52 is achieved in a four-step switching sequence. Any of the four combinations of switches 1 or 2 will produce an appropriate rotor position location. After the four switch combinations have been achieved, the switching cycle repeats itself. Each switching combination causes the motor to move one-fourth of a step.

A rotor similar to the one shown in Fig. 7-49 has 50 teeth. Using this rotor in the circuit of Fig. 7-52

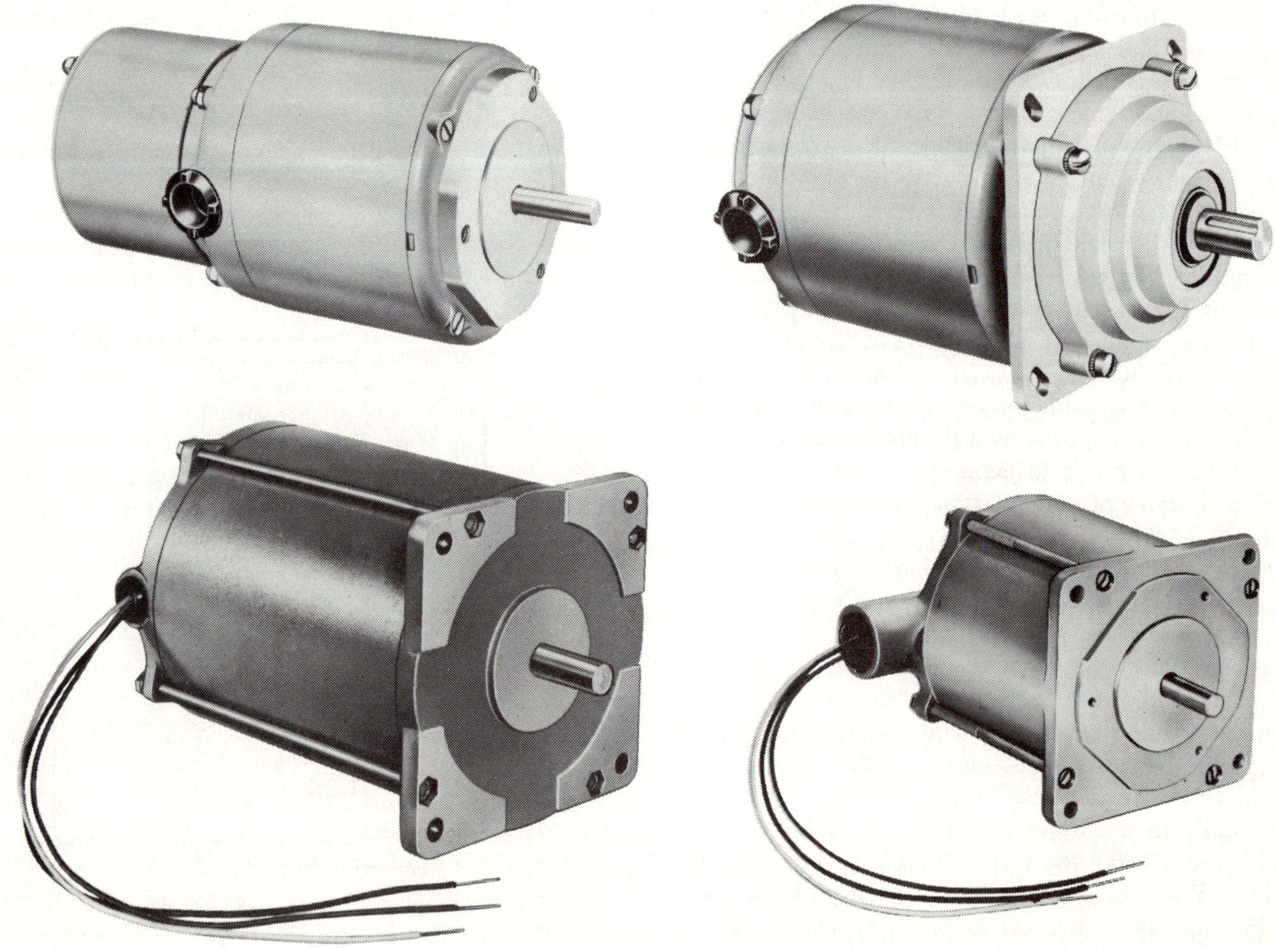

Courtesy The Superior Electric Co.

Fig. 7-53. An assortment of servomotors.

would permit four steps per tooth, or 200 steps per revolution. The amount of linear displacement, or step angle, of this motor is, therefore, determined by the number of teeth on the rotor and the switching sequence.

A stepping motor that takes 200 steps to produce one revolution will move 360°/200, or 1.8° per step. It is not unusual for stepping motors to use eight switching combinations to achieve one step. In this case, each switching combination could be used to produce 0.25° of linear displacement. Motors and switching circuits of this type permit a very precise type of controlled movement. Fig. 7-53 shows an assortment of slow-speed servomotors that can be used in either dc stepping applications or as ac synchronous motors.

SUMMARY

Mechanical power systems all contain an *energy source*, a *transmission path*, *control*, a *load device*, and possibly one or more *indicators* in order to operate. Systems of this type are designed to produce some unique type of mechanical motion. Hydraulic, pneumatic, and electrical systems can be designed to produce physical changes that have mechanical movement.

Fluid power is the generic term used to describe both hydraulic and pneumatic systems. The basic premise of a fluid power system is that pressure applied to a confined fluid is transmitted undiminished in all directions throughout the fluid. This basic premise is known today as *Pascal's law*.

Force is defined as something that modifies or produces motion. Weight expressions are used to denote force values. Applying a force to a specific unit area is called *pressure*. When an applied force causes an object to move, *work* is being done. *Power* refers to the length of time that work is being accomplished.

A unique difference between hydraulic and pneumatic systems is in the compressibility of fluids. Air is compressed to increase its pressure. Hydraulic fluid does not compress under normal operating conditions. Its pressure is developed by circulating fluid through system resistance.

The pump of a fluid system is designed to provide the system with fluid flow that will develop pressure. This device accepts fluid at an inlet port, forces it to move through a confined area, and expels it at an outlet port. Hydraulic pumps operate continuously while air compressors work on a demand basis. Positive-displacement pumps move certain volumes of fluid during each revolution. Nonpositive-displacement pumps have no set amount of fluid that will pass during one revolution. Reciprocating action, gears, and vanes are used in positive-displacement pumps. Centrifugal force is used to produce nonpositive pump operation.

Conditioning is needed to prolong component life in fluid systems. Filters, strainers, and heat exchangers are some of the common hydraulic conditioning components. Regulators, filters, and lubricators are used to condition air systems.

The transmission line of a fluid system is either flexible or rigid metal tubing. Rigid tubing is used in applications that are stationary and free of vibration. Flexible tubing is selected according to pressure, temperature range, and exposure resistance.

Fluid system control is designed to alter pressure, flow direction, and volume of flow. Pressure control is achieved by valves that relieve pressure, reduce its volume, bypass components, and operate them in a sequence. Direction control is designed to start, stop, or reverse system flow. One-, two-, three-, and four-way valves are some of the common types of directional control devices in operation today. Flow control is designed to alter the volume of fluid flow or flow rate. Flow is altered by restricting the path in some way.

The load of a fluid system that does useful work is commonly called an *actuator*. Linear motion is achieved by an actuating cylinder. Single-acting and double-acting cylinders are both used in industrial applications today. Applying pressure to one side of a piston causes it to move linearly. Applying pressure to the other side of the piston causes it to retract in a similar action.

Rotary actuators are designed to produce a limited amount of rotary motion in either direction. Single-vane motors are used to achieve this function.

Fluid motors are designed to produce continuous rotary motion. The power of a moving fluid or air flow causes vanes, gears, or pistons to move in a rotary direction. The power output of a fluid motor is commonly measured in horsepower.

Fluid system indicators are designed to measure pressure, temperature, and flow. Pressure is measured by physically changing the shape of an element. The physical change that results from this action is used to deflect a hand or move a marking stylus. Flow is measured by pressure differences, by developed voltage, or by striking a turbine blade or target. Clamp-on flow measurements are performed by determining the rate at which ultrasonic waves pass through the fluid.

Electromechanical power systems produce a very important part of the mechanical energy used by industry. Linear, reciprocating, and rotary motion are the resulting outputs of these systems.

Linear motion is achieved by solenoids, relays, contactors, and motor-driven actuators. Mechanical motion produced by these components is used to move objects, actuate electrical control devices, or to physically change certain operating conditions.

The rotary actuators discussed in this chapter are designed to produce controlled angular shaft displacements instead of continuous rotory motion. Synchro systems include a generator/motor unit connected in a way that will permit mechanical shaft changes to be developed by electromagnetic field changes. Turning the shaft of the generator a certain number of degrees will cause the motor or receiver unit to move an equal amount.

Servo systems are a specific type of rotary machine designed to change the mechanical position of an object or alter its speed. A command given to the input is applied to the error detector. Data from the output is compared with the command information to determine the type of control needed to honor the command. Correction signals are amplified and applied to the servomotor, which produces shaft displacement changes.

Servomotors are designed to achieve a precise degree of rotary motion. An ac synchronous motor is often used as a servomotor. Special permanent-magnet rotor construction is used to develop precise shaft displacements. An RC network is used to produce a 90° phase change that is used to produce rotation. Typical operating speeds are 28, 72, and 200 r/min. These devices are quick starting, easily stopped, and can reverse rotational direction without hesitation.

Dc stepping motors are unique rotary actuators. The shaft of this type of motor is designed to move a specific number of degrees when a switching combination is achieved. With two switches, the sequence repeats itself after four combinations. The amount of linear displacement produced is based on the number of switching combinations needed to make one step.

STUDY QUESTIONS

1. *Explain the physical makeup of a fluid power system in terms of source, transmission path, control, load, and indicator.*
2. *What are some of the unique differences in hydraulic and pneumatic systems?*
3. *Define the terms force, work, pressure, and power. How are these factors measured?*
4. *How does Pascal's law apply to fluid power systems?*
5. *What causes flow resistance in a fluid system?*
6. *How does the flow of hydraulic fluid produce pressure?*
7. *What causes pressure drop? Of what significance is it?*
8. *Explain the basic operating principles that apply to the following pumps: gear, reciprocating, vane, and centrifugal.*
9. *Why is conditioning equipment important?*
10. *What are the differences in hydraulic and pneumatic conditioning?*
11. *Explain how control applies pressure, direction, and flow.*
12. *What are some specific applications of one-, two-, three-, and four-way valves?*
13. *What are the differences in actuators that produce linear, reciprocating, and rotary motion in fluid systems?*
14. *Briefly explain the difference between a single-acting and a double-acting cylinder.*
15. *What are the principle differences between a fluid pump and a motor?*
16. *What do indicators display in fluid systems?*
17. *What are some of the common pressure indicator elements? How do they change physically?*
18. *Compare the operational basics of fluid and electromechanical power systems.*
19. *What is the difference between an electrical load and an actuator?*
20. *What is a synchro system?*
21. *What is servomotor?*
22. *How does a dc stepping motor operate?*

PROBLEMS

1. If a cylinder has a cross-sectional area of 60 in^2, what force is developed when pressures of 30 lb/in^2, 100 lb/in^2, and 500 lb/in^2 are applied?
2. What pressure is developed when 10 newtons of force is applied to 90 cm^2?
3. A man moves metal boxes 4 ft. If he moves 1.0 ton in 1.0 hr how much power was used? Horsepower?
4. If 100 lb of force is applied to a piston of 2 in^2, what force will be developed by a 100-in^2 piston in the same system?
5. If the 2-in^2 piston of problem 4 moves 4 in, how far will the 100-in^2 piston move?
6. If a 24-in diameter tank is filled to 25 ft, and a 2-in diameter pipe is filled to 25 ft, what would be the pressure developed at the bottom of the tank and pipe? Why?
7. If the compression possibility of hydraulic fluid is ½% per volume per 100 lb/in^2, how many cubic inches of compression would occur in a tank that is 4 ft tall and 24 in in diameter, when 1000 lb/in^2 is applied?
8. How many pounds of force would be developed by a 12-in diameter cylinder when 100 lb/in^2 of pressure is applied?
9. How many newtons of force would be developed by a 22-cm diameter cylinder when 400 pascals of pressure is applied?
10. How much torque in ft-lb is developed by a gear motor with a 1.0-ft arm radius and 1.25-in^3 displacement when 600 lb/in^2 is applied?

APPENDIX 1

Comparison Chart for U. S. and Metric Units

U.S.		METRIC	METRIC		U.S.
1 inch	= 25.4	millimeters	1 millimeter	= 0.03937	inch
1 foot	= 0.3048	meter	1 meter	= 3.2808	feet
1 yard	= 0.9144	meter	1 meter	= 1.0936	yards
1 mile	= 1.609	kilometers	1 kilometer	= 0.6214	mile
1 square inch	= 6.4516	square centimeters	1 square centimeter	= 0.155	square inch
1 square foot	= 0.0929	square meter	1 square meter	= 10.7639	square feet
1 square yard	= 0.836	square meter	1 square meter	= 1.196	square yards
1 acre	= 0.4047	hectare	1 hectare	= 2.471	acres
1 cubic inch	= 16.387	cubic centimeters	1 cubic centimeter	= 0.061	cubic inch
1 cubic foot	= 0.028	cubic meter	1 cubic meter	= 35.3147	cubic feet
1 cubic yard	= 0.764	cubic meter	1 cubic meter	= 1.308	cubic yards
1 quart (liq)	= 0.946	liter	1 liter	= 1.0567	quarts (liq)
1 gallon	= 0.00378	cubic meter	1 cubic meter	= 264.172	gallons
1 ounce (avdp)	= 28.349	grams	1 gram	= 0.035	ounce (avdp)
1 pound (avdp)	= 0.4536	kilogram	1 kilogram	= 2.2046	pounds (avdp)
1 horsepower	= 0.7457	kilowatt	1 kilowatt	= 1.341	horsepower

APPENDIX 2

Common Industrial Symbols

ELECTRICAL/ELECTRONIC SYMBOLS

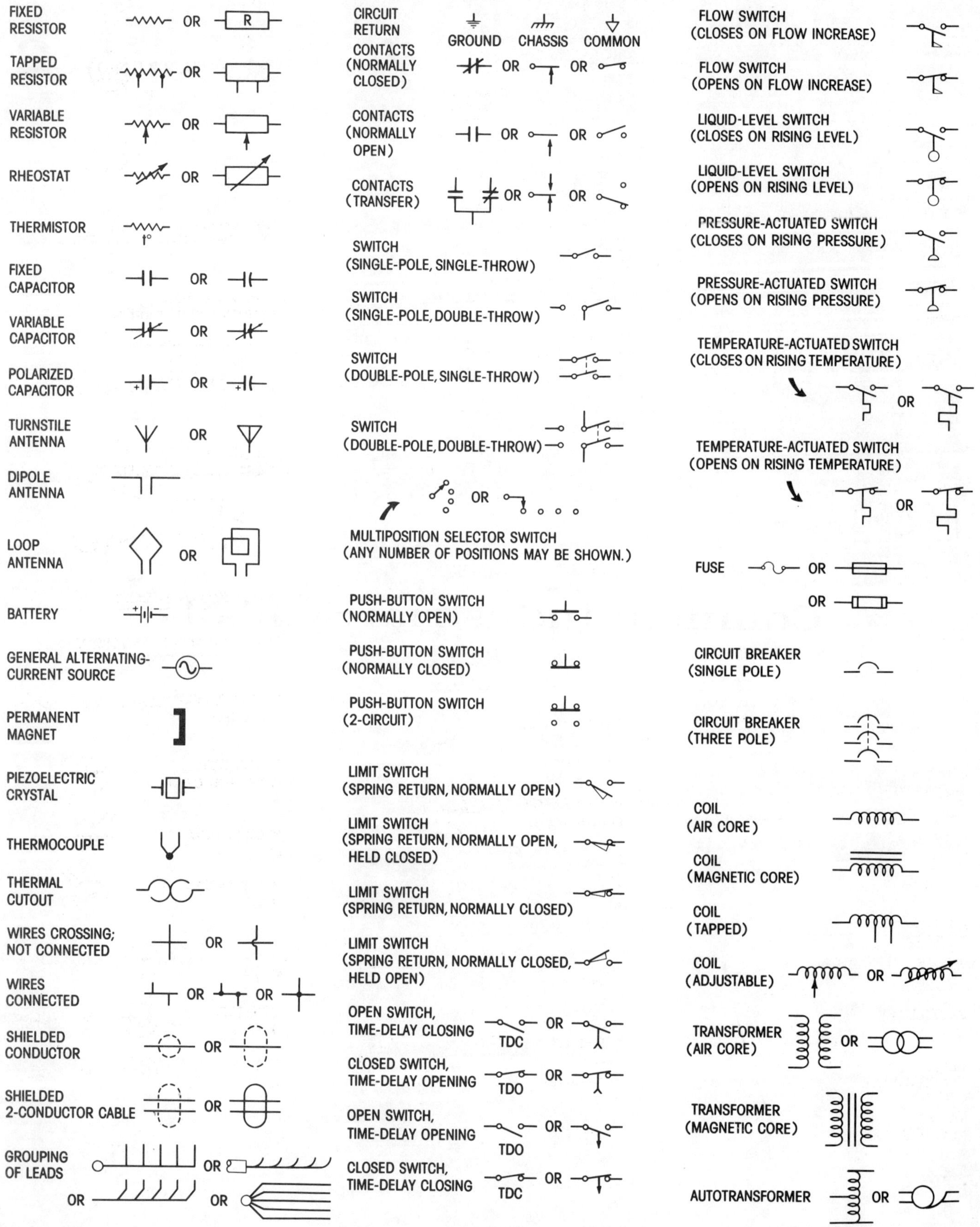

CURRENT TRANSFORMER

RELAY, TRANSFER CONTACTS

OR ... OR ®

CIRCUIT TERMINAL

TERMINAL BOARD OR STRIP ... OR

FEMALE CONNECTOR

MALE CONNECTOR

ENGAGED CONNECTORS

ENGAGED COAXIAL CONNECTORS (OUTSIDE CONDUCTOR CARRIED THROUGH)

JACK (2-CONDUCTOR)

PLUG (2-CONDUCTOR)

POWER SUPPLY CONNECTORS: NONPOLARIZED, 2-CONDUCTOR MALE

OR ... OR

POLARIZED, 3-CONDUCTOR FEMALE

OR ... OR

ELECTRON-TUBE SYMBOLS:

VACUUM-TYPE ENVELOPE

GAS-FILLED ENVELOPE

FILAMENT AND DIRECTLY HEATED CATHODE

INDIRECTLY HEATED CATHODE

COLD CATHODE

PHOTOCATHODE

POOL CATHODE

IONICALLY HEATED CATHODE

GRID

DEFLECTING ELECTRODE

ANODE OR PLATE

TARGET OR X-RAY ANODE

DYNODE

IGNITOR OR STARTER

VACUUM-TYPE DIODE ... OR

GAS-FILLED DIODE ... OR

COLD-CATHODE, GAS-FILLED DIODE

VACUUM-TYPE PHOTOTUBE

MULTIPLIER-TYPE PHOTOTUBE

X-RAY TUBE

VACUUM-TYPE TRIODE ... OR

GAS-FILLED TRIODE (THYRATRON) ... OR

VACUUM-TYPE TEDRODE

VACUUM-TYPE PENTODE

BEAM-POWER TUBE

MERCURY-POOL TUBE (IGNITRON)

CATHODE-RAY TUBE (ELECTROSTATIC DEFLECTION)

CATHODE-RAY TUBE (ELECTROMAGNETIC DEFLECTION)

SEMICONDUCTOR SYMBOLS:

DIODE (A) ANODE CATHODE (K)

CAPACITIVE DIODE (VARACTOR) ... OR

TEMPERATURE-DEPENDENT DIODE

PHOTOSENSITIVE DIODE

PHOTOEMISSIVE DIODE

ZENER DIODE ... OR

THYRECTOR DIODE ... OR

TUNNEL DIODE ... OR

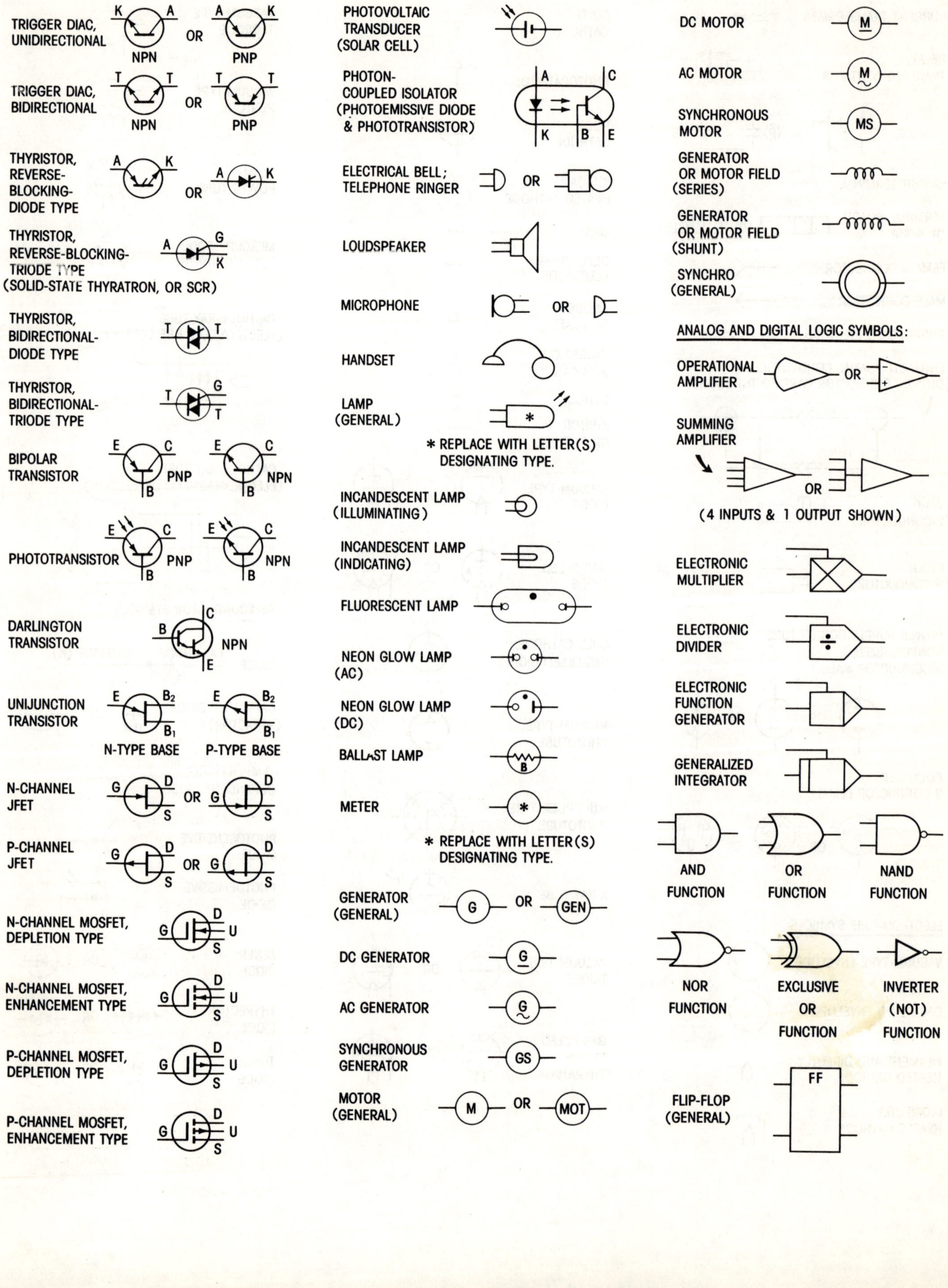
TRIGGER DIAC, UNIDIRECTIONAL
K A OR A K
NPN PNP
TRIGGER DIAC, BIDIRECTIONAL
T T OR T T
NPN PNP
THYRISTOR, REVERSE-BLOCKING-DIODE TYPE
A K OR A K
THYRISTOR, REVERSE-BLOCKING-TRIODE TYPE (SOLID-STATE THYRATRON, OR SCR)
A G K
THYRISTOR, BIDIRECTIONAL-DIODE TYPE
T T
THYRISTOR, BIDIRECTIONAL-TRIODE TYPE
T G T
BIPOLAR TRANSISTOR
E C B PNP
E C B NPN
PHOTOTRANSISTOR
E C B PNP
E C B NPN
DARLINGTON TRANSISTOR
C B E NPN
UNIJUNCTION TRANSISTOR
E B2 B1 E B2 B1
N-TYPE BASE P-TYPE BASE
N-CHANNEL JFET
G D S OR G D S
P-CHANNEL JFET
G D S OR G D S
N-CHANNEL MOSFET, DEPLETION TYPE
G D U S
N-CHANNEL MOSFET, ENHANCEMENT TYPE
G D U S
P-CHANNEL MOSFET, DEPLETION TYPE
G D U S
P-CHANNEL MOSFET, ENHANCEMENT TYPE
G D U S
PHOTOVOLTAIC TRANSDUCER (SOLAR CELL)
PHOTON-COUPLED ISOLATOR (PHOTOEMISSIVE DIODE & PHOTOTRANSISTOR)
A C K B E
ELECTRICAL BELL; TELEPHONE RINGER
OR
LOUDSPEAKER
MICROPHONE
OR
HANDSET
LAMP (GENERAL)
*
* REPLACE WITH LETTER(S) DESIGNATING TYPE.
INCANDESCENT LAMP (ILLUMINATING)
INCANDESCENT LAMP (INDICATING)
FLUORESCENT LAMP
NEON GLOW LAMP (AC)
NEON GLOW LAMP (DC)
BALLAST LAMP
B
METER
*
* REPLACE WITH LETTER(S) DESIGNATING TYPE.
GENERATOR (GENERAL)
G OR GEN
DC GENERATOR
G
AC GENERATOR
G
SYNCHRONOUS GENERATOR
GS
MOTOR (GENERAL)
M OR MOT
DC MOTOR
M
AC MOTOR
M
SYNCHRONOUS MOTOR
MS
GENERATOR OR MOTOR FIELD (SERIES)
GENERATOR OR MOTOR FIELD (SHUNT)
SYNCHRO (GENERAL)
ANALOG AND DIGITAL LOGIC SYMBOLS:
OPERATIONAL AMPLIFIER
OR
SUMMING AMPLIFIER
OR
(4 INPUTS & 1 OUTPUT SHOWN)
ELECTRONIC MULTIPLIER
ELECTRONIC DIVIDER
ELECTRONIC FUNCTION GENERATOR
GENERALIZED INTEGRATOR
AND FUNCTION
OR FUNCTION
NAND FUNCTION
NOR FUNCTION
EXCLUSIVE OR FUNCTION
INVERTER (NOT) FUNCTION
FLIP-FLOP (GENERAL)
FF

FLUID POWER SYMBOLS

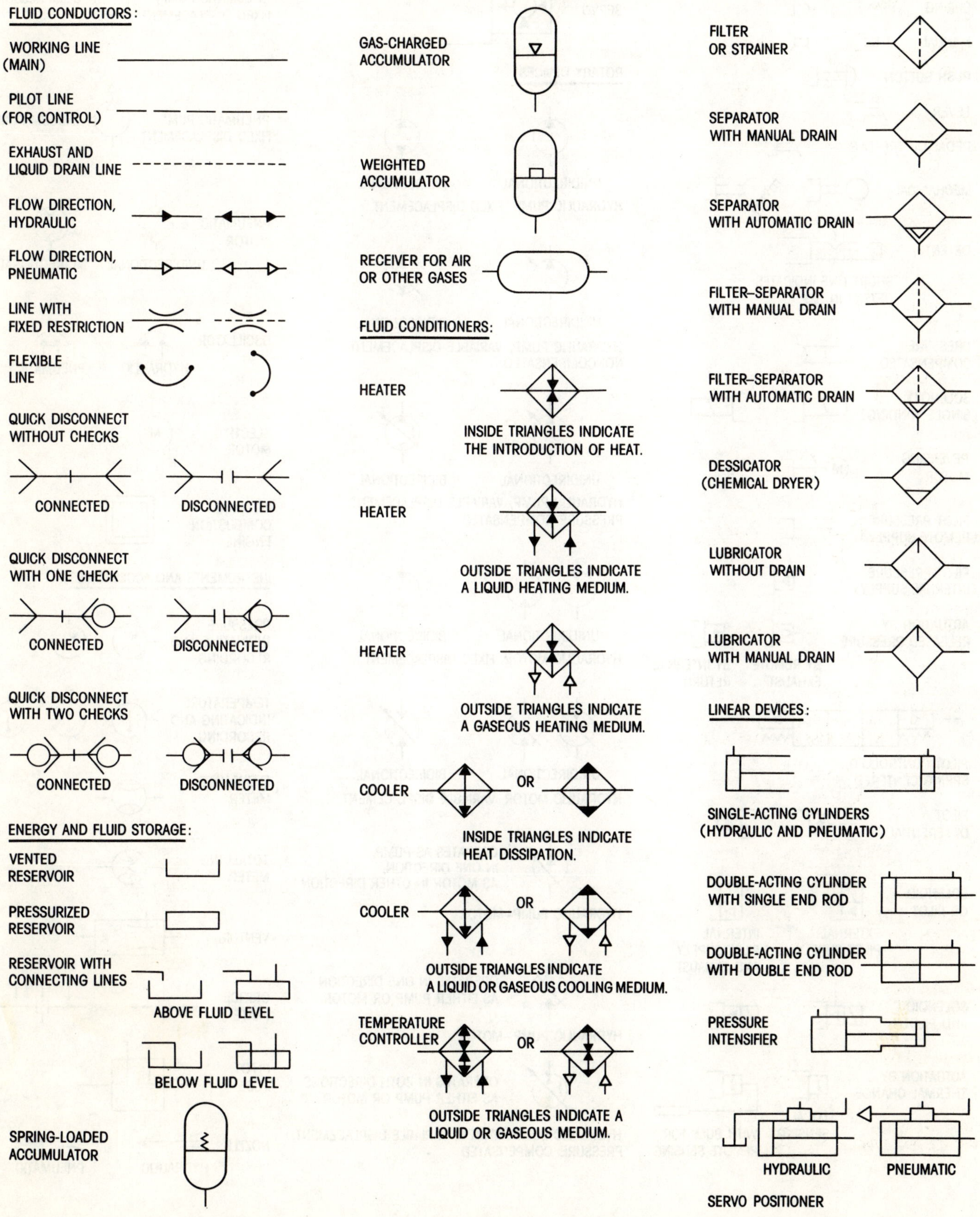

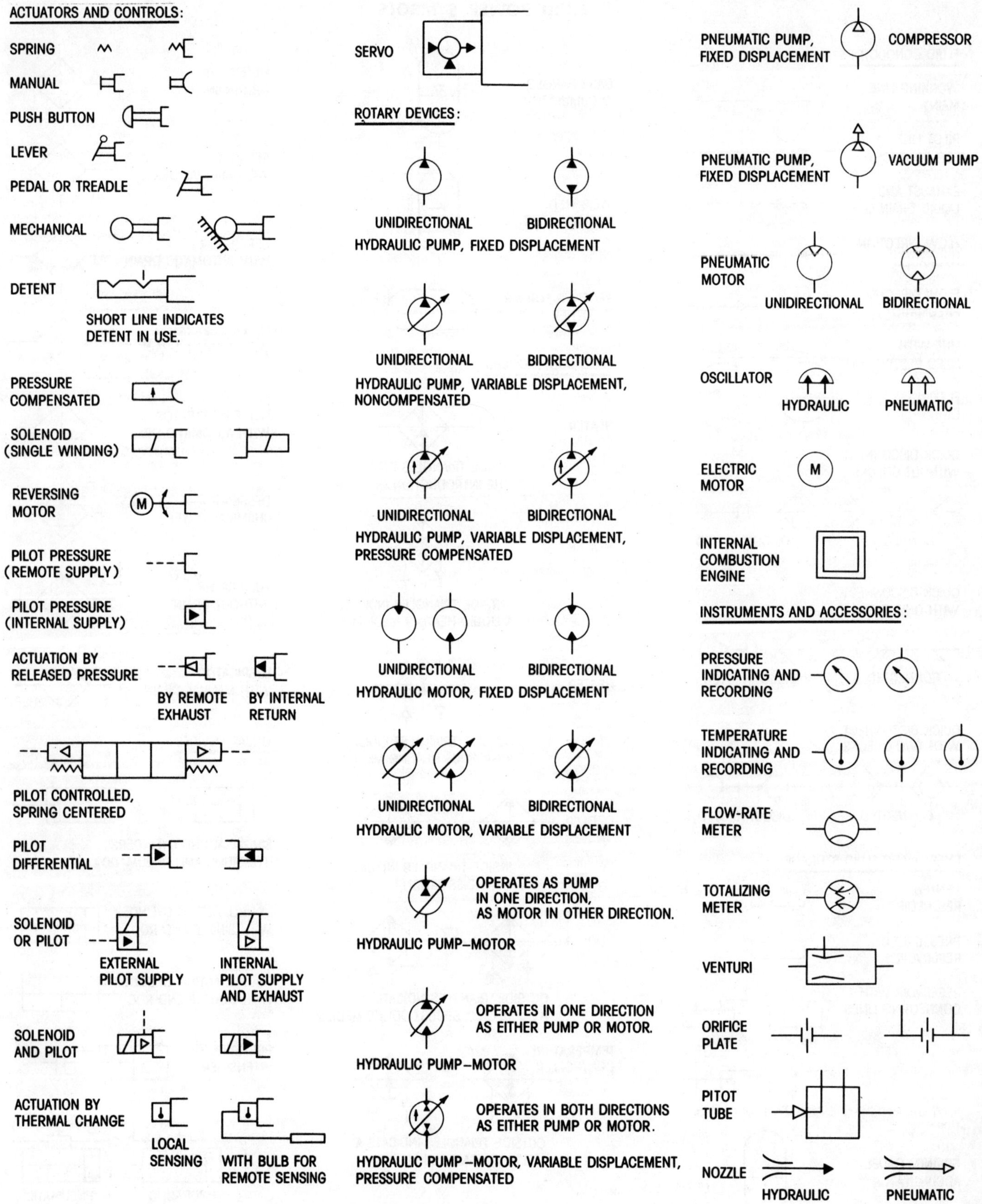

ACTUATORS AND CONTROLS:
SPRING
MANUAL
PUSH BUTTON
LEVER
PEDAL OR TREADLE
MECHANICAL
DETENT
SHORT LINE INDICATES DETENT IN USE.
PRESSURE COMPENSATED
SOLENOID (SINGLE WINDING)
REVERSING MOTOR
M
PILOT PRESSURE (REMOTE SUPPLY)
PILOT PRESSURE (INTERNAL SUPPLY)
ACTUATION BY RELEASED PRESSURE
BY REMOTE EXHAUST
BY INTERNAL RETURN
PILOT CONTROLLED, SPRING CENTERED
PILOT DIFFERENTIAL
SOLENOID OR PILOT
EXTERNAL PILOT SUPPLY
INTERNAL PILOT SUPPLY AND EXHAUST
SOLENOID AND PILOT
ACTUATION BY THERMAL CHANGE
LOCAL SENSING
WITH BULB FOR REMOTE SENSING
SERVO
ROTARY DEVICES:
UNIDIRECTIONAL
BIDIRECTIONAL
HYDRAULIC PUMP, FIXED DISPLACEMENT
UNIDIRECTIONAL
BIDIRECTIONAL
HYDRAULIC PUMP, VARIABLE DISPLACEMENT, NONCOMPENSATED
UNIDIRECTIONAL
BIDIRECTIONAL
HYDRAULIC PUMP, VARIABLE DISPLACEMENT, PRESSURE COMPENSATED
UNIDIRECTIONAL
BIDIRECTIONAL
HYDRAULIC MOTOR, FIXED DISPLACEMENT
UNIDIRECTIONAL
BIDIRECTIONAL
HYDRAULIC MOTOR, VARIABLE DISPLACEMENT
OPERATES AS PUMP IN ONE DIRECTION, AS MOTOR IN OTHER DIRECTION.
HYDRAULIC PUMP–MOTOR
OPERATES IN ONE DIRECTION AS EITHER PUMP OR MOTOR.
HYDRAULIC PUMP–MOTOR
OPERATES IN BOTH DIRECTIONS AS EITHER PUMP OR MOTOR.
HYDRAULIC PUMP–MOTOR, VARIABLE DISPLACEMENT, PRESSURE COMPENSATED
PNEUMATIC PUMP, FIXED DISPLACEMENT
COMPRESSOR
PNEUMATIC PUMP, FIXED DISPLACEMENT
VACUUM PUMP
PNEUMATIC MOTOR
UNIDIRECTIONAL
BIDIRECTIONAL
OSCILLATOR
HYDRAULIC
PNEUMATIC
ELECTRIC MOTOR
M
INTERNAL COMBUSTION ENGINE
INSTRUMENTS AND ACCESSORIES:
PRESSURE INDICATING AND RECORDING
TEMPERATURE INDICATING AND RECORDING
FLOW-RATE METER
TOTALIZING METER
VENTURI
ORIFICE PLATE
PITOT TUBE
NOZZLE
HYDRAULIC
PNEUMATIC

PRESSURE SWITCH

MUFFLER

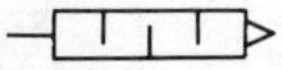

VALVES, BASIC SYMBOLS:

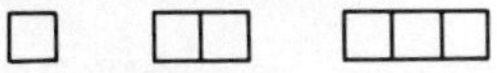

ENVELOPES

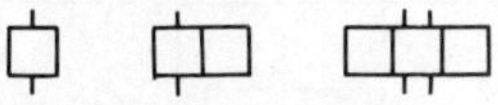

ENVELOPES WITH PORTS

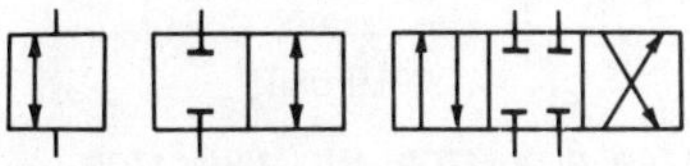

PORTS INTERNALLY BLOCKED

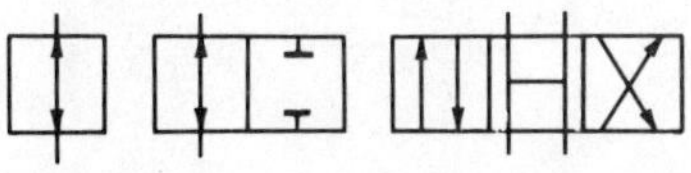

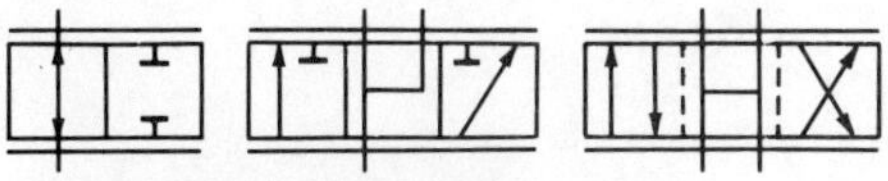

FLOW PATHS INTERNALLY OPEN

TWO-WAY VALVES (2-PORTED):

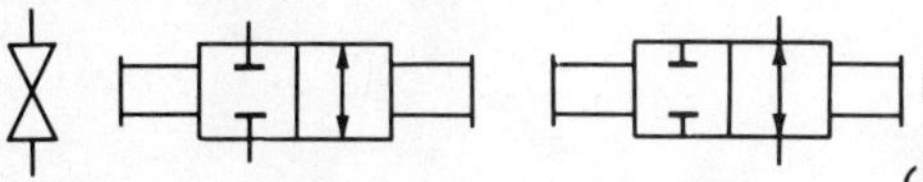

SIMPLIFIED OFF ON

MANUAL SHUTOFF ON-OFF VALVE

SIMPLIFIED

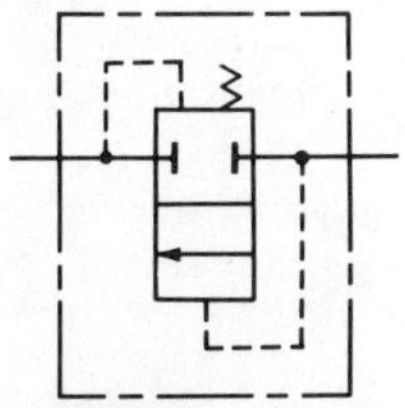

COMPOSITE
(FLOW BLOCKED TO RIGHT; PERMITTED TO LEFT.)

CHECK VALVE

CHECK VALVE, PILOT-OPERATED TO OPEN

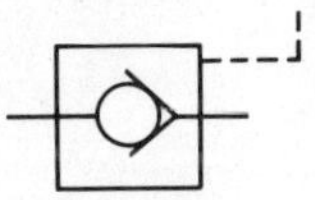

CHECK VALVE, PILOT-OPERATED TO CLOSE

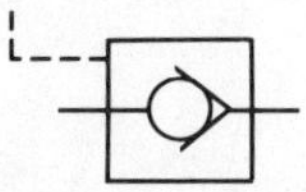

TWO-POSITION TWO-WAY VALVES

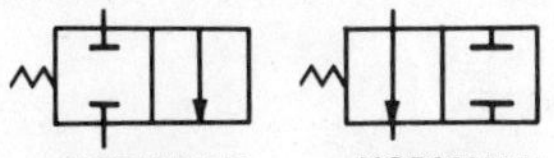

NORMALLY CLOSED NORMALLY OPEN

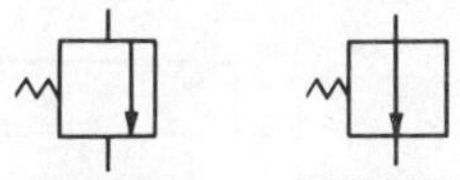

NORMALLY CLOSED NORMALLY OPEN

INFINITE-POSITION TWO-WAY VALVES

THREE-WAY VALVES:

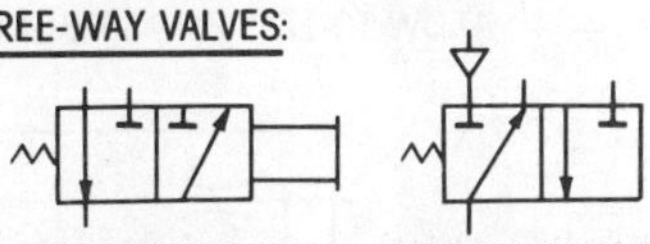

NORMALLY OPEN NORMALLY CLOSED

TWO-POSITION

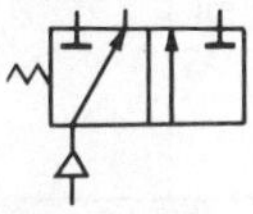

PRESSURE IS DISTRIBUTED FIRST TO ONE PORT, THEN THE OTHER.

TWO-POSITION DISTRIBUTOR

TWO-POSITION, TWO-PRESSURE

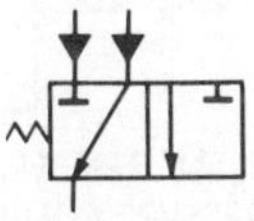

DOUBLE CHECK VALVE WITHOUT CROSS BLEED (ONE-WAY FLOW)

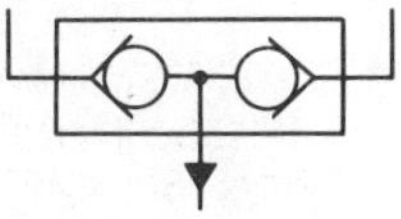

DOUBLE CHECK VALVE WITH CROSS BLEED (REVERSE FLOW PERMITTED)

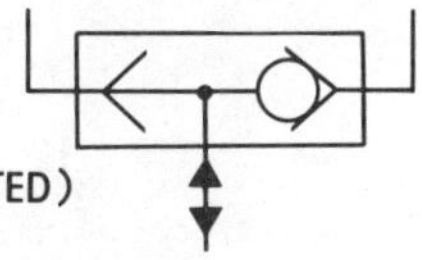

FOUR-WAY VALVES:

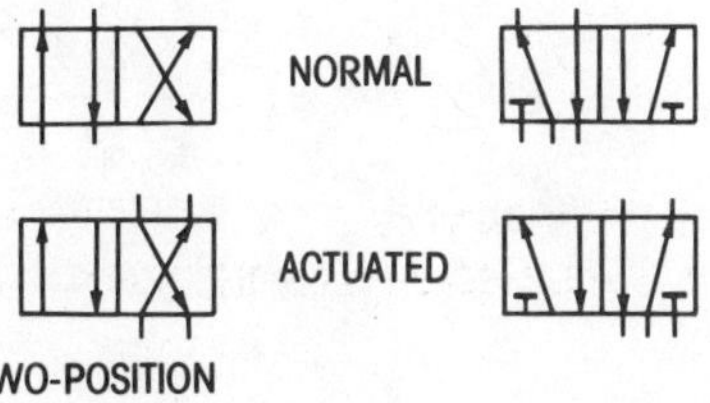

NORMAL

ACTUATED

TWO-POSITION

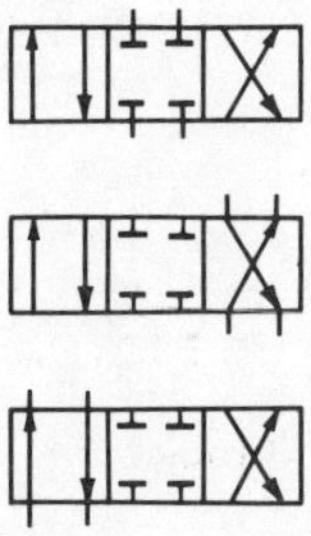

NORMAL

ACTUATED LEFT

ACTUATED RIGHT

THREE-POSITION

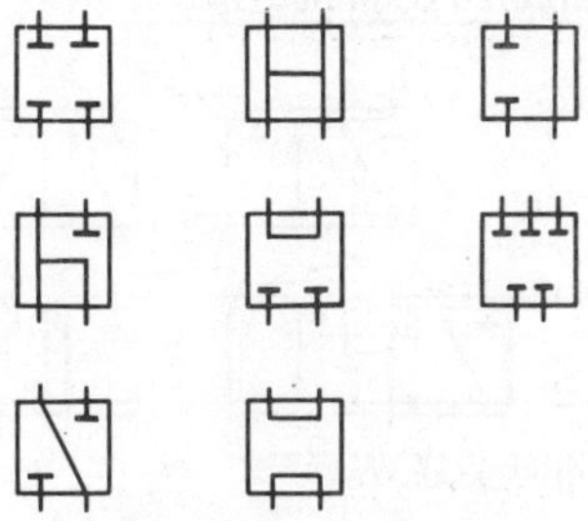

TYPICAL FLOW PATHS FOR CENTER CONDITION OF THREE-POSITION, FOUR-WAY VALVES

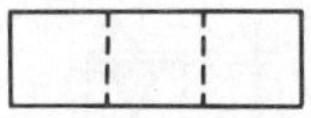

TRANSITION SYMBOL

TWO-POSITION, SNAP ACTION WITH TRANSITION

INFINITE POSITIONING (BETWEEN OPEN AND CLOSED):

NORMALLY CLOSED

NORMALLY OPEN

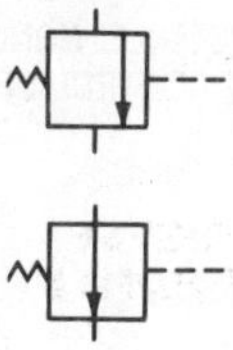

PRESSURE CONTROL VALVES:

PRESSURE RELIEF

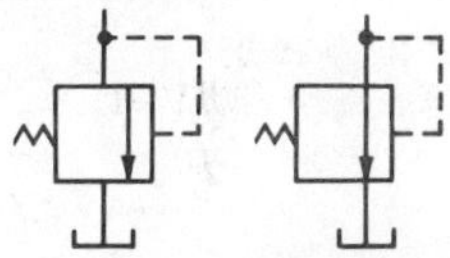

NORMAL ACTUATED (RELIEVING)

SEQUENCE

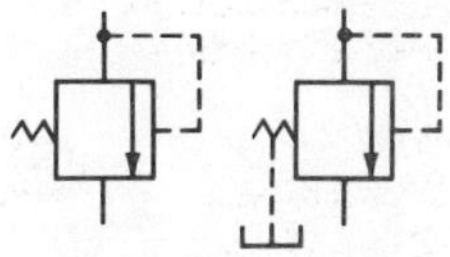

PRESSURE REDUCING

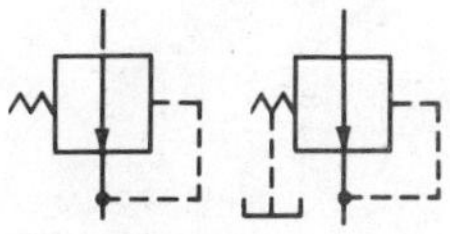

PRESSURE REDUCING AND RELIEVING

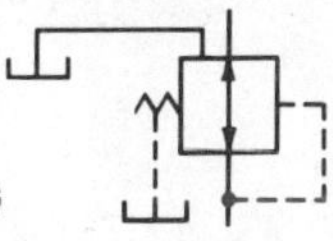

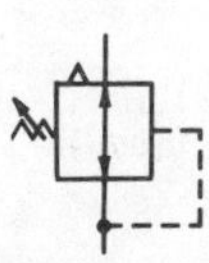

AIR LINE PRESSURE REGULATOR (ADJUSTABLE, RELIEVING)

INFINITE POSITIONING VALVES:

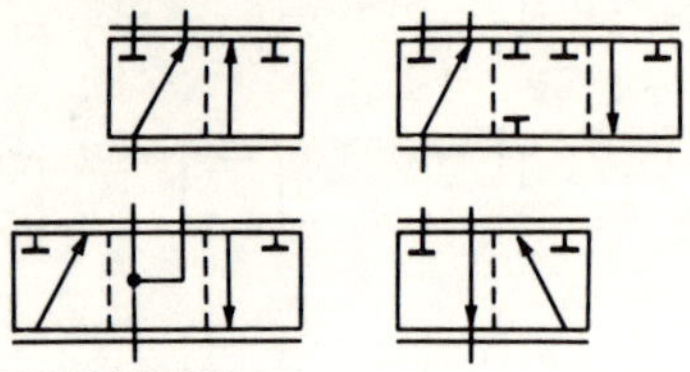

THREE-WAY VALVES

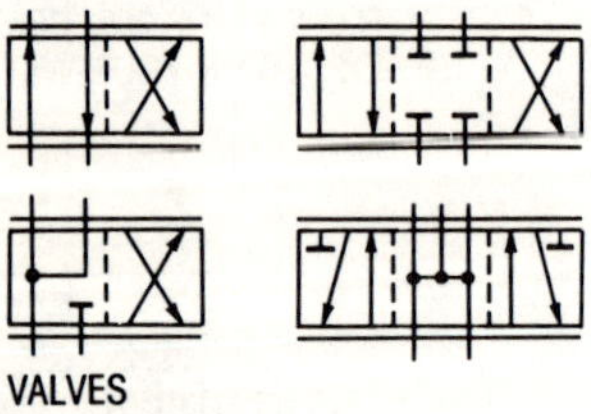

FOUR-WAY VALVES

FLOW-CONTROL VALVES:

ADJUSTABLE, NONCOMPENSATED
(FLOW CONTROL IN EACH DIRECTION)

ADJUSTABLE,
WITH BYPASS

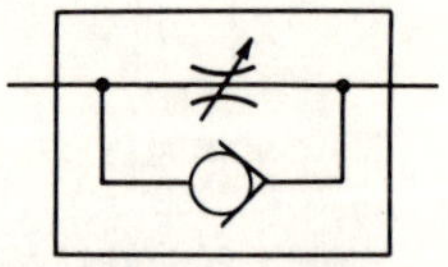

FLOW CONTROLLED TO RIGHT,
FLOW TO LEFT BYPASSES CONTROL.

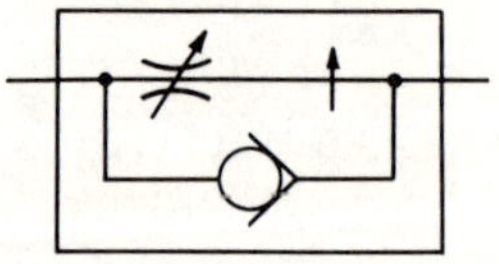

ADJUSTABLE AND PRESSURE
COMPENSATED, WITH BYPASS

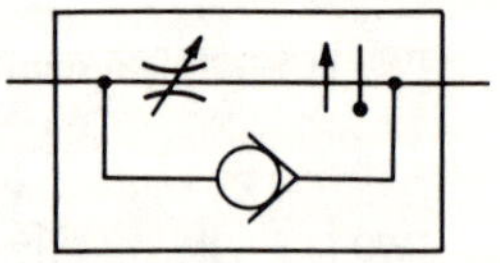

ADJUSTABLE, TEMPERATURE
AND PRESSURE COMPENSATED

AIR LINE ACCESSORIES:

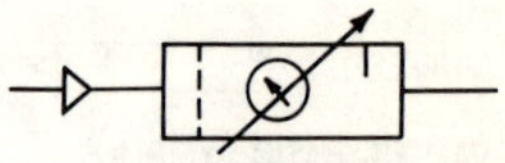

SIMPLIFIED

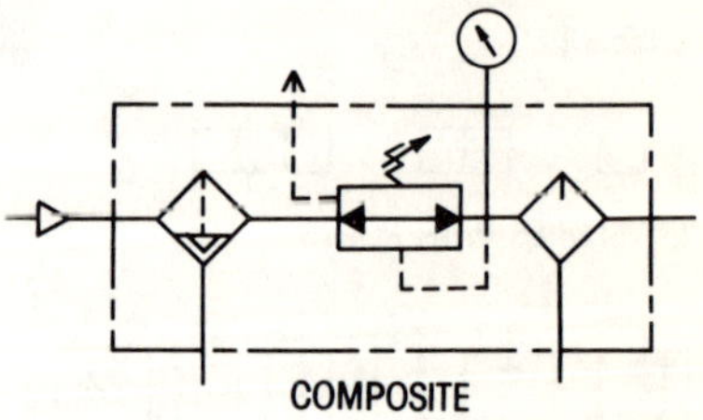

COMPOSITE

FILTER, REGULATOR, AND LUBRICATOR

APPENDIX 3

Trigonometric Functions

Trigonometry is a very valuable form of mathematics for anyone who studies electricity/electronics. Trigonometry deals with angles and triangles, particularly the right triangle, which has one angle of 90°. An electronic example of a right triangle is shown in Fig. A3-1. This example illustrates how resistance, reactance, and impedance are related in ac circuits. We know that resistance (R) and reactance (X) are 90° apart, so their angle of intersection forms a right angle. We can use the law of right triangles, known as the *Pythagorean theorem,* to solve for the value of any side. This theorem states that *in any right triangle, the square of the hypotenuse is equal to the sum of the squares of the other two sides.* With reference to Fig. A3-1, we can express the Pythagorean theorem mathematically as:

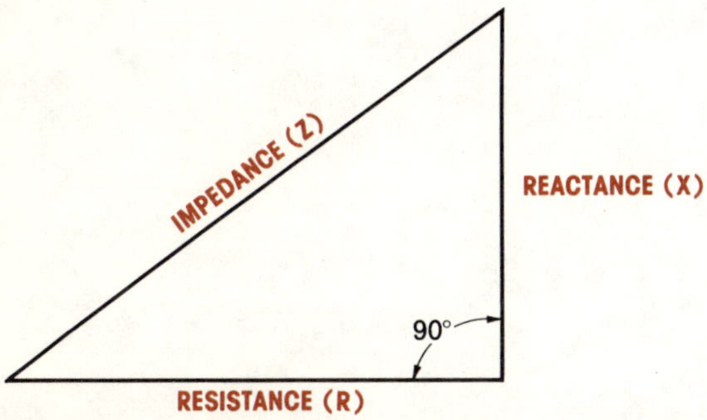

Fig. A3-1. Right triangle illustrating the trigonometric relationships between resistance, reactance, and impedance in ac circuits.

$$Z^2 = R^2 + X^2$$

or,

$$Z = \sqrt{R^2 + X^2}$$

By using trigonometric relationships, we can solve problems dealing with phase angles, power factor, and reactive power in ac circuits. The three most used trigonometric functions are the *sine,* the *cosine,* and the *tangent.* These functions show the ratios of the sides of the triangle, which determine the size of the angles. Fig. A3-2 illustrates how these ratios are expressed mathematically, and their values can be found in any table of trigonometric functions. Table A3-1 lists the natural sines, cosines, tangents, and cotangents of angles from 0° to 90°. In order to find these values for angles from 0° to 45°, use the headings at the top of the table and the degree listings from top to bottom in the left-hand column. For angles from 45° to 90°, use the headings at the bottom of the table and the degree listings from bottom to top in the right-hand column.

We can reverse this process to find the size of an angle when the ratios of the sides are known. The term *arc function* is used to indicate this process. For example, the notation "arc sin $x = \theta$" means that θ is the angle whose sine function is x. Thus, if we are given "arc sin $0.9455 = \theta$," we look through the listing of sine functions in Table A3-1 and find that angle θ is 71°.

Trigonometric ratios hold true for angles of any size; however, we use angles in the first quadrant of a standard graph (0° to 90°) as a reference, and in

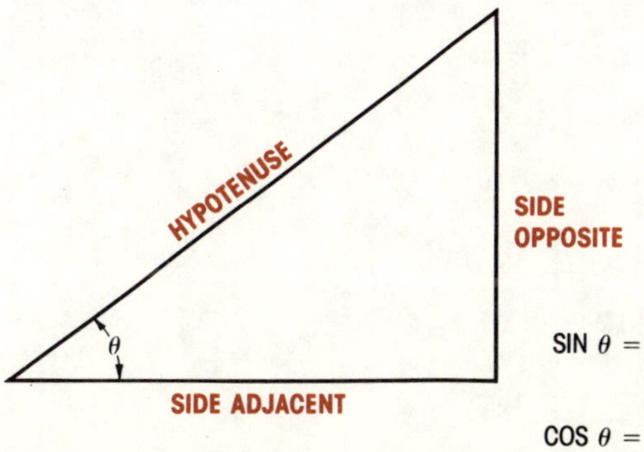

$$\text{SIN } \theta = \frac{\text{SIDE OPPOSITE}}{\text{HYPOTENUSE}}$$

$$\text{COS } \theta = \frac{\text{SIDE ADJACENT}}{\text{HYPOTENUSE}}$$

$$\text{TAN } \theta = \frac{\text{SIDE OPPOSITE}}{\text{SIDE ADJACENT}}$$

Fig. A3-2. Illustration of the trigonometric relationships of the sides of a right triangle to the angle θ.

order to solve for angles greater than 90° (second-, third-, and fourth-quadrant angles), we must convert them to first-quadrant angles (see Fig. A3-3). All first-quadrant angles have positive functions, while angles in the second, third, and fourth quadrants have two negative functions and one positive function.

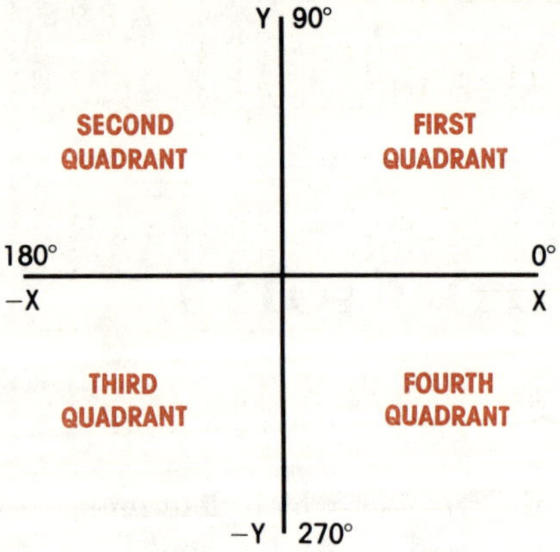

SECOND QUADRANT (90° TO 180°)

$\sin\theta = \sin(180^\circ - \theta)$
$\cos\theta = -\cos(180^\circ - \theta)$
$\tan\theta = -\tan(180^\circ - \theta)$

THIRD QUADRANT (180° TO 270°)

$\sin\theta = -\sin(\theta - 180^\circ)$
$\cos\theta = -\cos(\theta - 180^\circ)$
$\tan\theta = \tan(\theta - 180^\circ)$

FOURTH QUADRANT (270° TO 360°)

$\sin\theta = -\sin(360^\circ - \theta)$
$\cos\theta = \cos(360^\circ - \theta)$
$\tan\theta = -\tan(360^\circ - \theta)$

Fig. A3-3. Standard graph illustrating how trigonometric functions for angles greater than 90° (second-, third-, and fourth-quadrant angles) are derived by converting to first-quadrant angles.

Table A3-1. Natural Trigonometric Functions

Degrees	Sin	Cos	Tan	Cot	
0° 00′	0.0000	1.0000	0.0000	∞	90° 00′
10	.0029	1.0000	.0029	343.77	50
20	.0058	1.0000	.0058	171.89	40
30	.0087	1.0000	.0087	114.59	30
40	.0116	.9999	.0116	85.940	20
50	.0145	.9999	.0145	68.750	10
1° 00′	0.0175	0.9998	0.0175	57.290	89° 00′
10	.0204	.9998	.0204	49.104	50
20	.0233	.9997	.0233	42.964	40
30	.0262	.9997	.0262	38.188	30
40	.0291	.9996	.0291	34.368	20
50	.0320	.9995	.0320	31.242	10
2° 00′	0.0349	0.9994	0.0349	28.636	88° 00′
10	.0378	.9993	.0378	26.432	50
20	.0407	.9992	.0407	24.542	40
30	.0436	.9990	.0437	22.904	30
40	.0465	.9989	.0466	21.470	20
50	.0494	.9988	.0495	20.206	10
3° 00′	0.0523	0.9986	0.0524	19.081	87° 00′
10	.0552	.9985	.0553	18.075	50
20	.0581	.9983	.0582	17.169	40
30	.0610	.9981	.0612	16.350	30
40	.0640	.9980	.0641	15.605	20
50	.0669	.9978	.0670	14.924	10
4° 00′	0.0698	0.9976	0.0699	14.301	86° 00′
10	.0727	.9974	.0729	13.727	50
20	.0756	.9971	.0758	13.197	40
30	.0785	.9969	.0787	12.706	30
40	.0814	.9967	.0816	12.251	20
50	.0843	.9964	.0846	11.826	10
5° 00′	0.0872	0.9962	0.0875	11.430	85° 00′
10	.0901	.9959	.0904	11.059	50
20	.0929	.9957	.0934	10.712	40
30	.0958	.9954	.0963	10.385	30
40	.0987	.9951	.0992	10.078	20
50	.1016	.9948	.1022	9.7882	10
6° 00′	0.1045	0.9945	0.1051	9.5144	84° 00′
10	.1074	.9942	.1080	9.2553	50
20	.1103	.9939	.1110	9.0098	40
30	.1132	.9936	.1139	8.7769	30
40	.1161	.9932	.1169	8.5555	20
50	.1190	.9929	.1198	8.3450	10
7° 00′	0.1219	0.9925	0.1228	8.1443	83° 00′
10	.1248	.9922	.1257	7.9530	50
20	.1276	.9918	.1287	7.7704	40
30	.1305	.9914	.1317	7.5958	30
40	.1334	.9911	.1346	7.4287	20
50	.1363	.9907	.1376	7.2687	10
	Cos	Sin	Cot	Tan	Degrees

Degrees	Sin	Cos	Tan	Cot	
8° 00′	0.1392	0.9903	0.1405	7.1154	82° 00′
10	.1421	.9899	.1435	6.9682	50
20	.1449	.9894	.1465	6.8269	40
30	.1478	.9890	.1495	6.6912	30
40	.1507	.9886	.1524	6.5606	20
50	.1536	.9881	.1554	6.4348	10
9° 00′	0.1564	0.9877	0.1584	6.3138	81° 00′
10	.1593	.9872	.1614	6.1970	50
20	.1622	.9868	.1644	6.0844	40
30	.1650	.9863	.1673	5.9758	30
40	.1679	.9858	.1703	5.8708	20
50	.1708	.9853	.1733	5.7694	10
10° 00′	0.1736	0.9848	0.1763	5.6713	80° 00′
10	.1765	.9843	.1793	5.5764	50
20	.1794	.9838	.1823	5.4845	40
30	.1822	.9833	.1853	5.3955	30
40	.1851	.9827	.1883	5.3093	20
50	.1880	.9822	.1914	5.2257	10
11° 00′	0.1908	0.9816	0.1944	5.1446	79° 00′
10	.1937	.9811	.1974	5.0658	50
20	.1965	.9805	.2004	4.9894	40
30	.1994	.9799	.2035	4.9152	30
40	.2022	.9793	.2065	4.8430	20
50	.2051	.9787	.2095	4.7729	10
12° 00′	0.2079	0.9781	0.2126	4.7046	78° 00′
10	.2108	.9775	.2156	4.6382	50
20	.2136	.9769	.2186	4.5736	40
30	.2164	.9763	.2217	4.5107	30
40	.2193	.9757	.2247	4.4494	20
50	.2221	.9750	.2278	4.3897	10
13° 00′	0.2250	0.9744	0.2309	4.3315	77° 00′
10	.2278	.9737	.2339	4.2747	50
20	.2306	.9730	.2370	4.2193	40
30	.2334	.9724	.2401	4.1653	30
40	.2363	.9717	.2432	4.1126	20
50	.2391	.9710	.2462	4.0611	10
14° 00′	0.2419	0.9703	0.2493	4.0108	76° 00′
10	.2447	.9696	.2524	3.9617	50
20	.2476	.9689	.2555	3.9136	40
30	.2504	.9681	.2586	3.8667	30
40	.2532	.9674	.2617	3.8208	20
50	.2560	.9667	.2648	3.7760	10
15° 00′	0.2588	0.9659	0.2679	3.7321	75° 00′
10	.2616	.9652	.2711	3.6891	50
20	.2644	.9644	.2742	3.6470	40
30	.2672	.9636	.2773	3.6059	30
40	.2700	.9628	.2805	3.5656	20
50	.2728	.9621	.2836	3.5261	10
	Cos	Sin	Cot	Tan	Degrees

Table A3-1. Natural Trigonometric Functions (cont)

Degrees	Sin	Cos	Tan	Cot	
16° 00′	0.2756	0.9613	0.2867	3.4874	74° 00′
10	.2784	.9605	.2899	3.4495	50
20	.2812	.9596	.2931	3.4124	40
30	.2840	.9588	.2962	3.3759	30
40	.2868	.9580	.2994	3.3402	20
50	.2896	.9572	.3026	3.3052	10
17° 00′	0.2924	0.9563	0.3057	3.2709	73° 00′
10	.2952	.9555	.3089	3.2371	50
20	.2979	.9546	.3121	3.2041	40
30	.3007	.9537	.3153	3.1716	30
40	.3035	.9528	.3185	3.1397	20
50	.3062	.9520	.3217	3.1084	10
18° 00′	0.3090	0.9511	0.3249	3.0777	72° 00′
10	.3118	.9502	.3281	3.0475	50
20	.3145	.9492	.3314	3.0178	40
30	.3173	.9483	.3346	2.9887	30
40	.3201	.9474	.3378	2.9600	20
50	.3228	.9465	.3411	2.9319	10
19° 00′	0.3256	0.9455	0.3443	2.9042	71° 00′
10	.3283	.9446	.3476	2.8770	50
20	.3311	.9436	.3508	2.8502	40
30	.3338	.9426	.3541	2.8239	30
40	.3365	.9417	.3574	2.7980	20
50	.3393	.9407	.3607	2.7725	10
20° 00′	0.3420	0.9397	0.3640	2.7475	70° 00′
10	.3448	.9387	.3673	2.7228	50
20	.3475	.9377	.3706	2.6985	40
30	.3502	.9367	.3739	2.6746	30
40	.3529	.9356	.3772	2.6511	20
50	.3557	.9346	.3805	2.6279	10
21° 00′	0.3584	0.9336	0.3839	2.6051	69° 00′
10	.3611	.9325	.3872	2.5826	50
20	.3638	.9315	.3906	2.5605	40
30	.3665	.9304	.3939	2.5386	30
40	.3692	.9293	.3973	2.5172	20
50	.3719	.9283	.4006	2.4960	10
22° 00′	0.3746	0.9272	0.4040	2.4751	68° 00′
10	.3773	.9261	.4074	2.4545	50
20	.3800	.9250	.4108	2.4342	40
30	.3827	.9239	.4142	2.4142	30
40	.3854	.9228	.4176	2.3945	20
50	.3881	.9216	.4210	2.3750	10
23° 00′	0.3907	0.9205	0.4245	2.3559	67° 00′
10	.3934	.9194	.4279	2.3369	50
20	.3961	.9182	.4314	2.3183	40
30	.3987	.9171	.4348	2.2998	30
40	.4014	.9159	.4383	2.2817	20
50	.4041	.9147	.4417	2.2637	10
	Cos	Sin	Cot	Tan	Degrees

Degrees	Sin	Cos	Tan	Cot	
24° 00′	0.4067	0.9135	0.4452	2.2460	66° 00′
10	.4094	.9124	.4487	2.2286	50
20	.4120	.9112	.4522	2.2113	40
30	.4147	.9100	.4557	2.1943	30
40	.4173	.9088	.4592	2.1775	20
50	.4200	.9075	.4628	2.1609	10
25° 00′	0.4226	0.9063	0.4663	2.1445	65° 00′
10	.4253	.9051	.4699	2.1283	50
20	.4279	.9038	.4734	2.1123	40
30	.4305	.9026	.4770	2.0965	30
40	.4331	.9013	.4806	2.0809	20
50	.4358	.9001	.4841	2.0655	10
26° 00′	0.4384	0.8988	0.4877	2.0503	64° 00′
10	.4410	.8975	.4913	2.0353	50
20	.4436	.8962	.4950	2.0204	40
30	.4462	.8949	.4986	2.0057	30
40	.4488	.8936	.5022	1.9912	20
50	.4514	.8923	.5059	1.9768	10
27° 00′	0.4540	0.8910	0.5095	1.9626	63° 00′
10	.4566	.8897	.5132	1.9486	50
20	.4592	.8884	.5169	1.9347	40
30	.4617	.8870	.5206	1.9210	30
40	.4643	.8857	.5243	1.9074	20
50	.4669	.8843	.5280	1.8940	10
28° 00′	0.4695	0.8829	0.5317	1.8807	62° 00′
10	.4720	.8816	.5354	1.8676	50
20	.4746	.8802	.5392	1.8546	40
30	.4772	.8788	.5430	1.8418	30
40	.4797	.8774	.5467	1.8291	20
50	.4823	.8760	.5505	1.8165	10
29° 00′	0.4848	0.8746	0.5543	1.8040	61° 00′
10	.4874	.8732	.5581	1.7917	50
20	.4899	.8718	.5619	1.7796	40
30	.4924	.8704	.5658	1.7675	30
40	.4950	.8689	.5696	1.7556	20
50	.4975	.8675	.5735	1.7437	10
30° 00′	0.5000	0.8660	0.5774	1.7321	60° 00′
10	.5025	.8646	.5812	1.7205	50
20	.5050	.8631	.5851	1.7090	40
30	.5075	.8616	.5890	1.6977	30
40	.5100	.8601	.5930	1.6864	20
50	.5125	.8587	.5969	1.6753	10
31° 00′	0.5150	0.8572	0.6009	1.6643	59° 00′
10	.5175	.8557	.6048	1.6534	50
20	.5200	.8542	.6088	1.6426	40
30	.5225	.8526	.6128	1.6319	30
40	.5250	.8511	.6168	1.6212	20
50	.5275	.8496	.6208	1.6107	10
	Cos	Sin	Cot	Tan	Degrees

Table A3-1. Natural Trigonometric Functions (cont)

Degrees	Sin	Cos	Tan	Cot	
32° 00′	0.5299	0.8480	0.6249	1.6003	58° 00′
10	.5324	.8465	.6289	1.5900	50
20	.5348	.8450	.6330	1.5798	40
30	.5373	.8434	.6371	1.5697	30
40	.5398	.8418	.6412	1.5597	20
50	.5422	.8403	.6453	1.5497	10
33° 00′	0.5446	0.8387	0.6494	1.5399	57° 00′
10	.5471	.8371	.6536	1.5301	50
20	.5495	.8355	.6577	1.5204	40
30	.5519	.8339	.6619	1.5108	30
40	.5544	.8323	.6661	1.5013	20
50	.5568	.8307	.6703	1.4919	10
34° 00′	0.5592	0.8290	0.6745	1.4826	56° 00′
10	.5616	.8274	.6787	1.4733	50
20	.5640	.8258	.6830	1.4641	40
30	.5664	.8241	.6873	1.4550	30
40	.5688	.8225	.6916	1.4460	20
50	.5712	.8208	.6959	1.4370	10
35° 00′	0.5736	0.8192	0.7002	1.4281	55° 00′
10	.5760	.8175	.7046	1.4193	50
20	.5783	.8158	.7089	1.4106	40
30	.5807	.8141	.7133	1.4019	30
40	.5831	.8124	.7177	1.3934	20
50	.5854	.8107	.7221	1.3848	10
36° 00′	0.5878	0.8090	0.7265	1.3764	54° 00′
10	.5901	.8073	.7310	1.3680	50
20	.5925	.8056	.7355	1.3597	40
30	.5948	.8039	.7400	1.3514	30
40	.5972	.8021	.7445	1.3432	20
50	.5995	.8004	.7490	1.3351	10
37° 00′	.6018	.7986	.7536	1.3270	53° 00′
10	.6041	.7969	.7581	1.3190	50
20	.6065	.7951	.7627	1.3111	40
30	.6088	.7934	.7673	1.3032	30
40	.6111	.7916	.7720	1.2954	20
50	.6134	.7898	.7766	1.2876	10
38° 00′	0.6157	0.7880	0.7813	1.2799	52° 00′
10	.6180	.7862	.7860	1.2723	50
20	.6202	.7844	.7907	1.2647	40
30	.6225	.7826	.7954	1.2572	30
40	.6248	.7808	.8002	1.2497	20
50	.6271	.7790	.8050	1.2423	10
39° 00′	0.6293	0.7771	0.8098	1.2349	51° 00′
10	.6316	.7753	.8146	1.2276	50
20	.6338	.7735	.8195	1.2203	40
30	.6361	.7716	.8243	1.2131	30
40	.6383	.7698	.8292	1.2059	20
50	.6406	.7679	.8342	1.1988	10
	Cos	Sin	Cot	Tan	Degrees

Degrees	Sin	Cos	Tan	Cot	
40° 00′	0.6428	0.7660	0.8391	1.1918	50° 00′
10	.6450	.7642	.8441	1.1847	50
20	.6472	.7623	.8491	1.1778	40
30	.6494	.7604	.8541	1.1708	30
40	.6517	.7585	.8591	1.1640	20
50	.6539	.7566	.8642	1.1571	10
41° 00′	0.6561	0.7547	0.8693	1.1504	49° 00′
10	.6583	.7528	.8744	1.1436	50
20	.6604	.7509	.8796	1.1369	40
30	.6626	.7490	.8847	1.1303	30
40	.6648	.7470	.8899	1.1237	20
50	.6670	.7451	.8952	1.1171	10
42° 00′	0.6691	0.7431	0.9004	1.1106	48° 00′
10	.6713	.7412	.9057	1.1041	50
20	.6734	.7392	.9110	1.0977	40
30	.6756	.7373	.9163	1.0913	30
40	.6777	.7353	.9217	1.0850	20
50	.6799	.7333	.9271	1.0786	10
43° 00′	0.6820	0.7314	0.9325	1.0724	47° 00′
10	.6841	.7294	.9380	1.0661	50
20	.6862	.7274	.9435	1.0599	40
30	.6884	.7254	.9490	1.0538	30
40	.6905	.7234	.9545	1.0477	20
50	.6926	.7214	.9601	1.0416	10
44° 00′	0.6947	0.7193	0.9657	1.0355	46° 00′
10	.6967	.7173	.9713	1.0295	50
20	.6988	.7163	.9770	1.0235	40
30	.7009	.7133	.9827	1.0176	30
40	.7030	.7112	.9884	1.0117	20
50	.7050	.7092	.9942	1.0058	10
45° 00′	0.7071	0.7071	1.0000	1.0000	45° 00′
	Cos	Sin	Cot	Tan	Degrees

APPENDIX 4

Typical Industrial Diagrams

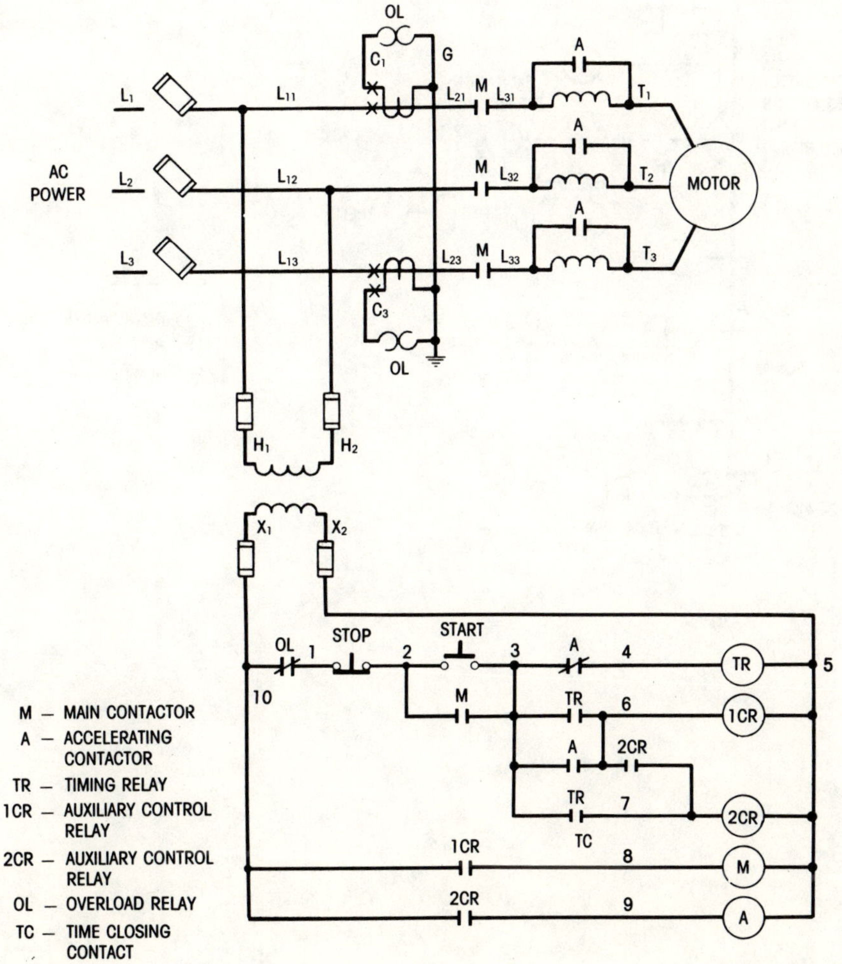

M – MAIN CONTACTOR
A – ACCELERATING CONTACTOR
TR – TIMING RELAY
1CR – AUXILIARY CONTROL RELAY
2CR – AUXILIARY CONTROL RELAY
OL – OVERLOAD RELAY
TC – TIME CLOSING CONTACT

Fig. A4-1. Typical industrial control schematic diagram.

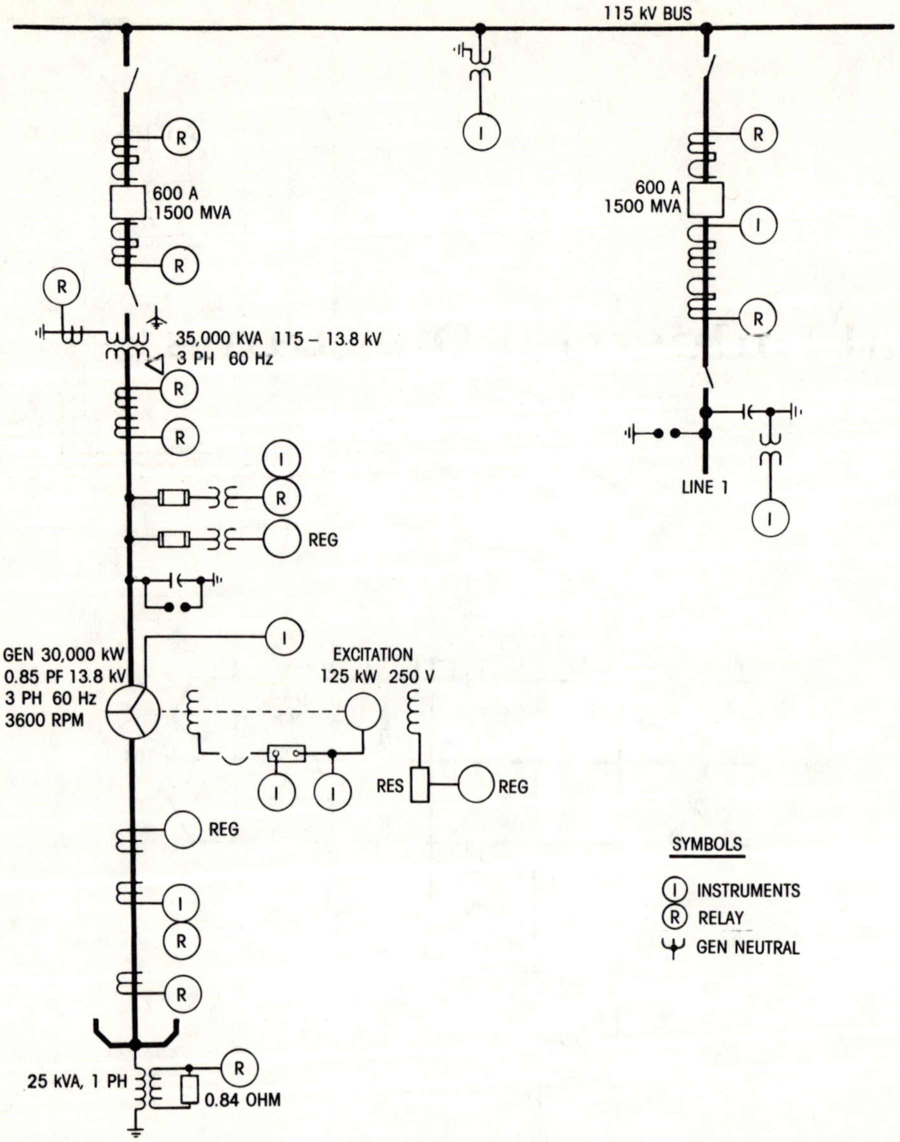

Fig. A4-2. Typical power switchgear single-line diagram.

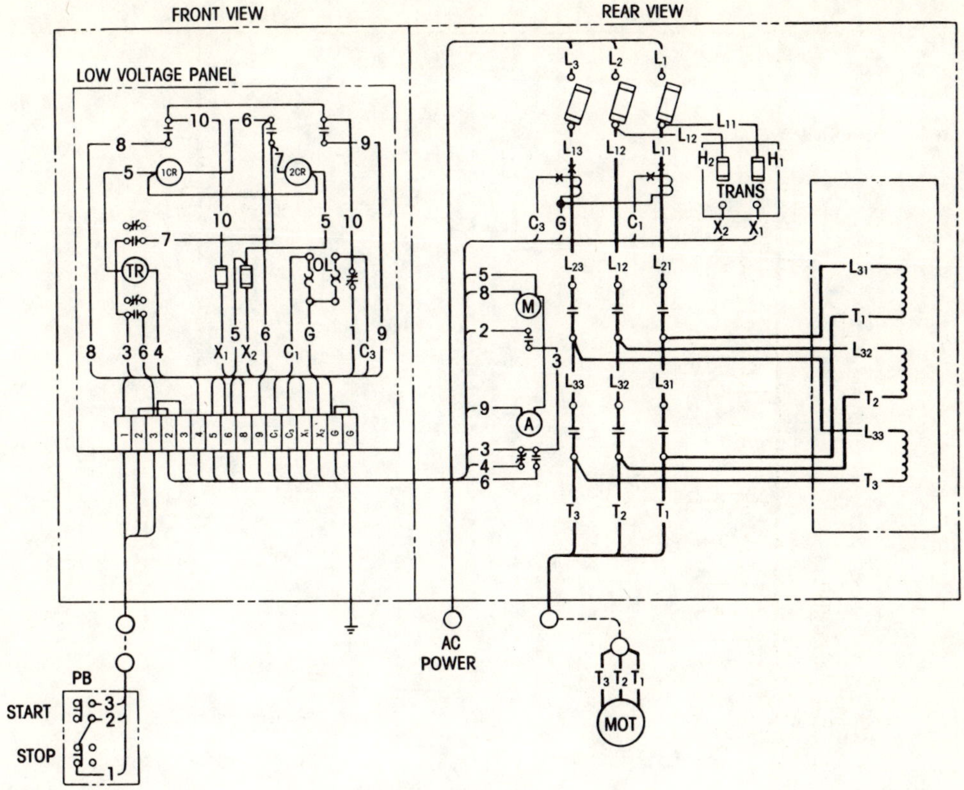

Fig. A4-3. Typical highway or cable connection diagram.

Index